Infrared Spectra of Inorganic and Coordination Compounds

Infrared Spectra of Inorganic and Coordination Compounds

SECOND EDITION

KAZUO NAKAMOTO

Wehr Professor of Chemistry
Marquette University

WILEY-INTERSCIENCE, a Division of JOHN WILEY & SONS
New York · London · Sydney · Toronto

Library of Congress Catalogue Card Number: 78-107588

SBN 471 62980 4

Printed in the United States of America

10 9 8 7 6 5 4 3 2 1

Dedicated to the memory of
Professor Ryutaro Tsuchida
who first stimulated my interest in
inorganic and coordination chemistry

Preface to the Second Edition

A reference book is doomed to become outdated rapidly because of the ever-increasing volume of new literature. I began to feel a need for a new edition of this book several years ago, but it was not until the spring of 1968 that I finished my survey of the literature and started to rewrite the chapters. This task was finally completed in the summer of 1969.

In the second edition, I have expanded Part I by adding several new sections concerning the theories of Raman spectra, crystal spectra, and infrared intensities. In Parts II and III, I have attempted to include all important and significant results published before the end of 1967, while omitting those proved to be erroneous by later studies. Some results obtained after 1967 were included rather arbitrarily simply because they were easily accessible to me. It was clearly not possible to cover all the work on infrared and Raman spectra of inorganic and coordination compounds. I have tried, however, to present topics and results that I consider important or interesting in a proper and reasonable balance. Finally, I have replaced several appendices by new ones that I consider to be more useful and convenient to the reader.

I express my sincere thanks to all persons who gave valuable comments on the first edition through private communications or book reviews. All these comments were taken into consideration in preparing the second edition. I am particularly indebted to Dr. Marcia Cordes who read the whole manuscript of the second edition. Thanks are also due to Professor I. Nakagawa (Part I), Dr. A. Fadini and Professor A. Müller (Part II), and Professor D. F. Schriver (Part III) who helped me in writing some sections. Most of the literature search and writing for the second edition were done at the Department of Chemistry, Illinois Institute of Technology, whose library facilities greatly helped me in the preparation of this volume. The second edition includes some recent results we obtained during its preparation. Most of these projects were supported by the ACS-PRF unrestricted research grant, and I would like to thank the American Chemical Society for giving me opportunities to pursue these, rather uncoordinated, research projects.

Milwaukee, Wisconsin
October, 1969

KAZUO NAKAMOTO

Preface to the First Edition

Since 1945, the volume of literature on the infrared spectra of inorganic and coordination compounds has grown with ever-increasing rapidity. As a result, it is becoming more and more difficult for any one individual to read all pertinent articles. Excellent books on the theory of vibrational spectra and on the infrared spectra of organic compounds have been published, but relatively few comprehensive reviews of the vibrational spectra of inorganic or coordination compounds are available. This situation has prompted me to write this book.

In reviewing the literature, I have attempted to interpret the experimental results in terms of normal vibrations, as far as it is possible. Consequently, I have been most interested in normal coordinate analysis, in fundamental frequencies, in band assignments and in structural considerations, while omitting any consideration of band intensities or rotational fine structure. The experimental aspects of vibrational spectra have also been omitted.

Part I is devoted to the minimum amount of theory necessary for an understanding of the concept of the normal vibration and of the method of normal coordinate analysis. For a more detailed discussion of the theory, the reader may consult the references cited in Part I. An application of the method of normal coordinate analysis to a complex system is given in detail in Appendix III. In Parts II and III, the observed fundamental vibrational frequencies of molecules are discussed in terms of their relation to molecular structure. Although most of the data are from infrared spectra, Raman spectral data have been quoted wherever pertinent or necessary. This volume is intended as a critical review of the infrared spectra of inorganic and coordination compounds; but it was clearly impossible to refer to all work which has been reported in these fields. I have, however, tried to give a broad and reasonably complete coverage of the fields. But in the discussion I have had to select examples which I considered more illustrative, more important or more interesting. Although this book is intended to cover all inorganic and coordination compounds, the border line between these and organic compounds is a difficult one to draw. Therefore I have chosen to omit most of the metallo-organic compounds.

I wish to express my sincere thanks to Professor Arthur E. Martell whose discussions inspired me to begin this book. A special word of gratitude is due to Professors S. Mizushima, T. Shimanouchi and I. Nakagawa whose publications and personal communications have greatly helped me in writing on normal coordinate analysis. I am also deeply indebted to Dr. J. L. Bethune, Rev. Paul J. McCarthy, S.J., and Sister M. Paulita, C.S.J., who read the whole manuscript and gave many valuable comments. Thanks are also due to Dr. Junnosuke Fujita and Mr. Yukiyoshi Morimoto, who assisted in the preparation of this book, and to Mrs. Helen Kwan and Mrs. Jeannette Lynch, who typed the manuscript.

Chicago, Illinois
December, 1962

KAZUO NAKAMOTO

NOTES ON TABLES

The following abbreviations and symbols are used in the tables: IR, infrared; R, Raman; p, polarized; dp, depolarized; ν, stretching; δ, deformation; ρ_w, wagging; ρ_r, rocking; ρ_t, twisting; π, out-of-plane bending; as, antisymmetric; s, symmetric; d, degenerate; GVF, generalized valence force field; and UBF, Urey-Bradley force field. Approximate normal modes of vibration corresponding to the vibrations mentioned above are given in Figs. III-2 and III-5.

Contents

Part I. Theory of Normal Vibration

Appendices

Infrared Spectra of Inorganic and Coordination Compounds

Theory of Normal Vibration

Part I

I-1. ORIGIN OF MOLECULAR SPECTRA

As a first approximation, it is possible to separate the energy of a molecule as three additive components associated with (1) the rotation of the molecule as a whole, (2) the vibrations of the constituent atoms and (3) the motion of the electrons in the molecule.* The translational energy of the molecule may be ignored in this discussion. The basis for this separation lies in the fact that the velocity of electrons is much greater than the vibrational velocity of nuclei, which is again much greater than the velocity of molecular rotation. If a molecule is placed in an electromagnetic field (e.g., light), a transfer of energy from the field to the molecule will occur only when Bohr's frequency condition is satisfied:

$$\Delta E = h\nu \tag{1.1}$$

where ΔE is the difference in energy between two quantized states, h is Planck's constant, and ν is the frequency of the light.† If

$$\Delta E = E'' - E' \tag{1.2}$$

where E'' is a quantized state of higher energy than E', the molecule *absorbs* radiation when it is excited from E' to E'' and *emits* radiation of the same frequency as given by Eq. 1.1 when it reverts from E'' to E'.

Because rotational levels are relatively close to each other, transitions between these levels occur at low frequencies (long wavelengths). In fact, pure rotational spectra appear in the range between 1 cm^{-1} (10^4 μ) and 10^2 cm^{-1} (10^2 μ). The separation of vibrational energy levels is greater, and the transitions occur at higher frequencies (shorter wavelengths) than do the rotational transitions. As a result, pure vibrational spectra are observed in

*Hereafter, the word *molecule* may also represent an *ion*.

†The frequency, ν, is converted to the wave number, $\tilde{\nu}$, or the wavelength, λ_ω, through the relation

$$\nu = c\tilde{\nu} = c/\lambda_\omega$$

where c is the velocity of light. For theoretical discussion, ν and $\tilde{\nu}$ are more convenient than λ_ω, since they are proportional to the energy of radiation. More explicit relations between these three units are given in the following table for the region in which vibrational spectra occur.

Frequency (sec^{-1})	Wave Number (cm^{-1})	Wavelength (μ)
$3 \cdot 10^{14}$	10^4	1
$3 \cdot 10^{13}$	10^3	10
$3 \cdot 10^{12}$	10^2	10^2

Although the dimensions of ν and $\tilde{\nu}$ differ from one another, it is conventional to use them interchangeably. For example, a phrase such as "a frequency shift of 25 cm^{-1}" is often employed. All the spectral data in this book are given in terms of $\tilde{\nu}$ (cm^{-1}).

the range between 10^2 cm^{-1} (10^2 μ) and 10^4 cm^{-1} (1 μ). Finally, electronic energy levels are usually far apart, and electronic spectra are observed in the range between 10^4 cm^{-1} (1 μ) and 10^5 cm^{-1} (10^{-1} μ). Thus pure rotational, vibrational, and electronic spectra are usually observed in the microwave and far-infrared, the infrared, and the visible and ultraviolet regions, respectively. This division into three regions, however, is to some extent arbitrary, for pure rotational spectra may appear in the near infrared region ($1.5 \sim 0.5 \times 10^4$ cm^{-1}) if transitions to higher excited states are involved, and pure electronic transitions may appear in the near infrared region if the levels are closely spaced.

Figure I-1 illustrates transitions of the three types mentioned for a diatomic molecule. As the figure shows, rotational intervals tend to increase as the rotational quantum number J increases, whereas vibrational intervals tend to decrease as the vibrational quantum number v increases. The dotted line below each electronic level indicates the zero point energy that exists even at a temperature of absolute zero as a result of nuclear vibration. It should be emphasized that not all transitions between these levels are possible. In order to see whether the transition is *allowed* or *forbidden*, the relevant selection rule must be examined. This, in turn, is determined by the symmetry of the molecule. As will be seen later, vibrational problems like those mentioned above can be solved for polyatomic molecules in an elegant manner by the use of group theory.

Since this book is concerned only with vibrational spectra, no description of electronic and rotational spectra is given. Although vibrational spectra are observed experimentally as infrared or Raman spectra, the physical origins of these two types of spectra are different. Infrared spectra originate in transitions between two vibrational levels of the molecule in the electronic ground state and are usually observed as *absorption spectra* in the infrared region. On the other hand, Raman spectra originate in the electronic polarization caused by ultraviolet or visible light. If a molecule is irradiated by monochromatic light of frequency ν,* then, because of electronic polarization induced in the molecule by this incident light, light of frequency ν (Rayleigh scattering) as well as of $\nu \pm \nu_i$ (Raman scattering) is emitted (ν_i represents a vibrational frequency). Thus the vibrational frequencies are observed as Raman shifts from the incident frequency ν in the ultraviolet or visible region. In the past, the mercury line at 4358A was used exclusively to excite Raman scattering. This line, however, is absorbed by colored substances such as transition metal complexes. Modern Raman instruments are equipped with gas lasers such as He—Ne(6328A, red) and Ar—ion(4579, 4658, 4765, 4880, 4915, and 5145A, blue-green); the former may be used for yellow or red and the latter for blue and green compounds.

* In principle, light of any frequency can be used unless the incident light is absorbed by the molecule.

It is to be expected from Fig. I-1 that electronic spectra are very complicated because they are accompanied by vibrational as well as rotational fine structure. The rotational fine structure in the electronic spectrum can be observed if a molecule is simple and the spectrum is obtained in the gaseous state under high resolution. The vibrational fine structure of the electronic

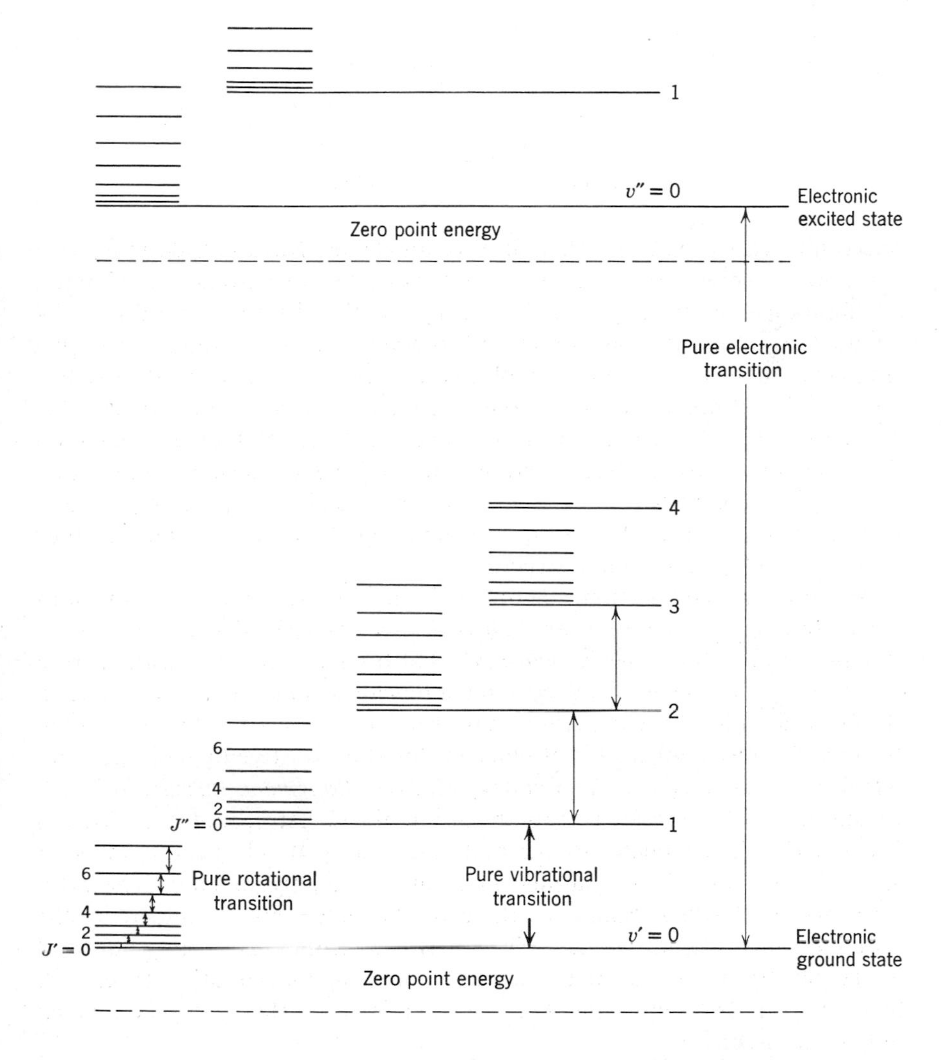

Fig. I-1. Energy levels of a diatomic molecule (the actual spacings of electronic levels are much larger and those of rotational levels are much smaller than those shown in the figure).

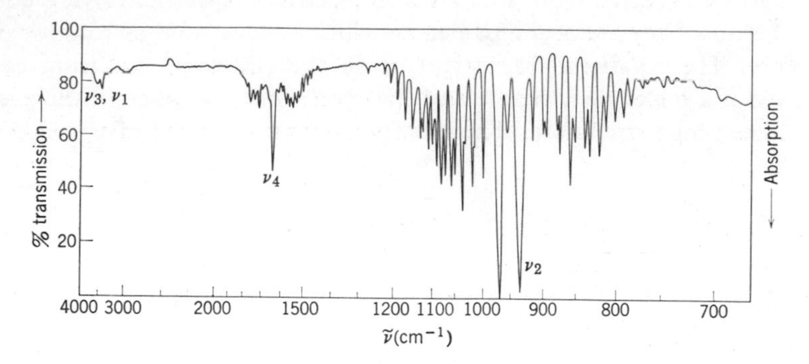

Fig. I-2. Rotational fine structure of gaseous NH_3.

spectrum is easier to observe than the rotational fine structure, and can provide structural and bonding information about molecules in electronic excited states.

Vibrational spectra are accompanied by rotational transitions. Figure I-2 shows the rotational fine structure observed for the gaseous ammonia molecule. In most polyatomic molecules, however, such a rotational fine structure is not observed because the rotational levels are closely spaced due to relatively large moments of inertia. Vibrational spectra obtained in solution do not exhibit rotational fine structure, since molecular collisions occur before a rotation is completed and the levels of the individual molecules are perturbed differently. Since Raman spectra are often obtained in liquid state, they do not exhibit rotational fine structure.

According to the selection rule for the harmonic oscillator, any transitions corresponding to $\Delta v = \pm 1$ are allowed (Sec. I-2). Under ordinary conditions, however, only the *fundamentals* that originate in the transition from $v = 0$ to $v = 1$ in the electronic ground state can be observed. This is due to the fact that most of the transitions have $v = 0$ in the initial state, for at room temperature the number of molecules in this state is exceedingly large compared with that in the excited states (*Maxwell-Boltzmann distribution law*). In addition to the selection rule for the harmonic oscillator, another restriction results from the symmetry of the molecule (Sec. I-9). Thus the number of allowed transitions in polyatomic molecules is greatly reduced. The *overtones and combination bands** of these fundamentals are forbidden by the selection rule of the harmonic oscillator. However, they are weakly observed in the spectrum because of the anharmonicity of the vibration (Sec. I-2). Since they are less important than the fundamentals, they will be discussed only when necessary.

*Overtones represent multiples of some fundamental, whereas combination bands arise from the sum or difference of two different fundamentals.

I-2. VIBRATION OF A DIATOMIC MOLECULE

Through quantum mechanical considerations,[7] the vibration of two nuclei in a diatomic molecule can be reduced to the motion of a single particle of mass μ, whose displacement q from its equilibrium position is equal to the change of the internuclear distance. The mass μ is called the *reduced mass* and is represented by

$$\frac{1}{\mu} = \frac{1}{m_1} + \frac{1}{m_2} \tag{2.1}$$

where m_1 and m_2 are the masses of the two nuclei. The kinetic energy is then

$$T = \frac{1}{2}\mu\dot{q}^2 = \frac{1}{2\mu}p^2 \tag{2.2}$$

where p is the conjugate momentum, $\mu\dot{q}$. If a simple parabolic potential function such as that shown in Fig. I-3 is assumed, the system represents a *harmonic oscillator*, and the potential energy is simply given by

$$V = \tfrac{1}{2}Kq^2 \tag{2.3}$$

Here K is the force constant for the vibration. Then the Schrödinger wave

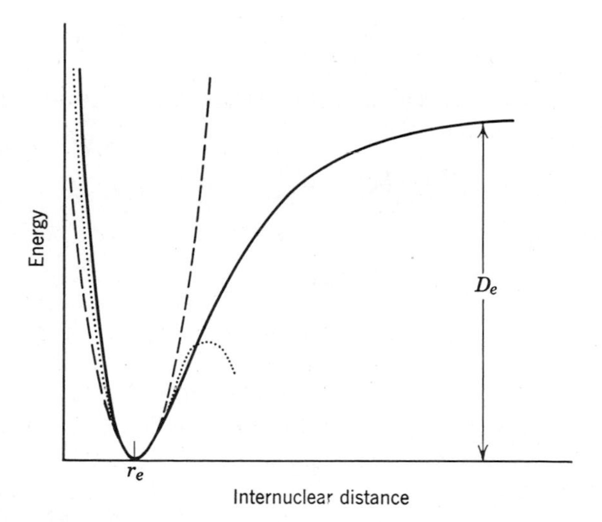

Fig. I-3. Potential curve for a diatomic molecule. Actual potential, solid line; parabola, broken line; cubic parabola, dotted line.

equation becomes

$$\frac{d^2\psi}{dq^2} + \frac{8\pi^2\mu}{h^2}\left(E - \frac{1}{2}Kq^2\right)\psi = 0 \tag{2.4}$$

If this equation is solved with the condition that ψ must be single-valued, finite, and continuous, the eigenvalues are

$$E_v = h\nu(v + \tfrac{1}{2}) = hc\tilde{\nu}(v + \tfrac{1}{2}) \tag{2.5}$$

with the frequency of vibration

$$\nu = \frac{1}{2\pi}\sqrt{\frac{K}{\mu}} \quad \text{or} \quad \tilde{\nu} = \frac{1}{2\pi c}\sqrt{\frac{K}{\mu}} \tag{2.6}$$

Here v is the vibrational quantum number, and it can have the values 0, 1, 2, 3, $\cdots$.

The corresponding eigenfunctions are

$$\psi_v = \frac{(\alpha/\pi)^{1/4}}{\sqrt{2^v v!}} e^{-\alpha q^2/2} H_v(\sqrt{\alpha}q) \tag{2.7}$$

where $\alpha = 2\pi\sqrt{\mu K}/h = 4\pi^2\mu\nu/h$, and $H_v(\sqrt{\alpha}q)$ is a Hermite polynomial of the vth degree. Thus the eigenvalues and the corresponding eigenfunctions are

$$\begin{array}{ll} E_0 = \tfrac{1}{2}h\nu & \psi_0 = (\alpha/\pi)^{1/4}e^{-\alpha q^2/2} \\ E_1 = \tfrac{3}{2}h\nu & \psi_1 = (\alpha/\pi)^{1/4}2^{1/2}qe^{-\alpha q^2/2} \\ \cdots\cdots & \cdots\cdots\cdots\cdots \\ \cdots\cdots & \cdots\cdots\cdots\cdots \end{array} \tag{2.8}$$

As Fig. I-3 shows, actual potential curves can be approximated more exactly by adding a cubic term:[9]

$$V = \tfrac{1}{2}Kq^2 - Gq^3 \qquad (K \gg G) \tag{2.9}$$

Then the eigenvalues are

$$E_v = hc\omega_e(v + \tfrac{1}{2}) - hc\omega_e x_e(v + \tfrac{1}{2})^2 + \cdots \tag{2.10}$$

where ω_e is the wave number corrected for *anharmonicity*, and $\omega_e x_e$ indicates the magnitude of anharmonicity. Table II-1*a* of Part II indicates the observed wave numbers and ω_e values for a number of diatomic molecules. Equation 2.10 shows that the energy levels of the anharmonic oscillator are not equidistant, and the separation decreases slowly as v increases. This anharmonicity is responsible for the appearance of overtones and combination vibrations, that are forbidden in the harmonic oscillator.[4] Since the anharmonicity correction has not been made for most polyatomic molecules, in large part because of the complexity of the calculation, the frequencies given in Parts II and III are not corrected for anharmonicity (except for those given in Table II-1*a*).

According to Eq. 2.6, the frequency of the vibration in a diatomic molecule is proportional to the square root of K/μ. If K is approximately the same for a series of diatomic molecules, the frequency is inversely proportional to the square root of μ. This point is illustrated by the series H_2, HD, and D_2 shown in Table II-1*a*. If μ is approximately the same for a series of diatomic molecules, the frequency is proportional to the square root of K. This point is illustrated by the series HF, HCl, HBr, and HI. These simple rules, obtained for a diatomic molecule, are helpful to the understanding of the vibrational spectra of polyatomic molecules.

Figure I-4 indicates the relation between the force constant K, calculated from Eq. 2.6, and the dissociation energy in a series of hydrogen halides. Evidently, the bond becomes stronger as the force constant becomes larger. It should be noted, however, that a general theoretical relation between these two quantities is difficult to derive even for a diatomic molecule.* The force constant is a measure of the curvature of the potential well near the

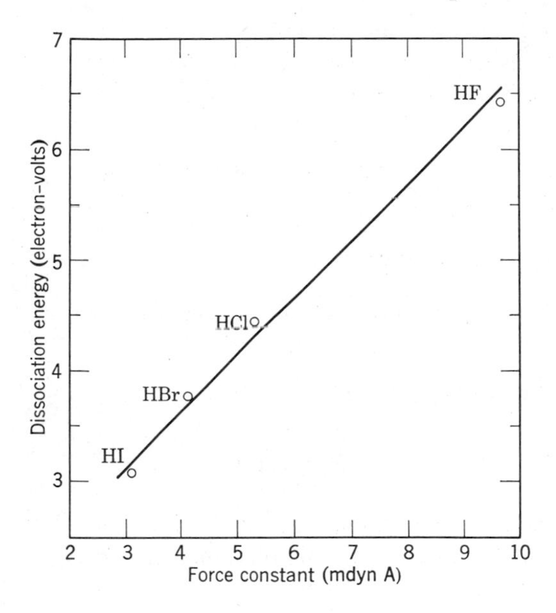

Fig. I-4. Relation between force constant and dissociation energy in hydrogen halides.

* Quantum mechanical expression of the potential function involves a Coulomb integral, an exchange integral and an overlap integral.[7] Calculation of K from Eq. 2.11, therefore, is not simple.

equilibrium position:

$$K = \left(\frac{d^2V}{dq^2}\right)_{q\to 0} \tag{2.11}$$

whereas the dissociation energy D_e is given by the depth of the potential energy curve (Fig. I-3). Thus a large force constant means sharp curvature of the potential well near the bottom but does not necessarily indicate a deep potential well. Usually, however, a larger force constant is an indication of a stronger bond in a series of molecules belonging to the same type (Fig. I-4).

I-3. NORMAL COORDINATES AND NORMAL VIBRATIONS

In diatomic molecules, the vibration of the nuclei occurs only along the line connecting two nuclei. In polyatomic molecules, however, the situation is much more complicated because all the nuclei perform their own harmonic oscillations. It can be shown, however, that any of these extremely complicated vibrations of the molecule may be represented as a superposition of a number of *normal vibrations*.

Let the displacement of each nucleus be expressed in terms of rectangular coordinate systems with the origin of each system at the equilibrium position of each nucleus. Then the kinetic energy of an N-atom molecule would be expressed as

$$T = \frac{1}{2}\sum_N m_N\left[\left(\frac{d\,\Delta x_N}{dt}\right)^2 + \left(\frac{d\,\Delta y_N}{dt}\right)^2 + \left(\frac{d\,\Delta z_N}{dt}\right)^2\right] \tag{3.1}$$

If generalized coordinates such as

$$q_1 = \sqrt{m_1}\,\Delta x_1, \quad q_2 = \sqrt{m_1}\,\Delta y_1, \quad q_3 = \sqrt{m_1}\,\Delta z_1, \quad q_4 = \sqrt{m_2}\,\Delta x_2, \cdots \tag{3.2}$$

are used, the kinetic energy is simply written

$$T = \frac{1}{2}\sum_i^{3N} \dot{q}_i^{\,2} \tag{3.3}$$

The potential energy of the system is a complex function of all the coordinates involved. For small values of the displacements, it may be expanded in a Taylor's series as

$$V(q_1, q_2, \cdots, q_{3N}) = V_0 + \sum_i^{3N}\left(\frac{\partial V}{\partial q_i}\right)_0 q_i + \frac{1}{2}\sum_{i,j}^{3N}\left(\frac{\partial^2 V}{\partial q_i\,\partial q_j}\right)_0 q_i q_j + \cdots \tag{3.4}$$

where the derivatives are evaluated at $q_i = 0$, the equilibrium position. The constant term V_0 can be taken as zero if the potential energy at $q_i = 0$ is taken as a standard. The $(\partial V/\partial q_i)_0$ terms also become zero, since V must be a minimum at $q_i = 0$. Thus V may be represented by

$$V = \frac{1}{2}\sum_{i,j}^{3N}\left(\frac{\partial^2 V}{\partial q_i\,\partial q_j}\right)_0 q_i q_j = \frac{1}{2}\sum_{i,j}^{3N} b_{ij} q_i q_j \qquad (3.5)$$

neglecting higher order terms.

If the potential energy given by Eq. 3.5 did not include any cross products such as $q_i q_j$, the problem could be solved directly by using Newton's equation:

$$\frac{d}{dt}\left(\frac{\partial T}{\partial \dot{q}_i}\right) + \frac{\partial V}{\partial q_i} = 0 \qquad i = 1, 2, \cdots, 3N \qquad (3.6)$$

From Eqs. 3.3 and 3.5, Eq. 3.6 is written

$$\ddot{q}_i + \sum_j b_{ij} q_j = 0 \qquad j = 1, 2, \cdots, 3N \qquad (3.7)$$

If $b_{ij} = 0$ for $i \neq j$, Eq. 3.7 becomes

$$\ddot{q}_i + b_{ii} q_i = 0 \qquad (3.8)$$

and the solution is given by

$$q_i = q_i^0 \sin\left(\sqrt{b_{ii}}\,t + \delta_i\right) \qquad (3.9)$$

where q_i^0 and δ_i are the amplitude and the phase constant, respectively.

Since, in general, this simplification is not applicable, the coordinates q_i must be transformed into a set of new coordinates Q_i through the relations

$$\begin{aligned} q_1 &= \sum_i B_{1i} Q_i \\ q_2 &= \sum_i B_{2i} Q_i \\ &\cdots\cdots\cdots \\ q_k &= \sum_i B_{ki} Q_i \end{aligned} \qquad (3.10)$$

The Q_i are called *normal coordinates* for the system. By appropriate choice of the coefficients B_{ki}, both the potential and the kinetic energies can be written

$$T = \tfrac{1}{2}\sum_i \dot{Q}_i^2 \qquad (3.11)$$

$$V = \tfrac{1}{2}\sum_i \lambda_i Q_i^2 \qquad (3.12)$$

without any cross products.

If Eqs. 3.11 and 3.12 are combined with Newton's equation (3.6), there results

$$\ddot{Q}_i + \lambda_i Q_i = 0 \qquad (3.13)$$

The solution of this equation is given by

$$Q_i = Q_i^0 \sin(\sqrt{\lambda_i} t + \delta_i) \tag{3.14}$$

and the frequency is

$$\nu_i = \frac{1}{2\pi}\sqrt{\lambda_i} \tag{3.15}$$

Such a vibration is called a *normal vibration.*

For the general N-atom molecule, it is obvious that the number of the normal vibrations is only $3N - 6$, since six coordinates are required to describe the translational and rotational motion of the molecule as a whole. Linear molecules have $3N - 5$ normal vibrations as no rotational freedom exists around the molecular axis. Thus the general form of the molecular vibration is a superposition of the $3N - 6$(or $3N - 5$) normal vibrations given by Eq. 3.14.

The physical meaning of the normal vibration may be demonstrated in the following way. As shown in Eq. 3.10, the original displacement coordinate is related to the normal coordinate by

$$q_k = \sum_i B_{ki} Q_i \tag{3.10}$$

Since all the normal vibrations are independent of each other, consideration may be limited to a special case in which only one normal vibration, subscripted by 1, is excited (i.e., $Q_1^0 \neq 0$, $Q_2^0 = Q_3^0 = \cdots = 0$). Then it follows from Eqs. 3.10 and 3.14 that

$$\begin{aligned} q_k = B_{k1} Q_1 &= B_{k1} Q_1^0 \sin(\sqrt{\lambda_1}\, t + \delta_1) \\ &= A_{k1} \sin(\sqrt{\lambda_1}\, t + \delta_1) \end{aligned} \tag{3.16}$$

This relation holds for all k. Thus it is seen that the excitation of one normal vibration of the system causes vibrations, given by Eq. 3.16, of all the nuclei in the system. In other words, in the normal vibration, all the nuclei move with the same frequency and in phase.

This is true for any other normal vibration. Thus Eq. 3.16 may be written in the more general form

$$q_k = A_k \sin(\sqrt{\lambda}\, t + \delta) \tag{3.17}$$

If Eq. 3.17 is combined with Eq. 3.7, there results

$$-\lambda A_k + \sum_j b_{kj} A_j = 0 \tag{3.18}$$

This is a system of first order simultaneous equations with respect to A. In order for all the A's to be nonzero,

$$\begin{vmatrix} b_{11}-\lambda & b_{12} & b_{13} & \cdots \\ b_{21} & b_{22}-\lambda & b_{23} & \cdots \\ b_{31} & b_{32} & b_{33}-\lambda & \cdots \\ \vdots & \vdots & \vdots & \end{vmatrix} = 0 \tag{3.19}$$

The order of this secular equation is equal to $3N$. Suppose one root, λ_1, is found for Eq. 3.19. If it is inserted in Eq. 3.18, A_{k1}, A_{k2}, $\cdots$ are obtained for all the nuclei. The same is true for the other roots of Eq. 3.19. Thus the most general solution may be written as a superposition of all the normal vibrations:

$$q_k = \sum_l B_{kl} Q_l^0 \sin(\sqrt{\lambda_l}\, t + \delta_l) \tag{3.20}$$

The general discussion developed above may be understood more easily if we apply it to a simple molecule such as CO_2, which is constrained to move only in one direction. If the mass and the displacement of each atom are defined as shown below,

m_1 m_2 m_3 x
Δx_1 Δx_2 Δx_3

the potential energy is given by

$$V = (\tfrac{1}{2})k\{(\Delta x_1 - \Delta x_2)^2 + (\Delta x_2 - \Delta x_3)^2\} \tag{3.21}$$

Considering that $m_1 = m_3$, we find that the kinetic energy is written as

$$T = (\tfrac{1}{2})m_1(\Delta\dot{x}_1{}^2 + \Delta\dot{x}_3{}^2) + (\tfrac{1}{2})m_2\, \Delta\dot{x}_2{}^2 \tag{3.22}$$

Using the generalized coordinates defined by Eq. 3.2, we may rewrite these energies as

$$2V = k\left\{\left(\frac{q_1}{\sqrt{m_1}} - \frac{q_2}{\sqrt{m_2}}\right)^2 + \left(\frac{q_2}{\sqrt{m_2}} - \frac{q_3}{\sqrt{m_1}}\right)^2\right\} \tag{3.23}$$

$$2T = \sum \dot{q}_i{}^2 \tag{3.24}$$

From the comparison of Eq. 3.23 with Eq. 3.5, we obtain

$$b_{11} = \frac{k}{m_1} \qquad b_{22} = \frac{2k}{m_2}$$

$$b_{12} = b_{21} = -\frac{k}{\sqrt{m_1 m_2}} \qquad b_{23} = b_{32} = -\frac{k}{\sqrt{m_1 m_2}}$$

$$b_{13} = b_{31} = 0 \qquad\qquad b_{33} = \frac{k}{m_1}$$

If these terms are inserted in Eq. 3.19, we obtain the following result

$$\begin{vmatrix} \dfrac{k}{m_1} - \lambda & -\dfrac{k}{\sqrt{m_1 m_2}} & 0 \\ -\dfrac{k}{\sqrt{m_1 m_2}} & \dfrac{2k}{m_2} - \lambda & -\dfrac{k}{\sqrt{m_1 m_2}} \\ 0 & -\dfrac{k}{\sqrt{m_1 m_2}} & \dfrac{k}{m_1} - \lambda \end{vmatrix} = 0 \tag{3.25}$$

By solving this secular equation, we obtain three roots:

$$\lambda_1 = \frac{k}{m_1}$$

$$\lambda_2 = k\mu \qquad \text{where } \mu = \frac{2m_1 + m_2}{m_1 m_2}$$

$$\lambda_3 = 0$$

Equation 3.18 gives the following three equations:

$$-\lambda A_1 + b_{11} A_1 + b_{12} A_2 + b_{13} A_3 = 0$$
$$-\lambda A_2 + b_{21} A_1 + b_{22} A_2 + b_{23} A_3 = 0$$
$$-\lambda A_3 + b_{31} A_1 + b_{32} A_2 + b_{33} A_3 = 0$$

Using Eq. 3.17, these are rewritten as

$$(b_{11} - \lambda) q_1 + b_{12} q_2 + b_{13} q_3 = 0$$
$$b_{21} q_1 + (b_{22} - \lambda) q_2 + b_{23} q_3 = 0$$
$$b_{31} q_1 + b_{32} q_2 + (b_{33} - \lambda) q_3 = 0$$

If $\lambda_1 = k/m_1$ is inserted in the above simultaneous equations, we obtain

$$q_1 = -q_3, \quad q_2 = 0$$

Similar calculations give

$$q_1 = q_3, \quad q_2 = -2\sqrt{\frac{m_1}{m_2}}\, q_1 \qquad \text{for } \lambda_2 = k\mu$$

$$q_1 = q_3, \quad q_2 = \sqrt{\frac{m_2}{m_1}}\, q_1 \qquad \text{for } \lambda_3 = 0$$

The relative displacements are depicted in the following figure:

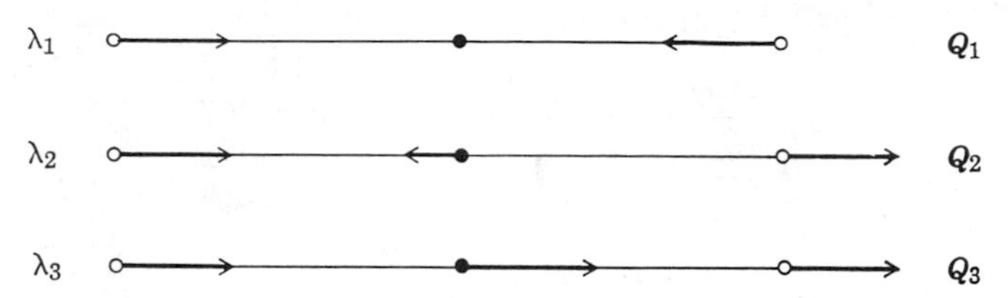

It is easy to see that λ_3 corresponds to the translational mode ($\Delta x_1 = \Delta x_2 = \Delta x_3$). The inclusion of λ_3 could be avoided if we consider the restriction that the center of gravity does not move; $m_1(\Delta x_1 + \Delta x_3) + m_2 \Delta x_2 = 0$.

The relationships between the generalized coordinates and the normal coordinates are given by Eq. 3.10. In the present case, it gives

$$q_1 = B_{11}Q_1 + B_{12}Q_2 + B_{13}Q_3$$

$$q_2 = B_{21}Q_1 + B_{22}Q_2 + B_{23}Q_3$$

$$q_3 = B_{31}Q_1 + B_{32}Q_2 + B_{33}Q_3$$

In the normal vibration whose normal coordinate is Q_1, $B_{11}:B_{21}:B_{31}$ gives the ratio of the displacements. From the previous calculation, it is obvious that $B_{11}:B_{21}:B_{31} = 1:0:-1$. Similarly, $B_{12}:B_{22}:B_{32} = 1:-2\sqrt{m_1/m_2}:1$ gives the ratio of the displacements in the normal vibration whose normal coordinate is Q_2. Thus the mode of a normal vibration can be drawn if the normal coordinate is translated into a set of rectangular coordinates, as is shown above.

So far, we have discussed only the vibrations whose displacements occur along the molecular axis. There are still, however, two other normal vibrations in which the displacements occur in the direction perpendicular to the molecular axis. They are not treated here, since the calculation is not simple. It is clear that the method described above will become more complicated as a molecule becomes larger. In this respect, the **GF** matrix method described in Sec. I-11 is important in the vibrational analysis of complex molecules.

By using the normal coordinates, the Schrödinger wave equation for the system can be written

$$\sum_i \frac{\partial^2 \psi_n}{\partial Q_i^2} + \frac{8\pi^2}{h^2}\left(E - \frac{1}{2}\sum_i \lambda_i Q_i^2\right)\psi_n = 0 \tag{3.26}$$

Since the normal coordinates are independent of each other, it is possible to write

$$\psi_n = \psi_1(Q_1) \cdot \psi_2(Q_2) \cdots \tag{3.27}$$

and solve the simpler one-dimensional problem.

If Eq. 3.27 is substituted in Eq. 3.26, there results

$$\frac{d^2\psi_i}{dQ_i^2} + \frac{8\pi^2}{h^2}\left(E_i - \frac{1}{2}\lambda_i Q_i^2\right)\psi_i = 0 \tag{3.28}$$

where

$$E = E_1 + E_2 + \cdots$$

with

$$E_i = h\nu_i\left(v_i + \frac{1}{2}\right)$$

$$\nu_i = \frac{1}{2\pi}\sqrt{\lambda_i} \tag{3.29}$$

I-4. SYMMETRY ELEMENTS AND POINT GROUPS[14-20]

As noted before, polyatomic molecules have $3N - 6$ or, if linear, $3N - 5$ normal vibrations. For any given molecule, however, only those vibrations that are permitted by the selection rule for that molecule appear in the infrared and Raman spectra. Since the selection rule is determined by the symmetry of the molecule, this must first be studied.

The spatial geometrical arrangement of the nuclei constituting the molecule determines the symmetry of the molecule. If a coordinate transformation (a reflection or a rotation or a combination of both) produces a configuration of the nuclei indistinguishable from the original one, this transformation is called a *symmetry operation*, and the molecule is said to have a corresponding *symmetry element*. Molecules may have the following symmetry elements.

(1) Identity, I

This is a symmetry element possessed by every molecule no matter how unsymmetrical it is, the corresponding operation being to leave the molecule unchanged. The inclusion of this element is necessitated by mathematical reasons which will be discussed in Sec. I-6.

(2) A Plane of Symmetry, σ

If reflection of a molecule with respect to some plane produces a configuration indistinguishable from the original one, the plane is called a plane of symmetry.

(3) A Center of Symmetry, i

If reflection at the center, that is, inversion, produces a configuration indistinguishable from the original one, the center is called a center of symmetry. This operation changes the signs of all the coordinates involved: $x_i \rightarrow -x_i$, $y_i \rightarrow -y_i$, $z_i \rightarrow -z_i$.

(4) A *p*-fold Axis of Symmetry, C_p

If rotation through an angle $360°/p$ about an axis produces a configuration indistinguishable from the original one, the axis is called a *p*-fold axis of symmetry, C_p. For example, a two-fold axis, C_2, implies that a rotation of 180° about the axis reproduces the original configuration. Molecules may have a two-, three-, four-, five-, or six-fold, or higher axis. Linear molecules have an infinite-fold (denoted by ∞-fold) axis of symmetry, C_∞, since a rotation of $360°/\infty$, that is, an infinitely small angle, transforms the molecule into one indistinguishable from the original.

(5) A *p*-fold Rotation-Reflection Axis, S_p

If rotation by $360°/p$ about the axis followed by reflection at a plane perpendicular to the axis produces a configuration indistinguishable from the original one, the axis is called a *p*-fold rotation-reflection axis. Molecules may have a two-, three-, four-, five-, or six-fold, or higher, rotation-reflection axis. Symmetrical linear molecules have an S_∞ axis. It is easily seen that the presence of S_p always means the presence of C_p as well as σ when p is odd.

A molecule may have more than one of these symmetry elements. Combination of more and more of these elements produces systems of higher and higher symmetry. Not all combinations of symmetry elements, however, are possible. For example, it is highly improbable that a molecule will have a C_3 and C_4 axis in the same direction because this requires the existence of a twelve-fold axis in the molecule. It should also be noted that the presence of some symmetry elements often implies the presence of other elements. For example, if a molecule has two σ planes at right angles to each other, the line of intersection of these two planes must be a C_2 axis. A possible combination of symmetry operations whose axes intersect at a point is called a *point group.** It can be shown theoretically[8] that only a limited number of such point groups exist.

I-5. SYMMETRY OF NORMAL VIBRATIONS AND SELECTION RULES

Figure I-5 indicates the normal modes of vibration in CO_2 and H_2O molecules. In each normal vibration, the individual nuclei carry out a simple harmonic motion in the direction indicated by the arrow, and all the nuclei

* In this respect, point groups differ from space groups, which involve translations and rotations about nonintersecting axes.

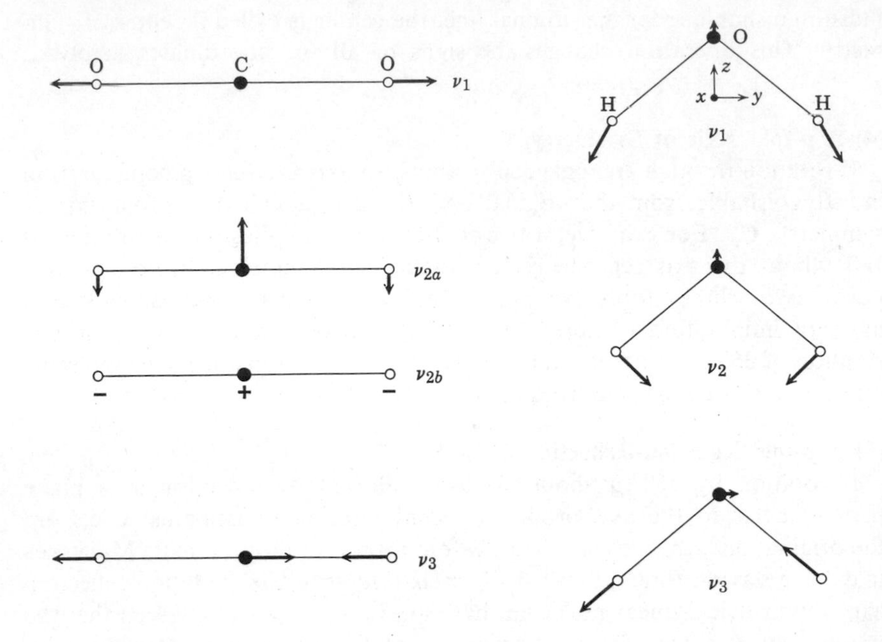

Fig. I-5. Normal modes of vibration in CO_2 and H_2O molecules (+ and − denote the vibrations going upward and downward, respectively, in the direction perpendicular to the paper plane).

have the same frequency of oscillation (i.e., the frequency of the normal vibration) and are moving in the same phase. Furthermore, the relative lengths of the arrows indicate the relative velocities and the amplitudes for each nucleus.* The ν_2 vibrations in CO_2 are worth comment, since they differ from the others in that two vibrations (ν_{2a} and ν_{2b}) have exactly the same frequency. Apparently, there are an infinite number of normal vibrations of this type, which differ only in their directions perpendicular to the molecular axis. Any of them, however, can be resolved into two vibrations such as ν_{2a} and ν_{2b}, which are perpendicular to each other. In this respect, the ν_2 vibrations in CO_2 are called *doubly degenerate vibrations*. Doubly degenerate vibrations occur only when a molecule has an axis higher than twofold. *Triply degenerate vibrations* also occur in molecules having more than one C_3 axis.

In order to determine the symmetry of a normal vibration, it is necessary to begin by considering the kinetic and potential energies of the system.

*In this respect, all the normal modes of vibration shown in this book are only approximate.

These have already been discussed in Sec. I–3.

$$T = \tfrac{1}{2}\sum_i \dot{Q}_i^2 \tag{5.1}$$

$$V = \tfrac{1}{2}\sum_i \lambda_i Q_i^2 \tag{5.2}$$

Consider a case in which a molecule performs only one normal vibration, Q_i. Then $T = \frac{1}{2}\dot{Q}_i^2$ and $V = \frac{1}{2}\lambda_i Q_i^2$. These energies must be invariant when a symmetry operation, R, changes Q_i to RQ_i. Thus,

$$T = \tfrac{1}{2}\dot{Q}_i^2 = \tfrac{1}{2}(R\dot{Q}_i)^2$$

$$V = \tfrac{1}{2}\lambda_i Q_i^2 = \tfrac{1}{2}\lambda_i(RQ_i)^2$$

In order that these relations hold, it is necessary that

$$(RQ_i)^2 = Q_i^2 \qquad \text{or} \qquad RQ_i = \pm Q_i \tag{5.3}$$

Thus the normal coordinate must change either into itself or its negative. If $Q_i = RQ_i$, the vibration is said to be *symmetric*. If $Q_i = -RQ_i$, it is said to be *antisymmetric*.

If the vibration is doubly degenerate, we have

$$T = \tfrac{1}{2}\dot{Q}_{ia}^2 + \tfrac{1}{2}\dot{Q}_{ib}^2$$

$$V = \tfrac{1}{2}\lambda_i(Q_{ia})^2 + \tfrac{1}{2}\lambda_i(Q_{ib})^2$$

In this case, a relation such as

$$(RQ_{ia})^2 + (RQ_{ib})^2 = Q_{ia}^2 + Q_{ib}^2 \tag{5.4}$$

must hold. As will be shown later, such a relationship is expressed more conveniently by using a matrix form:

$$R\begin{bmatrix} Q_{ia} \\ Q_{ib} \end{bmatrix} = \begin{bmatrix} A & B \\ C & D \end{bmatrix}\begin{bmatrix} Q_{ia} \\ Q_{ib} \end{bmatrix}$$

where the values of A, B, C, and D depend upon the symmetry operation, R. In any case, the normal vibration must either be symmetric or antisymmetric or degenerate for each symmetry operation.

The symmetry properties of the normal vibrations of the H_2O molecule shown in Fig. I-5 are classified as indicated in Table I-1. Here, +1 and −1

Table I-1

C_{2v}	I	$C_2(z)$	$\sigma_v(xz)$	$\sigma_v(yz)$
Q_1, Q_2	+1	+1	+1	+1
Q_3	+1	−1	−1	+1

σ_v: vertical plane of symmetry

denote symmetric and antisymmetric, respectively. In the ν_1 and ν_2 vibrations, all the symmetry properties are preserved during the vibration. Therefore they are *symmetric vibrations* and are called, in particular, *totally symmetric vibrations*. In the ν_3 vibration, however, symmetry elements such as C_2 and $\sigma_v(xz)$ are lost. Thus it is called a *nonsymmetric vibration*. If a molecule has a number of symmetry elements, the normal vibrations are classified as various species according to the number and the kind of symmetry elements preserved during the vibration.

In order to determine the activity of the vibrations in the infrared and Raman spectra, the selection rule must be applied to each normal vibration. From a quantum mechanical point of view,[7-13] *a vibration is active in the infrared spectrum if the dipole moment of the molecule is changed during the vibration, and is active in the Raman spectrum if the polarizability of the molecule is changed during the vibration.*

Since the concept of the polarizability is not as obvious as that of the dipole moment, the following brief description may be useful in discussing the Raman selection rules. When a molecule is placed in a static electric field, the nuclei are attracted toward the negative pole and the electrons toward the positive pole. The separation of charges caused by the external field gives the induced dipole moment. If E represents the strength of the electric field and P denotes the induced dipole moment, the relation

$$P = \alpha E \tag{5.5}$$

holds. Here, α is a proportionality constant called the polarizability. If we resolve P, α, and E in the x, y, and z directions, simple relationships such as

$$P_x = \alpha_x E_x,\ P_y = \alpha_y E_y,\ \text{and}\ P_z = \alpha_z E_z \tag{5.6}$$

are anticipated. In most molecules, however, Eq. 5.6 does not hold, since the direction of polarization does not coincide with the direction of the applied field. This is because the direction of chemical bonds in the molecule also affects the direction of polarization. Thus, instead of Eq. 5.6, we have the relationships:

$$\begin{aligned} P_x &= \alpha_{xx} E_x + \alpha_{xy} E_y + \alpha_{xz} E_z \\ P_y &= \alpha_{yx} E_x + \alpha_{yy} E_y + \alpha_{yz} E_z \\ P_z &= \alpha_{zx} E_x + \alpha_{zy} E_y + \alpha_{zz} E_z \end{aligned} \tag{5.7}$$

In a matrix form, Eq. 5.7 is written:

$$\begin{bmatrix} P_x \\ P_y \\ P_z \end{bmatrix} = \begin{bmatrix} \alpha_{xx} & \alpha_{xy} & \alpha_{xz} \\ \alpha_{yx} & \alpha_{yy} & \alpha_{yz} \\ \alpha_{zx} & \alpha_{zy} & \alpha_{zz} \end{bmatrix} \begin{bmatrix} E_x \\ E_y \\ E_z \end{bmatrix} \tag{5.8}$$

and the first matrix on the right-hand side is called the polarizability tensor.

It is a symmetric tensor; $\alpha_{xy} = \alpha_{yx}$, $\alpha_{yz} = \alpha_{zy}$ and $\alpha_{xz} = \alpha_{zx}$.

According to quantum mechanics, the vibration is Raman active if one of these six components of the polarizability changes during the vibration. Similarly, it is infrared active if one of the three components of the dipole moment (μ_x, μ_y, and μ_z) changes during the vibration. Changes in dipole moment or polarizability are not obvious from the inspection of normal modes of vibration in most polyatomic molecules. As will be shown later, application of group theory gives a clear-cut solution to this problem.

In simple molecules, however, the activity of a vibration may be determined by inspection of the normal mode. For example, it is obvious that the vibration in a homopolar diatomic molecule is not infrared active but is Raman active, whereas the vibration in a heteropolar diatomic molecule is both infrared and Raman active. It is also obvious that all the three vibrations of H_2O and ν_2 and ν_3 of CO_2 are infrared active. Except ν_1 of CO_2, the Raman activity is not easy to predict even for such simple molecules.

As stated before, the polarizability tensor is symmetric, and can be converted into a diagonal tensor if we choose a set of new axes, x', y', and z' that are perpendicular to each other:

$$\begin{bmatrix} P_{x'} \\ P_{y'} \\ P_{z'} \end{bmatrix} = \begin{bmatrix} \alpha_{x'x'} & 0 & 0 \\ 0 & \alpha_{y'y'} & 0 \\ 0 & 0 & \alpha_{z'z'} \end{bmatrix} \begin{bmatrix} E_{x'} \\ E_{y'} \\ E_{z'} \end{bmatrix} \tag{5.9}$$

These three axes are called the principal axes of polarizability. We can represent the polarizability by plotting $1/\sqrt{\alpha_i}$ in any direction from the origin. This gives a three-dimensional surface called the polarizability ellipsoid such as that shown in Fig. I-6. In terms of the polarizability ellipsoid, the Raman selection rule can be stated as follows: The vibration is Raman active if the ellipsoid changes in size, shape or orientation during the vibration. It is readily seen that this condition is satisfied in the ν_1 vibration of CO_2 and the three vibrations of H_2O.

The above definition of Raman activity is not complete, however. The vibration may be Raman inactive if the shape and size of the polarizability ellipsoid are identical before and after the nuclei pass through their equilibrium positions by equal distances. For example, Fig. I-6 shows that the size and shape of the ellipsoid are identical in two extreme positions in the ν_3 vibration of CO_2. If we consider a limiting case where the nuclei perform small displacements, there is effectively no change in the polarizability. Therefore the ν_3 vibration of CO_2 is not Raman active. The same is true for the ν_2 vibration.

It should be noted that in CO_2 the vibration symmetric with respect to the center of symmetry (ν_1) is Raman active and not infrared active, whereas the vibrations antisymmetric with respect to the center of symmetry (ν_2 and ν_3) are infrared active but not Raman active. In a polyatomic molecule having

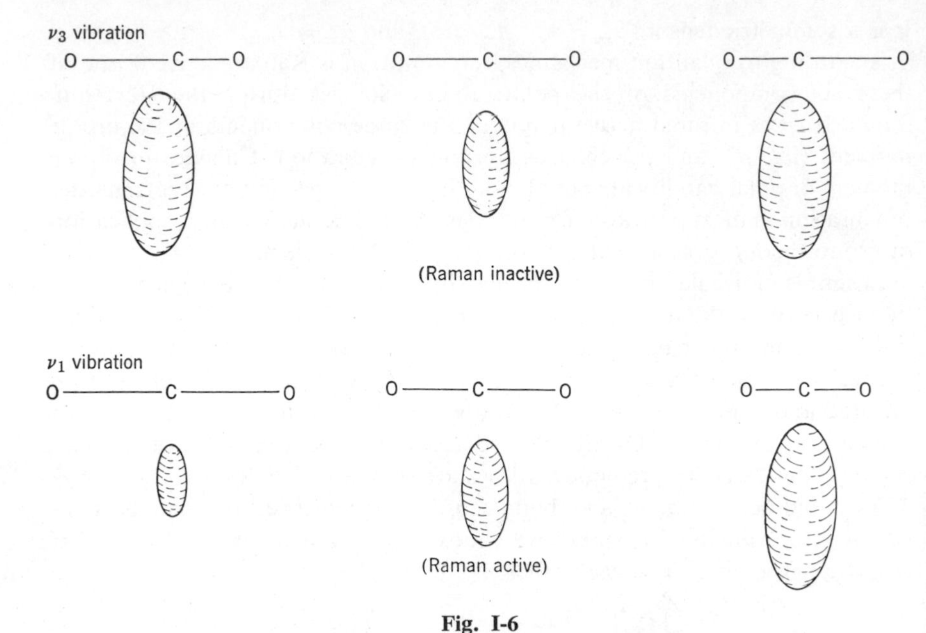

Fig. I-6

a center of symmetry, the vibrations symmetric with respect to the center of symmetry (g vibrations*) are Raman active and not infrared active, but the vibrations antisymmetric with respect to the center of symmetry (u vibrations*) are infrared active and not Raman active. This rule is called the *mutual exclusion rule.* It should be noted, however, that in polyatomic molecules having several symmetry elements besides the center of symmetry, the vibrations that should be active according to this rule may not necessarily be active, because of the presence of other symmetry elements. An example is seen in a square-planar XY_4 type of molecule of $\mathbf{D}_{4h}$ symmetry, where the A_{2g} vibrations are not Raman active and the A_{1u}, B_{1u}, and B_{2u} vibrations are not infrared active (see Sec. II-8(3) and Appendix I).

I-6. INTRODUCTION TO GROUP THEORY

In Sec. I-4, the symmetry and the point group allocation of a given molecule were discussed. In order to understand the symmetry and selection rules of normal vibrations in polyatomic molecules, however, a knowledge of group theory is required. The minimum amount of group theory needed for this purpose is given here†.

* The symbols g and u stand for "gerade" and "ungerade" in German, respectively.
† For details on group theory and matrix theory, see Refs. 15–20*a*.

Consider a pyramidal XY_3 molecule (Fig. I-7) for which the symmetry operations, I, C_3^+, C_3^-, σ_1, σ_2, and σ_3 are applicable. Here C_3^+ and C_3^- denote rotation through 120° in the clockwise and counterclockwise directions, respectively, and σ_1, σ_2, and σ_3 indicate the symmetry planes that pass through

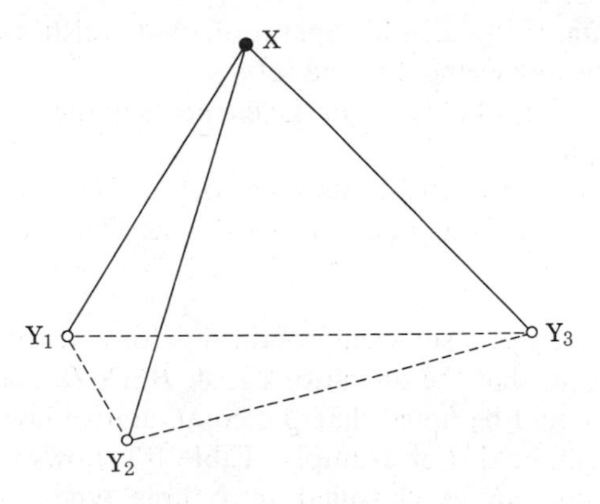

Fig. I-7

X and Y_1, X and Y_2, and X and Y_3, respectively. For simplicity, let these symmetry operations be denoted by I, A, B, C, D, and E, respectively. Other symmetry operations are possible, but they are all equivalent to some one of the operations mentioned. For instance, a clockwise rotation through 240° is identical with operation B. It may also be shown that two successive applications of any one of these operations is equivalent to some single operation of the group mentioned. Let operation C be applied to the original figure. This interchanges Y_2 and Y_3. If operation A is applied to the resulting figure, the net result is the same as application of the single operation D to the original figure. This is written $AC = D$. If all the possible multiplicative combinations are made, Table I-2 is obtained, in which the operation applied first is written across the top of the table.
This is called the *multiplication table* of the group.

TABLE I-2

	I	*A*	*B*	*C*	*D*	*E*
I	*I*	*A*	*B*	*C*	*D*	*E*
A	*A*	*B*	*I*	*D*	*E*	*C*
B	*B*	*I*	*A*	*E*	*C*	*D*
C	*C*	*E*	*D*	*I*	*B*	*A*
D	*D*	*C*	*E*	*A*	*I*	*B*
E	*E*	*D*	*C*	*B*	*A*	*I*

It is seen that a group consisting of the mathematical elements (symmetry operations) I, A, B, C, D, and E satisfies the following conditions:

1. The product of any two elements in the set is another element in the set.
2. The set contains the identity operation that satisfies the relation $IP = PI = P$, where P is any element in the set.
3. The associative law holds for all the elements in the set; that is, $(CB)A = C(BA)$, for example.
4. Every element in the set has its reciprocal X that satisfies the relation, $XP = PX = I$, where P is any element in the set. This reciprocal is usually denoted by P^{-1}.

These are necessary and sufficient conditions for a set of elements to form a *group*. It is evident that the operations I, A, B, C, D, and E form a group in this sense. It should be noted that the commutative law of multiplication does not necessarily hold. For example, Table I-2 shows that $CD \neq DC$.

The six elements can be classified into three types of operations: the identity operation, I; the rotations, C_3^+ and C_3^-; and the reflections, σ_1, σ_2, and σ_3. Each of these sets of operations is said to form a *class*. More precisely, two operations, P and Q, which satisfy the relation $X^{-1}PX = P$ or Q, where X is any operation of the group and X^{-1} is its reciprocal, are said to belong to the same class. It can easily be shown that C_3^+ and C_3^-, for example, satisfy the relation. Thus the six elements of the point group $\mathbf{C}_{3v}$ are usually abbreviated I, $2C_3$, and $3\sigma_v$.

The relations between the elements of the group are shown in the multiplication table in Table I-2. Such a tabulation of a group is, however, awkward to handle. The essential features of the table may be abstracted by replacing the elements by some analytical function that reproduces the multiplication table. Such an analytical expression may be composed of a simple integer, an exponential function, or a matrix. Any set of such expressions that satisfies the relations given by the multiplication table is called a *representation* of the group and is designated by Γ. The representations of the point group $\mathbf{C}_{3v}$ discussed above are indicated in Table I-3. It is easily proved that each representation in the table satisfies the multiplication table.

TABLE I-3

$\mathbf{C}_{3v}$	I	A	B	C	D	E
$A_1(\Gamma_1)$	1	1	1	1	1	1
$A_2(\Gamma_2)$	1	1	1	-1	-1	-1
$E(\Gamma_3)$	$\begin{pmatrix} 1 & 0 \\ 0 & 1 \end{pmatrix}$	$\begin{pmatrix} -\frac{1}{2} & \frac{\sqrt{3}}{2} \\ -\frac{\sqrt{3}}{2} & -\frac{1}{2} \end{pmatrix}$	$\begin{pmatrix} -\frac{1}{2} & -\frac{\sqrt{3}}{2} \\ \frac{\sqrt{3}}{2} & -\frac{1}{2} \end{pmatrix}$	$\begin{pmatrix} -1 & 0 \\ 0 & 1 \end{pmatrix}$	$\begin{pmatrix} \frac{1}{2} & -\frac{\sqrt{3}}{2} \\ -\frac{\sqrt{3}}{2} & -\frac{1}{2} \end{pmatrix}$	$\begin{pmatrix} \frac{1}{2} & \frac{\sqrt{3}}{2} \\ \frac{\sqrt{3}}{2} & -\frac{1}{2} \end{pmatrix}$

Beside the three representations in Table I-3, it is possible to write an infinite number of other representations of the group. If a set of six matrices of the type $S^{-1}R(K)S$ is chosen where $R(K)$ is a representation of the element K given in Table I-3, $S(|S| \neq 0)$ is any matrix of the same order as R, and S^{-1} is the reciprocal of S, this set also satisfies the relations given by the multiplication table. The reason is obvious from the relation

$$S^{-1}R(K)SS^{-1}R(L)S = S^{-1}R(K)R(L)S = S^{-1}R(KL)S$$

Such a transformation is called a *similarity transformation.* Thus it is possible to make an infinite number of representations by similarity transformations.

On the other hand, this statement suggests that a given representation may be broken into simpler ones. If each representation of the symmetry element K is transformed into the form

$$R(K) = \begin{vmatrix} Q_1(K) & 0 & 0 & 0 \\ 0 & Q_2(K) & 0 & 0 \\ 0 & 0 & Q_3(K) & 0 \\ 0 & 0 & 0 & Q_3(K) \end{vmatrix} \tag{6.1}$$

by a similarity transformation, $Q_1(K)$, $Q_2(K)$, $\cdots$ are simpler representations. In such a case, $R(K)$ is called *reducible.* If a representation cannot be simplified any further, it is said to be *irreducible.* The representations Γ_1, Γ_2, and Γ_3 in Table I-3 are all irreducible representations. It can be shown generally that the number of irreducible representations is equal to the number of classes. Thus only three irreducible representations exist for the point group $\mathbf{C}_{3v}$. These representations are entirely independent of each other. Furthermore, the sum of the squares of the dimensions (l) of the irreducible representations of a group is always equal to the total number of the symmetry elements, namely, the *order of the group* (h). Thus

$$\sum l_i^2 = l_1^2 + l_2^2 + \cdots = h \tag{6.2}$$

In the point group $\mathbf{C}_{3v}$, it is seen that

$$1^2 + 1^2 + 2^2 = 6$$

A point group is classified into *species* according to its irreducible representations. In the point group $\mathbf{C}_{3v}$, the species having the irreducible representations Γ_1, Γ_2, and Γ_3 are called the A_1, A_2, and E species, respectively.*

The sum of the diagonal elements of a matrix is called the *character* of the matrix and is denoted by χ. It is to be noted in Table I-3 that the character of each of the elements belonging to the same class is the same. Thus, using

* For the labeling of the irreducible representations (species), see Appendix I.

the character, Table I-3 can be simplified to Table I-4. Such a table is called

TABLE I-4. THE CHARACTER TABLE OF THE POINT GROUP $\mathbf{C}_{3v}$

$\mathbf{C}_{3v}$	I	$2C_3(z)$	$3\sigma_v$
$A_1(\chi_1)$	1	1	1
$A_2(\chi_2)$	1	1	−1
$E(\chi_3)$	2	−1	0

the *character table* of the point group $\mathbf{C}_{3v}$. The *character* of a matrix is not changed by a similarity transformation. This can be proved as follows. If a similarity transformation is expressed by $T = S^{-1}RS$, then

$$\chi_T = \sum_i (S^{-1}RS)_{ii} = \sum_{i,j,k} (S^{-1})_{ij} R_{jk} S_{ki} = \sum_{j,k,i} S_{ki}(S^{-1})_{ij} R_{jk}$$
$$= \sum_{j,k} \delta_{kj} R_{jk} = \sum_k R_{kk} = \chi_R$$

where δ_{kj} is Kronecker's delta (zero for $k \neq j$ and 1 for $k = j$). Thus any reducible representation can be reduced to its irreducible representations by a similarity transformation that leaves the character unchanged. Therefore the character of the reducible representation, $\chi(K)$, is written

$$\chi(K) = \sum_m a_m \chi_m(K) \tag{6.3}$$

where $\chi_m(K)$ is the character of $Q_m(K)$, and a_m is a positive integer that indicates the number of times $Q_m(K)$ appears in the matrix of Eq. 6.1. Hereafter the character will be used rather than the corresponding representation because a 1 : 1 correspondence exists between these two, and the former is sufficient for vibrational problems.

It is important to note that the following relation holds in Table I-4:

$$\sum_K \chi_i(K)\chi_j(K) = h\delta_{ij} \tag{6.4}$$

If Eq. 6.3 is multiplied by $\chi_i(K)$ on both sides, and the summation is taken over all the symmetry operations, then

$$\sum_K \chi(K)\chi_i(K) = \sum_K \sum_m a_m \chi_m(K)\chi_i(K)$$
$$= \sum_m \sum_K a_m \chi_m(K)\chi_i(K)$$

For a fixed m, we have

$$\sum_K a_m \chi_m(K)\chi_i(K) = a_m \sum_K \chi_m(K)\chi_i(K) = a_m h\delta_{im}$$

If we consider the sum of such a term over m, only the sum in which $m = i$ remains. Thus

$$\sum_k \chi(K)\, \chi_m(K) = h a_m$$

or

$$a_m = \frac{1}{h} \sum_K \chi(K) \chi_m(K) \tag{6.5}$$

This formula is written more conveniently as

$$\boxed{a_m = \frac{1}{h} \sum n \chi(K) \chi_m(K)} \tag{6.6}$$

where n is the number of symmetry elements in any one class, and the summation is made over the different classes. As Sec. I-7 will show, this formula is very useful in determining the number of normal vibrations belonging to each species.

I-7. THE NUMBER OF NORMAL VIBRATIONS FOR EACH SPECIES

As shown in Sec. I-5, the $3N - 6$ (or $3N - 5$) normal vibrations of an N-atom molecule can be classified into various species according to their symmetry properties. Using group theory, it is possible to find the number of normal vibrations belonging to each species.* The principle of the method is that all the representations are irreducible if normal coordinates are used as the basis for the representations. For example, the representations for the symmetry operations based on three normal coordinates, Q_1, Q_2, and Q_3, which correspond to the ν_1, ν_2, and ν_3 vibrations in the H_2O molecule of Fig. I-5 are

$$I \begin{bmatrix} Q_1 \\ Q_2 \\ Q_3 \end{bmatrix} = \begin{bmatrix} 1 & 0 & 0 \\ 0 & 1 & 0 \\ 0 & 0 & 1 \end{bmatrix} \begin{bmatrix} Q_1 \\ Q_2 \\ Q_3 \end{bmatrix} \qquad C_2(z) \begin{bmatrix} Q_1 \\ Q_2 \\ Q_3 \end{bmatrix} = \begin{bmatrix} 1 & 0 & 0 \\ 0 & 1 & 0 \\ 0 & 0 & -1 \end{bmatrix} \begin{bmatrix} Q_1 \\ Q_2 \\ Q_3 \end{bmatrix}$$

$$\sigma_v(xz) \begin{bmatrix} Q_1 \\ Q_2 \\ Q_3 \end{bmatrix} = \begin{bmatrix} 1 & 0 & 0 \\ 0 & 1 & 0 \\ 0 & 0 & -1 \end{bmatrix} \begin{bmatrix} Q_1 \\ Q_2 \\ Q_3 \end{bmatrix} \qquad \sigma_v(yz) \begin{bmatrix} Q_1 \\ Q_2 \\ Q_3 \end{bmatrix} = \begin{bmatrix} 1 & 0 & 0 \\ 0 & 1 & 0 \\ 0 & 0 & 1 \end{bmatrix} \begin{bmatrix} Q_1 \\ Q_2 \\ Q_3 \end{bmatrix}$$

Let a representation be written with the $3N$ rectangular coordinates of an N-atom molecule as its basis. If it is decomposed into its irreducible components, the basis for these irreducible representations must be the normal

* The results of the following calculations are tabulated in Appendix II.

coordinates, and the number of appearances of the same irreducible representation must be equal to the number of normal vibrations belonging to the species represented by this irreducible representation. As stated previously, however, the $3N$ rectangular coordinates involve 6 (or 5) coordinates, which correspond to the translational and rotational motions of the molecule as a whole. Therefore the representation that have such coordinates as their basis must be subtracted from the result obtained above. Use of the character of the representation, rather than the representation itself, yields the same result.

For example, consider a pyramidal XY_3 molecule that has six normal vibrations. At first, the representations for the various symmetry operations must be written with the 12 rectangular coordinates in Fig. I-8 as their basis. Consider pure rotation, C_p^+. If the clockwise rotation of the point (x, y, z) around the z axis by the angle θ brings it to the point denoted by the coordinates (x', y', z'), the relations between these two coordinates are given by

$$\begin{aligned} x' &= x \cos\theta + y \sin\theta \\ y' &= -x \sin\theta + y \cos\theta \\ z' &= z \end{aligned} \tag{7.1}$$

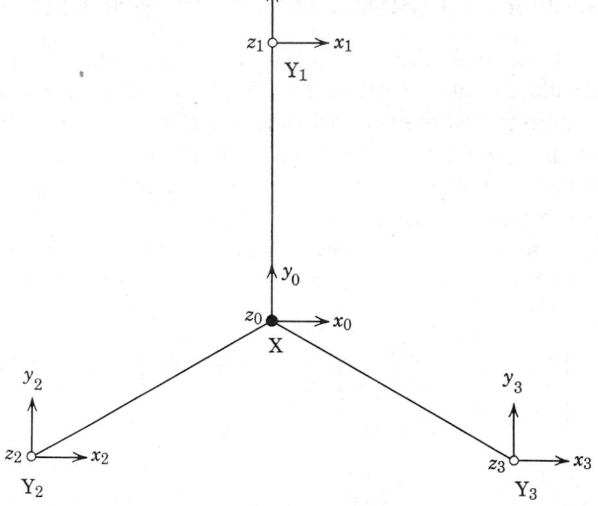

Fig. I-8 Rectangular coordinates in a pyramidal XY_3 molecule (z axis is perpendicular to the paper plane).

By using matrix notation, this can be written

$$\begin{bmatrix} x' \\ y' \\ z' \end{bmatrix} = C_\theta^+ \begin{bmatrix} x \\ y \\ z \end{bmatrix} = \begin{bmatrix} \cos\theta & \sin\theta & 0 \\ -\sin\theta & \cos\theta & 0 \\ 0 & 0 & 1 \end{bmatrix} \begin{bmatrix} x \\ y \\ z \end{bmatrix} \tag{7.2}$$

Then the character of the matrix is given by

$$\chi(C_\theta^+) = 1 + 2\cos\theta \tag{7.3}$$

The same result is obtained for $\chi(C_\theta^-)$. If this symmetry operation is applied to all the coordinates of the XY_3 molecule, the result is

$$C_\theta \begin{bmatrix} x_0 \\ y_0 \\ z_0 \\ x_1 \\ y_1 \\ z_1 \\ x_2 \\ y_2 \\ z_2 \\ x_3 \\ y_3 \\ z_3 \end{bmatrix} = \left[\begin{array}{c|c|c|c} \mathbf{A} & 0 & 0 & 0 \\ \hline 0 & 0 & 0 & \mathbf{A} \\ \hline 0 & \mathbf{A} & 0 & 0 \\ \hline 0 & 0 & \mathbf{A} & 0 \end{array}\right] \begin{bmatrix} x_0 \\ y_0 \\ z_0 \\ x_1 \\ y_1 \\ z_1 \\ x_2 \\ y_2 \\ z_2 \\ x_3 \\ y_3 \\ z_3 \end{bmatrix} \tag{7.4}$$

where **A** denotes the small square matrix given by Eq. 7.2. Thus the character of this representation is simply given by Eq. 7.3. It should be noted in Eq. 7.4 that only the small matrix **A**, related to the nuclei unchanged by the symmetry operation, appears as a diagonal element. Thus a more general form of the character of the representation for rotation around the axis by θ is

$$\boxed{\chi(R) = N_R(1 + 2\cos\theta)} \tag{7.5}$$

where N_R is the number of nuclei unchanged by the rotation. In the present case, $N_R = 1$ and $\theta = 120°$. Therefore

$$\chi(C_3) = 0 \tag{7.6}$$

Identity (I) can be regarded as a special case of Eq. 7.5 in which $N_R = 4$ and $\theta = 0°$. The character of the representation is

$$\chi(I) = 12 \tag{7.7}$$

Pure rotation and identity are called *proper rotation.*

It is evident from Fig. I-8 that a symmetry plane such as σ_1 changes the coordinates from (x_i, y_i, z_i) to $(-x_i, y_i, z_i)$. The corresponding representation is therefore written

$$\sigma_1 \begin{bmatrix} x \\ y \\ z \end{bmatrix} = \begin{bmatrix} -1 & 0 & 0 \\ 0 & 1 & 0 \\ 0 & 0 & 1 \end{bmatrix} \begin{bmatrix} x \\ y \\ z \end{bmatrix} \tag{7.8}$$

The result of such an operation on all the coordinates is

$$\sigma_1 \begin{bmatrix} x_0 \\ y_0 \\ z_0 \\ x_1 \\ y_1 \\ z_1 \\ x_2 \\ y_2 \\ z_2 \\ x_3 \\ y_3 \\ z_3 \end{bmatrix} = \left[\begin{array}{c|c|c|c} \mathbf{B} & 0 & 0 & 0 \\ \hline 0 & \mathbf{B} & 0 & 0 \\ \hline 0 & 0 & 0 & \mathbf{B} \\ \hline 0 & 0 & \mathbf{B} & 0 \end{array}\right] \begin{bmatrix} x_0 \\ y_0 \\ z_0 \\ x_1 \\ y_1 \\ z_1 \\ x_2 \\ y_2 \\ z_2 \\ x_3 \\ y_3 \\ z_3 \end{bmatrix}$$

where **B** denotes the small square matrix of Eq. 7.8. Thus the character of this representation is calculated as $2 \times 1 = 2$. It is noted again that the matrix on the diagonal is nonzero only for the nuclei unchanged by the operation.

More generally, a reflection at a plane (σ) is regarded as $\sigma = i \times C_2$. Thus the general form of Eq. 7.8 may be written

$$\begin{bmatrix} -1 & 0 & 0 \\ 0 & -1 & 0 \\ 0 & 0 & -1 \end{bmatrix} \begin{bmatrix} \cos\theta & \sin\theta & 0 \\ -\sin\theta & \cos\theta & 0 \\ 0 & 0 & 1 \end{bmatrix} = \begin{bmatrix} -\cos\theta & -\sin\theta & 0 \\ \sin\theta & -\cos\theta & 0 \\ 0 & 0 & -1 \end{bmatrix}$$

Then,

$$\chi(\sigma) = -(1 + 2\cos\theta)$$

As a result, the character of the large matrix shown in Eq. 7.9 is given by

$$\boxed{\chi(R) = -N_R(1 + 2\cos\theta)} \tag{7.10}$$

In the present case, $N_R = 2$ and $\theta = 180°$. This gives

$$\chi(\sigma_v) = 2 \tag{7.11}$$

Symmetry operations such as i and S_p are regarded as

$$\begin{aligned} i &= i \times I & \theta &= 0° \\ S_3 &= i \times C_6 & \theta &= 60° \\ S_4 &= i \times C_4 & \theta &= 90° \\ S_6 &= i \times C_3 & \theta &= 120° \end{aligned}$$

Therefore, the characters of these symmetry operations can be calculated by Eq. 7.10 with the values of θ defined above. Operations such as σ, i, and S_p are called *improper rotations*. Thus the character of the representation based on 12 rectangular coordinates is

I	$2C_3$	$3\sigma_v$
12	0	2

(7.12)

In order to determine the number of normal vibrations belonging to each species, the $\chi(R)$ thus obtained must be resolved into the $\chi_i(R)$ of the irreducible representations of each species in Table I-4. First, however, the characters corresponding to the translational and rotational motions of the molecule must be subtracted from the result shown in Eq. 7.12.

The characters for the translational motion of the molecule in the x, y, and z directions (denoted by T_x, T_y, and T_z) are the same as those obtained in Eqs. 7.5 and 7.10. They are

$$\boxed{\chi_t(R) = \pm(1 + 2 \cos \theta)} \tag{7.13}$$

where the + and − signs are for proper and improper rotations, respectively. The characters for the rotations around the x, y, and z axes (denoted by R_x, R_y, and R_z) are given by

$$\boxed{\chi_r(R) = +(1 + 2 \cos \theta)} \tag{7.14}$$

for both proper and improper rotations. This is due to the fact that a rotation of the vectors in the plane perpendicular to the x, y, and z axes can be regarded as a rotation of the components of angular momentum, M_x, M_y, and M_z, about the given axes. If p_x, p_y, and p_z are the components of linear momentum in the x, y, and z directions, the following relations hold:

$$M_x = yp_z - zp_y$$

$$M_y = zp_x - xp_z$$

$$M_z = xp_y - yp_x$$

Since (x, y, z) and (p_x, p_y, p_z) transform as shown in Eq. 7.2, it follows that

$$C_\theta \begin{bmatrix} M_x \\ M_y \\ M_z \end{bmatrix} = \begin{bmatrix} \cos\theta & \sin\theta & 0 \\ -\sin\theta & \cos\theta & 0 \\ 0 & \theta & 1 \end{bmatrix} \begin{bmatrix} M_x \\ M_y \\ M_z \end{bmatrix}$$

Then a similar relation holds for R_x, R_y, and R_z:

$$C_\theta \begin{bmatrix} R_x \\ R_y \\ R_z \end{bmatrix} = \begin{bmatrix} \cos\theta & \sin\theta & 0 \\ -\sin\theta & \cos\theta & 0 \\ 0 & 0 & 1 \end{bmatrix} \begin{bmatrix} R_x \\ R_y \\ R_z \end{bmatrix}$$

Thus the characters for the proper rotations are given by Eq. 7.14. The same result is obtained for the improper rotation if the latter is regarded as $i \times$ (proper rotation). Therefore the character for the vibration is obtained from

$$\boxed{\chi_v(R) = \chi(R) - \chi_t(R) - \chi_r(R)} \tag{7.15}$$

It is convenient to tabulate the foregoing calculations as in Table I-5.

TABLE I-5

Symmetry operation	I	$2C_3$	$3\sigma_v$
Kind of rotation	proper		improper
θ	0°	120°	180°
$\cos\theta$	1	$-\frac{1}{2}$	-1
$1+2\cos\theta$	3	0	-1
N_R	4	1	2
χ, $\pm N_R(1+2\cos\theta)$	12	0	2
χ_t, $\pm(1+2\cos\theta)$	3	0	1
χ_r, $+(1+2\cos\theta)$	3	0	-1
χ_v, $\chi-\chi_t-\chi_r$	6	0	2

By using the formula in Eq. 6.6 and the character of the irreducible representations in Table I-4, a_m can be calculated as follows:

$$a_m(A_1) = \tfrac{1}{6}[(1)(6)(1) + (2)(0)(1) \quad + (3)(2)(1)] \quad = 2$$

$$a_m(A_2) = \tfrac{1}{6}[(1)(6)(1) + (2)(0)(1) \quad + (3)(2)(-1)] = 0$$

$$a_m(E) = \tfrac{1}{6}[(1)(6)(2) + (2)(0)(-1) + (3)(2)(0)] \quad = 2$$

and

$$\chi_v = 2\chi_{A_1} + 2\chi_E \tag{7.16}$$

In other words, the six normal vibrations of a pyramidal XY_3 molecule are classified into two A_1 and two E species.

This procedure is applicable to any molecule. As another example, a similar calculation is shown in Table I-6 for an octahedral XY_6 molecule. By use of Eq. 6.6 and the character table in Appendix I, the a_m are obtained as

$$a_m(A_{1g}) = \tfrac{1}{48}\,[(1)(15)(1) + (8)(0)(1) + (6)(1)(1) + (6)(1)(1) + (3)(-1)(1) + (1)(-3)(1) + (6)(-1)(1) + (8)(0)(1) + (3)(5)(1) + (6)(3)(1)]$$
$$= 1$$

$$a_m(A_{1u}) = \tfrac{1}{48}\,[(1)(15)(1) + (8)(0)(1) + (6)(1)(1) + (6)(1)(1) + (3)(-1)(1) + (1)(-3)(-1) + (6)(-1)(-1) + (8)(0)(-1) + (3)(5)(-1) + (6)(3)(-1)]$$
$$= 0$$

....

and therefore

$$\chi_v = \chi_{A_{1g}} + \chi_{E_g} + 2\chi_{F_{1u}} + \chi_{F_{2g}} + \chi_{F_{2u}}$$

TABLE I-6

Symmetry operation	I	$8C_3$	$6C_2$	$6C_4$	$3C_4^2 \equiv C''_2$	$S_2 \equiv i$	$6S_4$	$8S_6 \equiv C_3 i$	$3\sigma_h$	$6\sigma_d$
Kind of rotation	proper					improper				
θ	0°	120°	180°	90°	−1	0°	90°	120°	180°	180°
$\cos\theta$	1	$-\frac{1}{2}$	−1	0	−1	1	0	$-\frac{1}{2}$	−1	−1
$1 + 2\cos\theta$	3	0	−1	1	−1	3	1	0	−1	−1
N_R	7	1	1	3	3	1	1	1	5	3
χ, $\pm N_R(1 + 2\cos\theta)$	21	0	−1	3	−3	−3	−1	0	5	3
χ_t, $\pm(1 + 2\cos\theta)$	3	0	−1	1	−1	−3	−1	0	1	1
χ_r, $+(1 + 2\cos\theta)$	3	0	−1	1	−1	3	1	0	−1	−1
χ_v, $\chi - \chi_t - \chi_v$	15	0	1	1	−1	−3	−1	0	5	3

σ_h = horizontal plane of symmetry; σ_d = diagonal plane of symmetry

I-8. INTERNAL COORDINATES

In Sec. I-3, the potential and the kinetic energies were expressed in terms of rectangular coordinates. If these energies are expressed in terms of *internal coordinates* such as increments of the bond length and bond angle, the corresponding force constants have a clearer physical meaning than those expressed in terms of rectangular coordinates, since these force constants are characteristic of the bond stretching and the angle deformation involved. The number of internal coordinates must be equal to, or greater than, $3N - 6$ (or $3N - 5$), the degrees of vibrational freedom of an N-atom molecule. If more than

$3N - 6$ (or $3N - 5$) coordinates are selected as the internal coordinates, this means that these coordinates are not independent of each other. Figure I-9 illustrates the internal coordinates for various types of molecules.

In linear XYZ (*a*), bent XY_2 (*b*), and pyramidal XY_3 (*c*) molecules, the number of internal coordinates is the same as the number of normal vibrations. In a nonplanar X_2Y_2 molecule (*d*) such as H_2O_2, the number of internal coordinates is the same as the number of vibrations if the twisting angle around the central bond ($\Delta\tau$) is considered. In a tetrahedral XY_4 molecule (*e*), however, the number of internal coordinates exceeds the number of normal vibrations by one. This is due to the fact that the six angle coordinates around the central atom are not independent of each other. That is, they must satisfy the relation

$$\Delta\alpha_{12} + \Delta\alpha_{23} + \Delta\alpha_{31} + \Delta\alpha_{41} + \Delta\alpha_{42} + \Delta\alpha_{43} = 0 \tag{8.1}$$

This is called a *redundant condition*. In a planar XY_3 molecule (*f*), the number of internal coordinates is seven when the coordinate, $\Delta\theta$, which represents the deviation from planarity is considered. Since the number of vibrations is six, one redundant condition such as

$$\Delta\alpha_{12} + \Delta\alpha_{23} + \Delta\alpha_{31} = 0 \tag{8.2}$$

must be involved. Such redundant conditions always exist for the angle coordinates around the central atom. In an octahedral XY_6 molecule (*g*), the number of internal coordinates exceeds the number of normal vibrations by three. This means that, of the twelve angle coordinates around the central atom, three redundant conditions are involved:

$$\begin{aligned} \Delta\alpha_{12} + \Delta\alpha_{26} + \Delta\alpha_{64} + \Delta\alpha_{41} &= 0 \\ \Delta\alpha_{15} + \Delta\alpha_{56} + \Delta\alpha_{63} + \Delta\alpha_{31} &= 0 \\ \Delta\alpha_{23} + \Delta\alpha_{34} + \Delta\alpha_{45} + \Delta\alpha_{52} &= 0 \end{aligned} \tag{8.3}$$

The redundant conditions are more complex in ring compounds. For example, the number of internal coordinates in a triangular X_3 molecule (*h*) exceeds the number of vibrations by three. One of these redundant conditions (A_1' species) is

$$\Delta\alpha_1 + \Delta\alpha_2 + \Delta\alpha_3 = 0 \tag{8.4}$$

The other two redundant conditions (E' species) involve bond stretching and angle deformation coordinates such as

$$\begin{aligned} (2\,\Delta r_1 - \Delta r_2 - \Delta r_3) + \frac{r}{\sqrt{3}}(\Delta\alpha_1 + \Delta\alpha_2 - 2\Delta\alpha_3) &= 0 \\ (\Delta r_2 - \Delta r_3) - \frac{r}{\sqrt{3}}(\Delta\alpha_1 - \Delta\alpha_2) &= 0 \end{aligned} \tag{8.5}$$

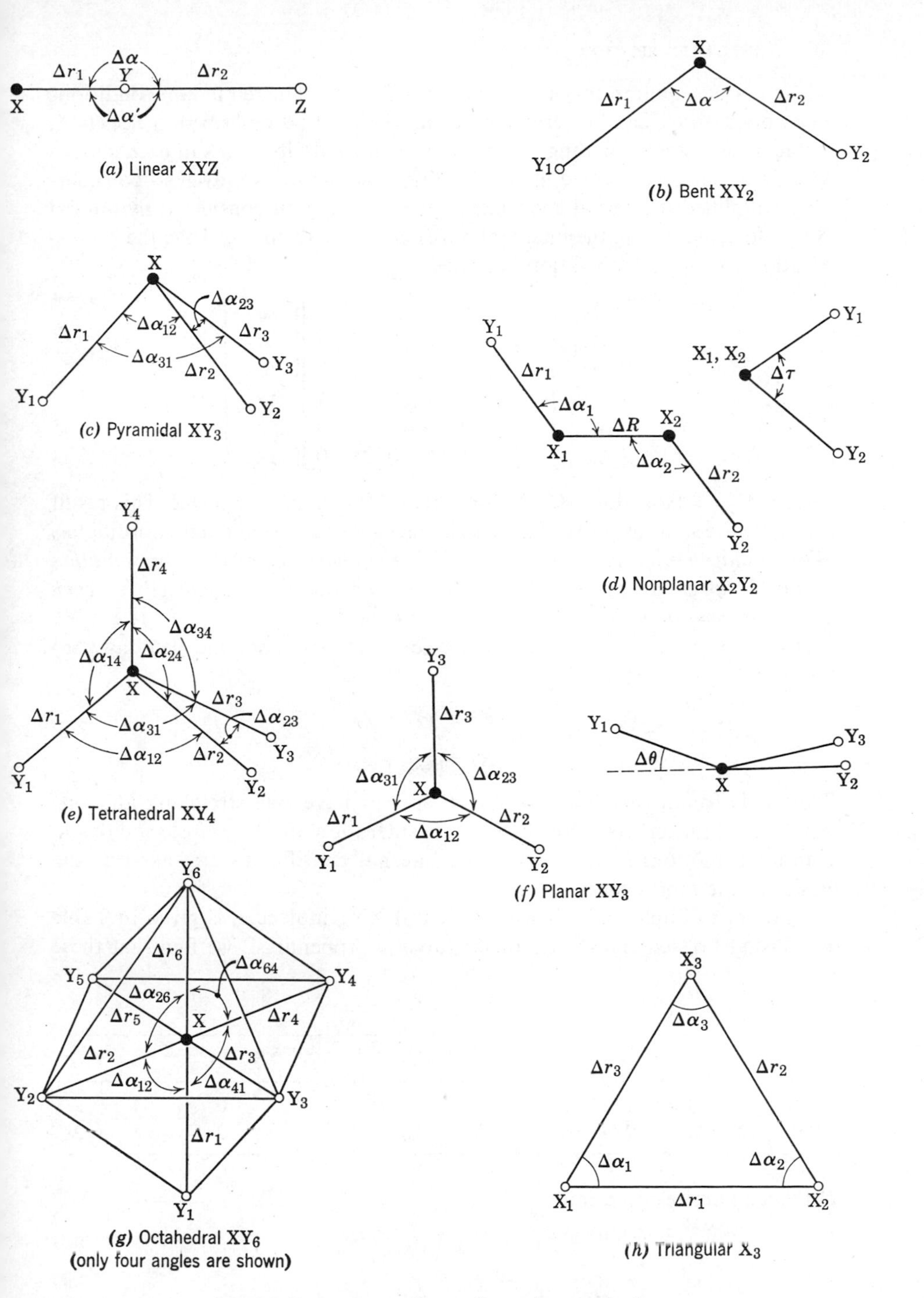

Fig. I-9 Internal coordinates for various molecules.

where r is the equilibrium length of the X—X bond. The redundant conditions mentioned above can be derived by using the method described in Sec. I-11.

The procedure for finding the number of normal vibrations in each species has already been described in Sec. I-7. This procedure is, however, considerably simplified if internal coordinates are used. Again consider a pyramidal XY_3 molecule. Using the internal coordinates shown in Fig. I-9c, the representation for the C_3^+ operation becomes

$$C_3^+ \begin{bmatrix} \Delta r_1 \\ \Delta r_2 \\ \Delta r_3 \\ \Delta\alpha_{12} \\ \Delta\alpha_{23} \\ \Delta\alpha_{31} \end{bmatrix} = \begin{bmatrix} 0 & 0 & 1 & 0 & 0 & 0 \\ 1 & 0 & 0 & 0 & 0 & 0 \\ 0 & 1 & 0 & 0 & 0 & 0 \\ 0 & 0 & 0 & 0 & 0 & 1 \\ 0 & 0 & 0 & 1 & 0 & 0 \\ 0 & 0 & 0 & 0 & 1 & 0 \end{bmatrix} \begin{bmatrix} \Delta r_1 \\ \Delta r_2 \\ \Delta r_3 \\ \Delta\alpha_{12} \\ \Delta\alpha_{23} \\ \Delta\alpha_{31} \end{bmatrix} \tag{8.6}$$

Thus $\chi(C_3^+) = 0$, as does $\chi(C_3^-)$. Similarly, $\chi(I) = 6$ and $\chi(\sigma_v) = 2$. This result is exactly the same as that obtained in Table I-5 using rectangular coordinates. *When using internal coordinates, however, the character of the representation is simply given by the number of internal coordinates unchanged by each symmetry operation.*

If this procedure is made separately for stretching (Δr) and bending ($\Delta\alpha$) coordinates, it is readily seen that

$$\begin{aligned} \chi^r(R) &= \chi_{A_1} + \chi_E \\ \chi^\alpha(R) &= \chi_{A_1} + \chi_E \end{aligned} \tag{8.7}$$

Thus it is found that both A_1 and E species have one stretching and one bending vibration, respectively. No consideration of the translational and rotational motions is necessary if the internal coordinates are taken as the basis for the representation.

Another example, one for an octahedral XY_6 molecule, is given in Table I-7. Using Eq. 6.6 and the character table in Appendix I, we find that these

TABLE I-7

	I	$8C_3$	$6C_2$	$6C_4$	$3C_4^2 \equiv C_2''$	$S_2 \equiv i$	$6S_4$	$8S_6 \equiv C_{3i}$	$3\sigma_h$	$6\sigma_d$
$\chi^r(R)$	6	0	0	2	2	0	0	0	4	2
$\chi^\alpha(R)$	12	0	2	0	0	0	0	0	4	2

characters are resolved into

$$\chi^r(R) = \chi_{A_{1g}} + \chi_{E_g} + \chi_{F_{1u}} \tag{8.8}$$

$$\chi^\alpha(R) = \chi_{A_{1g}} + \chi_{E_g} + \chi_{F_{1u}} + \chi_{F_{2g}} + \chi_{F_{2u}} \tag{8.9}$$

A comparison of this result with that obtained in Sec. I-7 immediately suggests that three redundant conditions are included in these bending vibrations (one in A_{1g} and one in E_g). Therefore $\chi^{\alpha}(R)$ for genuine vibrations becomes

$$\chi^{\alpha}(R) = \chi_{F_{1u}} + \chi_{F_{2g}} + \chi_{F_{2u}} \tag{8.10}$$

Thus it is concluded that six stretching and nine bending vibrations are distributed as indicated in Eqs. 8.8 and 8.10, respectively. Although the method given above is simpler that that of Sec. I-7, caution must be exercised with respect to the bending vibrations whenever redundancy is involved. In such a case, a comparison of the results obtained from both methods is useful in finding the species of redundancy.

I-9. SELECTION RULES FOR INFRARED AND RAMAN SPECTRA

According to quantum mechanics,[7-13] the selection rule for the infrared spectrum is determined by the integral:

$$[\mu]_{v'v''} = \int \psi_{v'}(Q_a)\mu\psi_{v''}(Q_a)\, dQ_a \tag{9.1}$$

Here μ is the dipole moment in the electronic ground state, ψ is the vibrational eigenfunction given by Eq. 2.7, and v' and v'' are the vibrational quantum numbers before and after the transition, respectively. The activity of the normal vibration whose normal coordinate is Q_a is being determined. By resolving the dipole moment into the three components in the x, y, and z directions, we obtain the result

$$\begin{aligned}
[\mu_x]_{v'v''} &= \int \psi_{v'}(Q_a)\mu_x\psi_{v''}(Q_a)\, dQ_a \\
[\mu_y]_{v'v''} &= \int \psi_{v'}(Q_a)\mu_y\psi_{v''}(Q_a)\, dQ_a \\
[\mu_z]_{v'v''} &= \int \psi_{v'}(Q_a)\mu_z\psi_{v''}(Q_a)\, dQ_a
\end{aligned} \tag{9.2}$$

If one of these integrals is not zero, the normal vibration associated with Q_a is infrared active. If all the integrals are zero, the vibration is infrared inactive.

Similarly, the selection rule for the Raman spectrum is determined by the integral:

$$[\alpha]_{v'v''} = \int \psi_{v'}(Q_a)\alpha\psi_{v''}(Q_n)\, dQ_a \tag{9.3}$$

As is shown in Sec. I-5, α consists of six components, α_{xx}, α_{yy}, α_{zz}, α_{xy}, α_{yz}, and α_{xz}. Thus Eq. 9.3 may be resolved into six components:

$$[\alpha_{xx}]_{v'v''} = \int \psi_{v'}(Q_a)\alpha_{xx}\psi_{v''}(Q_a)\, dQ_a$$

$$[\alpha_{yy}]_{v'v''} = \int \psi_{v'}(Q_a)\alpha_{yy}\psi_{v''}(Q_a)\, dQ_a \tag{9.4}$$

$$\cdots\cdots\cdots\cdots\cdots\cdots$$

$$\cdots\cdots\cdots\cdots\cdots\cdots$$

If one of these integrals is not zero, the normal vibration associated with Q_a is Raman active. If all the integrals are zero, the vibration is Raman inactive.

It is possible to decide whether the integrals of Eqs. 9.2 and 9.4 are zero or nonzero from a consideration of symmetry. As stated in Sec. I-1, the vibrations of interest are the fundamentals in which transitions occur from $v' = 0$ to $v'' = 1$. It is evident from the form of the vibrational eigenfunction (Eq. 2.8) that $\psi_0(Q_a)$ is invariant under any symmetry operation, whereas the symmetry of $\psi_1(Q_a)$ is the same as that of Q_a. Thus the integral does not vanish when the symmetry of μ_x, for example, is the same as that of Q_a. If the symmetry properties of μ_x and Q_a differ in even one symmetry element of the group, the integral becomes zero. In other words, for the integral to be nonzero, Q_a must belong to the same species as μ_x. More generally, the normal vibration associated with Q_a becomes infrared active when at least one of the components of the dipole moment belongs to the same species as Q_a. Similar conclusions are obtained for the Raman spectrum.

Since the species of the normal vibration can be determined by the methods described in Secs. I-7 and I-8, it is necessary only to determine the species of the components of the dipole moment and polarizability of the molecule. This can be done as follows. The components of the dipole moment, μ_x, μ_y, and μ_z, transform as do those of translational motion, T_x, T_y, and T_z, respectively. These have been discussed in Sec. I-7. Thus the character of the dipole moment is given by Eq. 7.13, which is

$$\boxed{\chi_\mu(R) = \pm(1 + 2\cos\theta)} \tag{9.5}$$

where + and − have the same meaning as before. In a pyramidal XY_3 molecule, Eq. 9.5 gives

	I	$2C_3$	$3\sigma_v$
$\chi_\mu(R)$	3	0	1

Using Eq. 6.6, this is resolved into $A_1 + E$. It is obvious that μ_z belongs to A_1. Then μ_x and μ_y must belong to E. In fact, the pair, μ_x and μ_y, transforms as follows.

$$I\begin{bmatrix}\mu_x\\ \mu_y\end{bmatrix} = \begin{bmatrix}1 & 0\\ 0 & 1\end{bmatrix}\begin{bmatrix}\mu_x\\ \mu_y\end{bmatrix} \qquad C_3^+\begin{bmatrix}\mu_x\\ \mu_y\end{bmatrix} = \begin{bmatrix}-\dfrac{1}{2} & \dfrac{\sqrt{3}}{2}\\ -\dfrac{\sqrt{3}}{2} & -\dfrac{1}{2}\end{bmatrix}\begin{bmatrix}\mu_x\\ \mu_y\end{bmatrix}$$

$$\chi(I) = 2 \qquad\qquad \chi(C_3^+) = -1$$

$$\sigma_1\begin{bmatrix}\mu_x\\ \mu_y\end{bmatrix} = \begin{bmatrix}-1 & 0\\ 0 & 1\end{bmatrix}\begin{bmatrix}\mu_x\\ \mu_y\end{bmatrix}$$

$$\chi(\sigma_1) = 0$$

Thus it is found that μ_z belongs to A_1 and (μ_x, μ_y) belong to E.

The character of the representation of the polarizability is given by

$$\boxed{\chi_\alpha(R) = 2\cos\theta(1 + 2\cos\theta)} \qquad (9.6)$$

for both proper and improper rotations. This can be derived as follows. The polarizability in the x, y, and z directions is related to that in X, Y, and Z coordinates by

$$\begin{bmatrix}\alpha_{XX} & \alpha_{XY} & \alpha_{XZ}\\ \alpha_{YX} & \alpha_{YY} & \alpha_{YZ}\\ \alpha_{ZX} & \alpha_{ZY} & \alpha_{ZZ}\end{bmatrix} = \begin{bmatrix}C_{Xx} & C_{Xy} & C_{Xz}\\ C_{Yx} & C_{Yy} & C_{Yz}\\ C_{Zz} & C_{Zy} & C_{Zz}\end{bmatrix}\begin{bmatrix}\alpha_{xx} & \alpha_{xy} & \alpha_{xz}\\ \alpha_{yx} & \alpha_{yy} & \alpha_{yz}\\ \alpha_{zx} & \alpha_{zy} & \alpha_{zz}\end{bmatrix}\begin{bmatrix}C_{Xx} & C_{Yx} & C_{Zx}\\ C_{Xy} & C_{Yy} & C_{Zy}\\ C_{Xz} & C_{Yz} & C_{Zz}\end{bmatrix}$$

where C_{Xx}, and so forth, denote the direction cosines between the two axes subscripted. If a rotation through θ around the Z axis superimposes the X, Y, and Z axes on the x, y, and z axes, the preceding relation becomes

$$C_\theta\begin{bmatrix}\alpha_{xx} & \alpha_{xy} & \alpha_{xz}\\ \alpha_{yx} & \alpha_{yy} & \alpha_{yz}\\ \alpha_{zx} & \alpha_{zy} & \alpha_{zz}\end{bmatrix} =$$

$$\begin{bmatrix}\cos\theta & \sin\theta & 0\\ -\sin\theta & \cos\theta & 0\\ 0 & 0 & 1\end{bmatrix}\begin{bmatrix}\alpha_{xx} & \alpha_{xy} & \alpha_{xz}\\ \alpha_{yx} & \alpha_{yy} & \alpha_{yz}\\ \alpha_{zx} & \alpha_{zy} & \alpha_{zz}\end{bmatrix}\begin{bmatrix}\cos\theta & -\sin\theta & 0\\ \sin\theta & \cos\theta & 0\\ 0 & 0 & 1\end{bmatrix}$$

This can be written

$$C_\theta \begin{bmatrix} \alpha_{xx} \\ \alpha_{yy} \\ \alpha_{zz} \\ \alpha_{xy} \\ \alpha_{xz} \\ \alpha_{yz} \end{bmatrix} =$$

$$\begin{bmatrix} \cos^2\theta & \sin^2\theta & 0 & 2\sin\theta\cos\theta & 0 & 0 \\ \sin^2\theta & \cos^2\theta & 0 & -2\sin\theta\cos\theta & 0 & 0 \\ 0 & 0 & 1 & 0 & 0 & 0 \\ -\sin\theta\cos\theta & \sin\theta\cos\theta & 0 & 2\cos^2\theta - 1 & 0 & 0 \\ 0 & 0 & 0 & 0 & \cos\theta & \sin\theta \\ 0 & 0 & 0 & 0 & -\sin\theta & \cos\theta \end{bmatrix} \begin{bmatrix} \alpha_{xx} \\ \alpha_{yy} \\ \alpha_{zz} \\ \alpha_{xy} \\ \alpha_{xz} \\ \alpha_{yz} \end{bmatrix}$$

Thus the character of this representation is given by Eq. 9.6. The same results are obtained for improper rotations if they are regarded as the product, $i \times$ (proper rotation). In a pyramidal XY_3 molecule, Eq. 9.6 gives

	I	$2C_3$	$3\sigma_v$
$\chi_\alpha(R)$	6	0	2

Using Eq. 6.6, this is resolved into $2A_1 + 2E$. Again, it is immediately seen that the component α_{zz} belongs to A_1, and the pair α_{zx} and α_{zy} belongs to E since

$$\begin{bmatrix} zx \\ zy \end{bmatrix} = z \begin{bmatrix} x \\ y \end{bmatrix} \approx A_1 \times E = E$$

It is more convenient to consider the components $\alpha_{xx} + \alpha_{yy}$ and $\alpha_{xx} - \alpha_{yy}$ rather than α_{xx} and α_{yy}. If a vector of unit length is considered, the relation

$$x^2 + y^2 + z^2 = 1$$

holds. Since α_{zz} belongs to A_1, $\alpha_{xx} + \alpha_{yy}$ must belong to A_1. Then the pair $\alpha_{xx} - \alpha_{yy}$ and α_{xy} must belong to E. As a result, the character table of the point group $\mathbf{C}_{3v}$ is completed as in Table I-8. Thus it is concluded that, in the point

TABLE I-8. CHARACTER TABLE OF THE POINT GROUP $\mathbf{C}_{3v}$

$\mathbf{C}_{3v}$	I	$2C_3$	$3\sigma_v$		
A_1	+1	+1	+1	μ_z	$\alpha_{xx} + \alpha_{yy}$, α_{zz}
A_2	+1	+1	−1		
E	+2	−1	0	(μ_x, μ_y)*	$(\alpha_{xz}, \alpha_{yz})$,* $(\alpha_{xx} - \alpha_{yy}, \alpha_{xy})$*

* A doubly degenerate pair is represented by two terms in parentheses.

group $\mathbf{C}_{3v}$, both the A_1 and the E vibrations are infrared as well as Raman active, while the A_2 vibrations are inactive.

Complete character tables like Table I-8 have already been worked out for all the point groups. Therefore no elaborate treatment such as that described in this section is necessary in practice. Appendix I gives complete character tables for the point groups that appear frequently in this book. From these tables, the selection rules for the infrared and Raman spectra are obtained immediately: *The vibration is infrared or Raman active if it belongs to the same species as one of the components of the dipole moment or polarizability, respectively.* For example, the character table of the point group $\mathbf{O}_h$ signifies immediately that only the F_{1u} vibrations are infrared active and only the A_{1g}, E_g, and F_{2g} vibrations are Raman active, for the components of the dipole moment or the polarizability belong to these species in this point group. It is to be noted in these character tables that (1) a totally symmetric vibration is Raman active in any point group, and (2) the infrared and Raman active vibrations always belong to u and g types, respectively, in the point groups having a center of symmetry.

I-10. STRUCTURE DETERMINATION

Suppose that a molecule has several probable structures each of which belongs to a different point group. Then the number of infrared and Raman active fundamentals should be different for each structure. Therefore the most probable model can be selected by comparing the observed number of infrared and Raman active fundamentals with that predicted theoretically for each model.

Consider the XeF_4 molecule as an example. It may be tetrahedral or square-planar. Using the methods described in the previous sections, the number of infrared or Raman active fundamentals can be found easily for each structure. Tables I-9*a* and I-9*b* summarize the results. It is seen that the tetrahedral structure predicts two infrared active fundamentals (one stretching and one bending), whereas the square-planar structure predicts three infrared active fundamentals (one stretching and two bendings). The infrared spectrum of XeF_4 in the vapor phase exhibits one Xe—F stretching at 586 cm^{-1} and two FXeF bendings at 291 and 123 cm^{-1} (Ref. II-622). Thus the square-planar structure of $\mathbf{D}_{4h}$ symmetry is preferable to the tetrahedral structure of $\mathbf{T}_d$ symmetry. As is seen in Tables I-9*a* and I-9*b*, group theory predicts two Raman active Xe—F stretchings for both structures, but two Raman active bendings for the tetrahedral and one Raman active bending for the square-planar structure. The observed Raman spectrum (543, 502 and 235 cm^{-1}) again confirms the square-planar structure.

TABLE I-9*a*. NUMBER OF FUNDAMENTALS FOR TETRAHEDRAL XeF_4

$\mathbf{T}_d$	Activity	Number of Fundamentals	Xe–F Stretching	FXeF Bending
A_1	R	1	1	0
A_2	ia	0	0	0
E	R	1	0	1
F_1	ia	0	0	0
F_2	IR, R	2	1	1
Total	IR	2	1	1
	R	4	2	2

TABLE I-9*b*. NUMBER OF FUNDAMENTALS FOR SQUARE-PLANAR XeF_4

$\mathbf{D}_{4h}$	Activity	Number of Fundamentals	Xe–F Stretching	FXeF Bending
A_{1g}	R	1	1	0
A_{1u}	ia	0	0	0
A_{2g}	ia	0	0	0
A_{2u}	IR	1	0	1
B_{1g}	R	1	1	0
B_{1u}	ia	0	0	0
B_{2g}	R	1	0	1
B_{2u}	ia	1	0	1
E_g	R	0	0	0
E_u	IR	2	1	1
Total	IR	3	1	2
	R	3	2	1

This method is widely used for the elucidation of molecular structure of inorganic, organic, and coordination compounds. In Part II, the number of infrared and Raman active fundamentals is compared for XY_3 (planar, $\mathbf{D}_{3h}$, and pyramidal, $\mathbf{C}_{3v}$), XY_4 (square-planar, $\mathbf{D}_{4h}$, and tetrahedral, $\mathbf{T}_d$), XY_5 (trigonal bipyramidal, $\mathbf{D}_{3h}$, and tetragonal pyramidal $\mathbf{C}_{4v}$) and other molecules. Recently, the structures of various metal carbonyl compounds (Sec. III-7) have been determined by this simple technique.

It should be noted, however, that this method does not give a clear-cut answer if the predicted number of infrared and Raman active fundamentals is similar for various probable structures. Furthermore, a practical difficulty arises in determining the number of fundamentals from the observed spectrum, since the intensities of overtone and combination bands are sometimes comparable to those of fundamentals when they appear as satellite bands of the fundamental. This is particularly true when overtone and combination bands are enhanced anomalously by *Fermi resonance* (accidental degeneracy).[8] For example, the frequency of the first overtone of the ν_2 vibration of CO_2 (667 cm^{-1}) is very close to that of the ν_1 vibration (1337 cm^{-1}). Since these two vibrations belong to the same symmetry species (Σ_g^+), they interact with each other and give rise to two strong Raman lines at 1388 and 1286 cm^{-1}. Fermi resonances similar to that observed for CO_2 may occur for a number of other molecules. It is to be noted also that the number of observed bands depends on the resolving power of the instrument used (Sec. III-7). Finally it should be remembered that the molecular symmetry in the isolated state is not necessarily the same as that in the crystalline state (Sec. I-16). Therefore this method must be applied with caution to spectra obtained for compounds in the crystalline state.

I-11. PRINCIPLE OF THE GF MATRIX METHOD*

As described in Sec. I-3, the frequency of the normal vibration is determined by the kinetic and potential energies of the system. The kinetic energy is determined by the masses of the individual atoms and their geometrical arrangement in the molecule. On the other hand, the potential energy arises from interaction between the individual atoms and is described in terms of the force constants. Since the potential energy provides valuable information about the nature of interatomic forces, it is highly desirable to obtain the force constants from the observed frequencies. This is usually done by calculating the frequencies, assuming a suitable set of the force constants. If the agreement between the calculated and observed frequencies is satisfactory, this particular set of the force constants is adopted as a representation of the potential energy of the system.

To calculate the vibrational frequencies, it is necessary first to express both the potential and kinetic energies in terms of some common coordinates (Sec. I-3.) Internal coordinates (Sec. I-8) are more suitable for this purpose than rectangular coordinates, since (1) the force constants expressed in terms of internal coordinates have a clearer physical meaning than those expressed in terms of rectangular coordinates, and (2) a set of internal coordinates does not involve translational and rotational motion of the molecule as a whole.

Using the internal coordinates, R_i, the potential energy is written

$$2V = \tilde{\mathbf{R}}\mathbf{F}\mathbf{R} \tag{11.1}$$

* For details, see Refs. 12 and 13.

For a bent Y_1XY_2 molecule such as that in Fig. I-9*b*, **R** is a column matrix of the form

$$\mathbf{R} = \begin{bmatrix} \Delta r_1 \\ \Delta r_2 \\ \Delta\alpha \end{bmatrix}$$

$\tilde{\mathbf{R}}$ is its transpose:

$$\tilde{\mathbf{R}} = [\Delta r_1 \; \Delta r_2 \; \Delta\alpha]$$

and **F** is a matrix whose components are the force constants:

$$\mathbf{F} = \begin{bmatrix} f_{11} & f_{12} & r_1 f_{13} \\ f_{21} & f_{22} & r_2 f_{23} \\ r_1 f_{31} & r_2 f_{32} & r_1 r_2 f_{33} \end{bmatrix} \equiv \begin{bmatrix} F_{11} & F_{12} & F_{13} \\ F_{21} & F_{22} & F_{23} \\ F_{31} & F_{32} & F_{33} \end{bmatrix} \tag{11.2}^*$$

Here r_1 and r_2 are the equilibrium lengths of the X—Y_1 and X—Y_2 bonds, respectively.

The kinetic energy is not easily expressed in terms of the same internal coordinates. Wilson[33] has shown, however, that the kinetic energy can be written

$$2T = \tilde{\dot{\mathbf{R}}}\mathbf{G}^{-1}\dot{\mathbf{R}} \tag{11.3}\dagger$$

where $\mathbf{G}^{-1}$ is the reciprocal of the **G** matrix, which will be defined later.

If Eqs. 11.1 and 11.3 are combined with Newton's equation,

$$\frac{d}{dt}\left(\frac{\partial T}{\partial \dot{R}_k}\right) + \frac{\partial V}{\partial R_k} = 0 \tag{3.6}$$

the following secular equation, which is similar to Eq. 3.19, is obtained.

$$\begin{vmatrix} F_{11} - (G^{-1})_{11}\lambda & F_{12} - (G^{-1})_{12}\lambda & \cdots \\ F_{21} - (G^{-1})_{21}\lambda & F_{22} - (G^{-1})_{22}\lambda & \cdots \\ \vdots & \vdots & \end{vmatrix} \equiv |\mathbf{F} - \mathbf{G}^{-1}\lambda| = 0 \tag{11.4}$$

By multiplying by the determinant of **G**

$$\begin{vmatrix} G_{11} & G_{12} & \cdots \\ G_{21} & G_{22} & \cdots \\ \vdots & \vdots & \end{vmatrix} \equiv |\mathbf{G}| \tag{11.5}$$

* Here f_{11} and f_{22} are the stretching force constants of the X—Y_1 and X—Y_2 bonds, respectively. f_{33} is the bending force constant of the Y_1XY_2 angle. The other symbols represent interaction force constants between stretching and stretching or between stretching and bending vibrations. f_{13} (or f_{31}), f_{23} (or f_{32}), and f_{33} are multiplied by r_1, r_2, and r_1r_2, respectively, to make the dimensions of all the force constants the same.

† Appendix III gives the derivation of Eq. 11.3.

from the left of Eq. 11.4, the following equation is obtained.

$$\begin{vmatrix} \sum G_{1t}F_{t1} - \lambda & \sum G_{1t}F_{t2} & \cdots \\ \sum G_{2t}F_{t1} & \sum G_{2t}F_{t2} - \lambda & \cdots \\ \vdots & \vdots & . \end{vmatrix} \equiv |\mathbf{GF} - \mathbf{E}\lambda| = 0 \qquad (11.6)$$

Here **E** is the unit matrix, and λ is related to the wave number, $\tilde{\nu}$, by the relation $\lambda = 4\pi^2c^2\tilde{\nu}^2$.* The order of the equation is equal to the number of internal coordinates used.

The **F** matrix can be written by assuming a suitable set of force constants. If the **G** matrix is constructed by the following method, the vibrational frequencies are obtained by solving Eq. 11.6. The **G** matrix is defined as

$$\mathbf{G} = \mathbf{BM}^{-1}\tilde{\mathbf{B}} \qquad (11.7)$$

Here $\mathbf{M}^{-1}$ is a diagonal matrix whose components are μ_i, where μ_i is the reciprocal of the mass of the ith atom. For a bent XY_2 molecule,

$$\mathbf{M}^{-1} = \begin{bmatrix} \mu_1 & & & & 0 \\ & \mu_1 & & & \\ & & \mu_1 & & \\ & & & \ddots & \\ 0 & & & & \mu_3 \end{bmatrix}$$

where μ_3 and μ_1 are the reciprocals of the masses of the X and Y atoms, respectively. The **B** matrix is defined as

$$\mathbf{R} = \mathbf{BX} \qquad (11.8)$$

where **R** and **X** are column matrices whose components are the internal and rectangular coordinates, respectively. For a bent XY_2 molecule, Eq. 11.8 is written

$$\begin{bmatrix} \Delta r_1 \\ \Delta r_2 \\ \Delta\alpha \end{bmatrix} = \left[\begin{array}{ccc:ccc:ccc} -s & -c & 0 & 0 & 0 & 0 & s & c & 0 \\ 0 & 0 & 0 & s & -c & 0 & -s & c & 0 \\ -c/r & s/r & 0 & c/r & s/r & 0 & 0 & -2s/r & 0 \end{array}\right] \begin{bmatrix} \Delta x_1 \\ \Delta y_1 \\ \Delta z_1 \\ \cdots \\ \Delta x_2 \\ \Delta y_2 \\ \Delta z_2 \\ \cdots \\ \Delta x_3 \\ \Delta y_3 \\ \Delta z_3 \end{bmatrix} \qquad (11.9)$$

* Here λ should not be confused with λ_w (wavelength).

where $s = \sin(\alpha/2)$, $c = \cos(\alpha/2)$ and r is the equilibrium distance between X and Y. (See Fig. I-10.)

If unit vectors such as those in Fig. I-10 are considered, Eq. 11.9 can be written in a more compact form using vector notation:

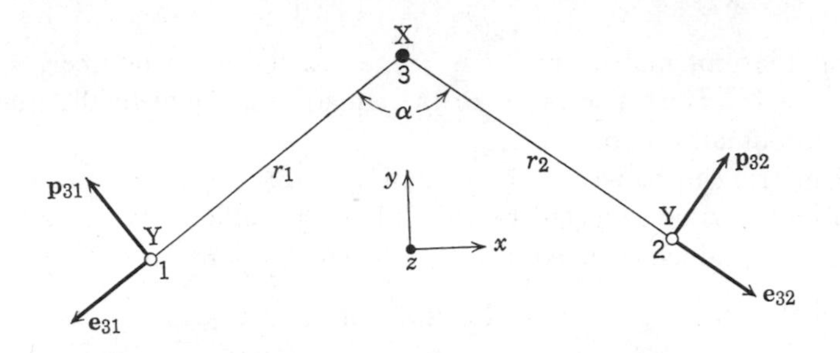

Fig. I-10. *Unit vectors in a bent XY_2 molecule.*

$$\begin{bmatrix} \Delta r_1 \\ \Delta r_2 \\ \Delta\alpha \end{bmatrix} = \begin{bmatrix} \mathbf{e}_{31} & 0 & -\mathbf{e}_{31} \\ 0 & \mathbf{e}_{32} & -\mathbf{e}_{32} \\ \mathbf{p}_{31}/r & \mathbf{p}_{32}/r & -(\mathbf{p}_{31}+\mathbf{p}_{32})/r \end{bmatrix} \begin{bmatrix} \boldsymbol{\rho}_1 \\ \boldsymbol{\rho}_2 \\ \boldsymbol{\rho}_3 \end{bmatrix}$$

Here $\boldsymbol{\rho}_1$, $\boldsymbol{\rho}_2$, and $\boldsymbol{\rho}_3$ are the displacement vectors of the atoms, 1, 2, and 3, respectively. This can be written simply as

$$\mathbf{R} = \mathbf{S} \cdot \boldsymbol{\rho} \tag{11.11}$$

where the dot represents the scalar product of the two vectors. Here, **S** is called the **S** matrix, and its components (**S** vector) can be written down according to the following formulas: (1) bond stretching,

$$\Delta r_1 = \Delta r_{31} = \mathbf{e}_{31} \cdot \boldsymbol{\rho}_1 - \mathbf{e}_{31} \cdot \boldsymbol{\rho}_3 \tag{11.12}$$

and (2) angle bending,

$$\Delta\alpha = \Delta\alpha_{132} = [\boldsymbol{\rho}_{31} \cdot \boldsymbol{\rho}_1 + \boldsymbol{\rho}_{32} \cdot \boldsymbol{\rho}_2 - (\boldsymbol{\rho}_{31} + \boldsymbol{\rho}_{32}) \cdot \boldsymbol{\rho}_3]/r \tag{11.13}$$

It is seen that the direction of the **S** vector is the direction in which a given displacement of ith atom will produce the greatest increase in Δr or $\Delta\alpha$. Formulas for obtaining the **S** vectors of other internal coordinates such as those of out-of-plane ($\Delta\theta$) and torsional ($\Delta\tau$) vibrations are also available.[12,13]

By using the **S** matrix, Eq. 11.7 is written

$$\mathbf{G} = \mathbf{S}\mathbf{m}^{-1}\tilde{\mathbf{S}} \tag{11.14}$$

For a bent XY_2 molecule, this becomes

$$\mathbf{G} = \begin{bmatrix} \mathbf{e}_{31} & 0 & -\mathbf{e}_{31} \\ 0 & \mathbf{e}_{32} & -\mathbf{e}_{32} \\ \mathbf{p}_{31}/r & \mathbf{p}_{32}/r & -(\mathbf{p}_{31}+\mathbf{p}_{32})/r \end{bmatrix} \begin{bmatrix} \mu_1 & 0 & 0 \\ 0 & \mu_1 & 0 \\ 0 & 0 & \mu_3 \end{bmatrix}$$

$$\times \begin{bmatrix} \mathbf{e}_{31} & 0 & \mathbf{p}_{31}/r \\ 0 & \mathbf{e}_{32} & \mathbf{p}_{32}/r \\ -\mathbf{e}_{31} & -\mathbf{e}_{32} & -(\mathbf{p}_{31}+\mathbf{p}_{32})/r \end{bmatrix}$$

$$= \begin{bmatrix} (\mu_3+\mu_1)\mathbf{e}_{31}{}^2 & \mu_3\,\mathbf{e}_{31}\cdot\mathbf{e}_{32} & \frac{\mu_1}{r}\mathbf{e}_{31}\cdot\mathbf{p}_{31}+\frac{\mu_3}{r}\mathbf{e}_{31}\cdot(\mathbf{p}_{31}+\mathbf{p}_{32}) \\ & (\mu_3+\mu_1)\mathbf{e}_{32}{}^2 & \frac{\mu_1}{r}\mathbf{e}_{32}\cdot\mathbf{p}_{32}+\frac{\mu_3}{r}\mathbf{e}_{32}\cdot(\mathbf{p}_{31}+\mathbf{p}_{32}) \\ & & \frac{\mu_1}{r^2}\mathbf{p}_{31}{}^2+\frac{\mu_1}{r^2}\mathbf{p}_{32}{}^2+\frac{\mu_3}{r}(\mathbf{p}_{31}+\mathbf{p}_{32})^2 \end{bmatrix}$$

Considering

$$\mathbf{e}_{31}\cdot\mathbf{e}_{31}=\mathbf{e}_{32}\cdot\mathbf{e}_{32}=\mathbf{p}_{31}\cdot\mathbf{p}_{31}=\mathbf{p}_{32}\cdot\mathbf{p}_{32}=1,\ \mathbf{e}_{31}\cdot\mathbf{p}_{31}=\mathbf{e}_{32}\cdot\mathbf{p}_{32}=0,$$
$$\mathbf{e}_{31}\cdot\mathbf{e}_{32}=\cos\alpha,\ \mathbf{e}_{31}\cdot\mathbf{p}_{32}=\mathbf{e}_{32}\cdot\mathbf{p}_{31}=-\sin\alpha,\text{ and}$$
$$(\mathbf{p}_{31}+\mathbf{p}_{32})^2=2(1-\cos\alpha)$$

we find that the **G** matrix is calculated as

$$\mathbf{G}=\begin{bmatrix} \mu_3+\mu_1 & \mu_3\cos\alpha & -\frac{\mu_3}{r}\sin\alpha \\ & \mu_3+\mu_1 & -\frac{\mu_3}{r}\sin\alpha \\ & & \frac{2\mu_1}{r^2}+\frac{2\mu_3}{r^2}(1-\cos\alpha) \end{bmatrix} \tag{11.15}$$

If the **G** matrix elements obtained are written for each combination of internal coordinates, there results

$$\begin{aligned} G(\Delta r_1, \Delta r_1) &= \mu_3+\mu_1 \\ G(\Delta r_2, \Delta r_2) &= \mu_3+\mu_1 \\ G(\Delta r_1, \Delta r_2) &= \mu_3\cos\alpha \\ G(\Delta\alpha, \Delta\alpha) &= \frac{2\mu_1}{r^2}+\frac{2\mu_3}{r^2}(1-\cos\alpha) \\ G(\Delta r_1, \Delta\alpha) &= -\frac{\mu_3}{r}\sin\alpha \\ G(\Delta r_2, \Delta\alpha) &= -\frac{\mu_3}{r}\sin\alpha \end{aligned} \tag{11.16}$$

If such calculations are made for several types of molecules, it is immediately seen that the **G** matrix elements themselves have many regularities. Decius[34] developed general formulas for writing **G** matrix elements.* Some of them are

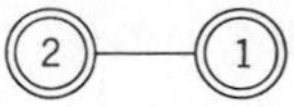

$$G_{rr}^{2} = \mu_1 + \mu_2$$

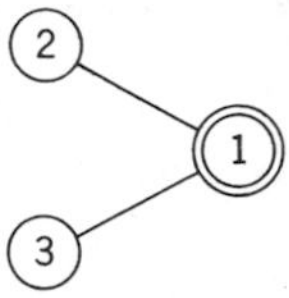

$$G_{rr}^{1} = \mu_1 \cos\phi$$

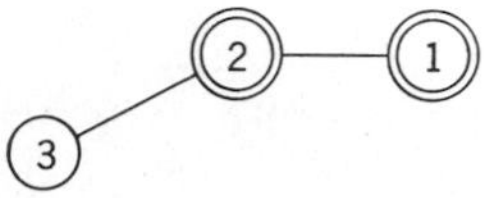

$$G_{r\phi}^{2} = -\rho_{23}\mu_2 \sin\phi$$

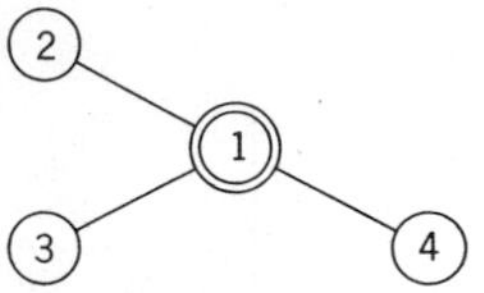

$$G_{r\phi}^{1}\begin{pmatrix}1\\1\end{pmatrix} = -(\rho_{13}\sin\phi_{213}\cos\psi_{234} + \rho_{14}\sin\phi_{214}\cos\psi_{243})\mu_1$$

$$G_{\phi\phi}^{3} = \rho_{12}^{2}\mu_1 + \rho_{23}^{2}\mu_3 + (\rho_{12}^{2} + \rho_{23}^{2} - 2\rho_{12}\rho_{23}\cos\phi)\mu_2$$

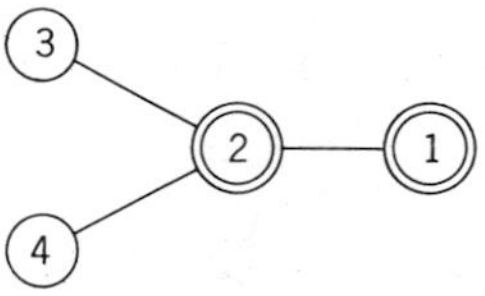

$$G_{\phi\phi}^{2}\begin{pmatrix}1\\1\end{pmatrix} = (\rho_{12}^{2}\cos\psi_{314})\mu_1 + [(\rho_{12} - \rho_{23}\cos\phi_{123} - \rho_{24}\cos\phi_{124})\rho_{12}\cos\psi_{314} + (\sin\phi_{123}\sin\phi_{124}\sin^2\psi_{314} + \cos\phi_{324}\cos\psi_{314})\rho_{23}\rho_{24}]\mu_2$$

Here the atoms surrounded by a double circle indicate the atoms common to both coordinates. The symbols μ and ρ denote the reciprocals of mass and bond distance, respectively. The solid angle, $\psi_{\alpha\beta\gamma}$, in Fig. I-11 is defined as

$$\cos\psi_{\alpha\beta\gamma} = \frac{\cos\phi_{\alpha\delta\gamma} - \cos\phi_{\alpha\delta\beta}\cos\phi_{\beta\delta\gamma}}{\sin\phi_{\alpha\delta\beta}\sin\phi_{\beta\delta\gamma}} \quad (11.17)$$

The correspondence between the Decius formulas and the results obtained in Eq. 11.16 is evident.

* See also Ref. 35.

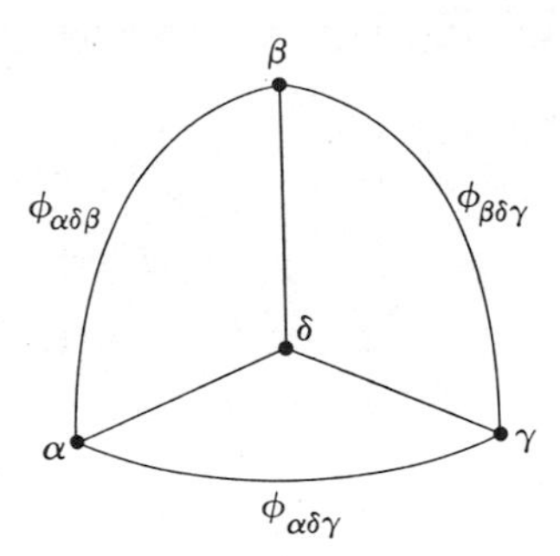

Fig. I-11

With the Decius formulas, the **G** matrix elements of a pyramidal XY_3 molecule have been calculated and are shown in Table I-10.

TABLE I-10

	Δr_1	Δr_2	Δr_3	$A\alpha_{23}$	$\Delta\alpha_{31}$	$\Delta\alpha_{12}$
Δr_1	A	B	B	C	D	D
Δr_2	—	A	B	D	C	D
Δr_3	—	—	A	D	D	C
$\Delta\alpha_{23}$	—	—	—	E	F	F
$\Delta\alpha_{31}$	—	—	—	—	E	F
$\Delta\alpha_{12}$	—	—	—	—	—	E

$$A = G_{rr}{}^2 = \mu_X + \mu_Y$$

$$B = G_{rr}{}^1 = \mu_X \cos\alpha$$

$$C = G_{r\phi}{}^1\binom{1}{1} = -\frac{2}{r}\frac{\cos\alpha(1-\cos\alpha)\mu_X}{\sin\alpha}$$

$$D = G_{r\phi}{}^2 = -\frac{\mu_X}{r}\sin\alpha$$

$$E = G_{\phi\phi}{}^3 = \frac{2}{r^2}[\mu_Y + \mu_X(1-\cos\alpha)]$$

$$F = G_{\phi\phi}{}^2\binom{1}{1} = \frac{\mu_Y}{r^2}\frac{\cos\alpha}{1+\cos\alpha} + \frac{\mu_X}{r^2}\frac{(1+3\cos\alpha)(1-\cos\alpha)}{1+\cos\alpha}$$

I-12. UTILIZATION OF SYMMETRY PROPERTIES

Considering the equivalence of the two X—Y bonds of a bent XY_2 molecule, the **F** and **G** matrices obtained in Eqs. 11.2 and 11.15 are written

$$\mathbf{F} = \begin{bmatrix} f_{11} & f_{12} & rf_{13} \\ f_{12} & f_{11} & rf_{13} \\ rf_{13} & rf_{13} & r^2 f_{33} \end{bmatrix} \tag{12.1}$$

$$\mathbf{G} = \begin{bmatrix} \mu_3 + \mu_1 & \mu_3 \cos\alpha & -\dfrac{\mu_3}{r}\sin\alpha \\ \mu_3 \cos\alpha & \mu_3 + \mu_1 & -\dfrac{\mu_3}{r}\sin\alpha \\ -\dfrac{\mu_3}{r}\sin\alpha & -\dfrac{\mu_3}{r}\sin\alpha & \dfrac{2\mu_1}{r^2} + \dfrac{2\mu_3}{r^2}(1 - \cos\alpha) \end{bmatrix} \tag{12.2}$$

Both of these matrices are of the form

$$\begin{bmatrix} A & C & D \\ C & A & D \\ D & D & B \end{bmatrix} \tag{12.3}$$

The appearance of the same elements is evidently due to the equivalence of the two internal coordinates, Δr_1 and Δr_2. Such symmetrically equivalent sets of internal coordinates are seen in many other molecules, such as those in Fig. I-9. In these cases, it is possible to reduce the order of the **F** and **G** matrices (and hence the order of the secular equation resulting from them) by a coordinate transformation.

Let the internal coordinates **R** be transformed by

$$\mathbf{R}' = \mathbf{UR} \tag{12.4}$$

More explicitly,

$$\begin{bmatrix} R_1' \\ R_2' \\ \vdots \\ R_j' \end{bmatrix} = \begin{bmatrix} u_{11} & u_{12} & \cdots & u_{1k} \\ u_{21} & u_{22} & \cdots & u_{2k} \\ \cdots & \cdots & \cdots & \cdots \\ u_{j1} & u_{j2} & \cdots & u_{jk} \end{bmatrix} \begin{bmatrix} R_1 \\ R_2 \\ \vdots \\ R_k \end{bmatrix} \tag{12.5}$$

Here **U** is an orthogonal matrix ($\tilde{\mathbf{U}}\mathbf{U} = \mathbf{E}$) whose elements must satisfy the relations

$$\sum_k (u_{jk})^2 = 1 \quad \text{(normalization)} \tag{12.6}$$

$$\sum_k u_{jk} u_{lk} = 0 \quad \text{(orthogonality)} \tag{12.7}$$

Furthermore, the symmetry of the molecule must be taken into consideration in constructing the **U** matrix. The new sets of coordinates, $R'_1, \cdots, R'_j$, thus obtained are linear combinations of internal coordinates and are called (internal) *symmetry coordinates.* An appropriate choice of these u elements can usually be made through intuition and experience. If follows from Eqs. 12.6 and 12.7 that the relations

$$u_{ak} = \pm(1/q)^{1/2} \tag{12.8}$$

$$u_{ak}^2 + u_{bk}^2 = 2/q \tag{12.9}$$

$$u_{ak}^2 + u_{bk}^2 + u_{ck}^2 = 3/q \tag{12.10}$$

must hold for degeneracies 1, 2, and 3, respectively.[33] Here q is the total number of symmetrically equivalent internal coordinates, and (u_{ak}, u_{bk}) and (u_{ak}, u_{bk}, u_{ck}) represent the elements of the kth internal coordinates for doubly and triply degenerate combinations, respectively.

In a bent XY_2 molecule, Δr_1 and Δr_2 are equivalent. Then, from Eqs. 12.6, 12.7, and 12.8, the symmetry coordinates are

$$\begin{aligned} R'_1 &= (1/\sqrt{2})\,\Delta r_1 + (1/\sqrt{2})\,\Delta r_2 \\ R'_3 &= (1/\sqrt{2})\,\Delta r_1 - (1/\sqrt{2})\,\Delta r_2 \end{aligned} \tag{12.11}$$

Consider the symmetry property of the R'_3 coordinate. If the identity operation is performed, each internal coordinate is transformed into itself. Thus

$$(I)R'_3 = 1\,(/\sqrt{2})\,\Delta r_1 - (1/\sqrt{2})\,\Delta r_2 = (+1)R'_3$$

If the operation $C_2(z)$ is applied, Δr_1 is transformed into Δr_2, and Δr_2 into Δr_1. Thus

$$(C_2(z))R'_3 = (1/\sqrt{2})\,\Delta r_2 - (1/\sqrt{2})\,\Delta r_1 = (-1)R'_3$$

The operations $\sigma_v(xz)$ and $\sigma_v(yz)$ give the same results as $C_2(z)$ and I, respectively. Thus the characters for the operations I, $C_2(z)$, $\sigma_v(xz)$, and $\sigma_v(yz)$ are $+1, -1, -1, +1$, respectively. In other words, R'_3 belongs to the B_2 species of the point group $\mathbf{C}_{2v}$. Similarly, it can be shown that R_1 belongs to the A_1 species. It is evident that R'_1 and R'_3 represent the coordinates corresponding to the *symmetric and antisymmetric stretching vibrations*, respectively. Evidently, the bending coordinate, $\Delta\alpha$, belongs to the A_1 species. Therefore the complete **U** matrix of the bent XY_2 molecule is written

$$\begin{bmatrix} R'_1(A_1) \\ R'_2(A_1) \\ R'_3(B_2) \end{bmatrix} = \begin{bmatrix} 1/\sqrt{2} & 1/\sqrt{2} & 0 \\ 0 & 0 & 1 \\ 1/\sqrt{2} & -1/\sqrt{2} & 0 \end{bmatrix} \begin{bmatrix} \Delta r_1 \\ \Delta r_2 \\ \Delta\alpha \end{bmatrix} \tag{12.12}$$

If the $\mathbf{F}$ and $\mathbf{G}$ matrices of the type (12.3) are transformed by the relations

$$\mathbf{F}' = \mathbf{U}\mathbf{F}\tilde{\mathbf{U}}$$
$$\mathbf{G}' = \mathbf{U}\mathbf{G}\tilde{\mathbf{U}} \tag{12.13}$$

where $\mathbf{U}$ is given by Eq. 12.12, they become formally

$$\mathbf{F}', \mathbf{G}' = \left[\begin{array}{cc|c} A + C & \sqrt{2}\, D & 0 \\ \sqrt{2}\, D & B & 0 \\ \hline 0 & 0 & A - C \end{array}\right] \tag{12.14}$$

Or, more explicitly,

$$\mathbf{F}' = \left[\begin{array}{cc|c} f_{11} + f_{12} & r\sqrt{2} f_{13} & 0 \\ r\sqrt{2} f_{13} & r^2 f_{33} & 0 \\ \hline 0 & 0 & f_{11} - f_{12} \end{array}\right] \tag{12.15}$$

$$\mathbf{G}' = \left[\begin{array}{cc|c} \mu_3(1 + \cos\alpha) + \mu_1 & -\dfrac{\sqrt{2}}{r}\mu_3 \sin\alpha & 0 \\ -\dfrac{\sqrt{2}}{r}\mu_3 \sin\alpha & \dfrac{2\mu_1}{r^2} + \dfrac{2\mu_3}{r^2}(1 - \cos\alpha) & 0 \\ \hline 0 & 0 & \mu_3(1 - \cos\alpha) + \mu_1 \end{array}\right] \tag{12.16}$$

It can be shown, in general, that the transformed secular equation, $|\mathbf{G}'\mathbf{F}' - \mathbf{E}\lambda| = 0$, obtained by the orthogonal transformation above, has the same roots as the original one.* Since $|\mathbf{G}'\mathbf{F}' - \mathbf{E}\lambda| = 0$ can be resolved into one quadratic (A_1) and one first order (B_2) secular equation, in a bent XY_2 molecule, it is necessary only to solve these low order secular equations. The advantage of solving the former equation is therefore that it can be resolved into several equations of lower order.† In general, the burden of computation rapidly increases as the order of the secular equation becomes large. Therefore a coordinate transformation such as that shown above greatly reduces the labor involved in calculation.

* $\mathbf{G}'\mathbf{F}' = \mathbf{U}\mathbf{G}\tilde{\mathbf{U}}\mathbf{U}\mathbf{F}\tilde{\mathbf{U}} = \mathbf{U}\mathbf{G}\mathbf{F}\tilde{\mathbf{U}} = \mathbf{U}\mathbf{G}\mathbf{F}\mathbf{U}^{-1}$. Thus $\mathbf{G}'\mathbf{F}'$ gives the same eigenvalues as $\mathbf{G}\mathbf{F}$.

† As shown in Eqs. 11.1 and 11.3, the potential and kinetic energies in terms of internal coordinates involve cross products. This makes the order of the secular equation high. On the other hand, the equation could be resolved completely if normal coordinates were used (Sec. I-3). Although symmetry coordinates are not so satisfactory as normal coordinates, the order of the secular equation, $|\mathbf{G}'\mathbf{F}' - \mathbf{E}\lambda| = 0$, in terms of symmetry coordinates becomes lower than that of $|\mathbf{G}\mathbf{F} - \mathbf{E}\lambda| = 0$ in terms of internal coordinates.

In a pyramidal XY_3 molecule (Fig. I-9*c*), Δr_1, Δr_2, and Δr_3 are the equivalent set. So are $\Delta\alpha_{23}$, $\Delta\alpha_{31}$, and $\Delta\alpha_{12}$. It is already known from Eq. 8.7 that one A_1 and one E vibration is involved both in the stretching and in the bending vibrations. Considering the relations given by Eqs. 12.6, 12.7, 12.8, and 12.9, the symmetry coordinates for the three stretching vibrations are selected as follows:

$$\begin{aligned} R_1'(A_1) &= (1/\sqrt{3})\,\Delta r_1 + (1/\sqrt{3})\,\Delta r_2 + (1/\sqrt{3})\,\Delta r_3 \\ R_{2a}'(E) &= (2/\sqrt{6})\,\Delta r_1 - (1/\sqrt{6})\,\Delta r_2 - (1/\sqrt{6})\,\Delta r_3 \\ R_{2b}'(E) &= \qquad\qquad\quad (1/\sqrt{2})\,\Delta r_2 - (1/\sqrt{2})\,\Delta r_3 \end{aligned} \tag{12.17}$$

It is evident that R_1' transforms according to the character of the A_1 species of the point group $\mathbf{C}_{3v}$. That the pair R_{2a}' and R_{2b}' belongs to the E species can be proved as follows. For example, for the C_3^+ operation

$$\begin{aligned} (C_3^+)R_{2a}' &= (2/\sqrt{6})\,\Delta r_3 - (1/\sqrt{6})\,\Delta r_1 - (1/\sqrt{6})\,\Delta r_2 = AR_{2a}' + BR_{2b}' \\ (C_3^+)R_{2b}' &= \qquad\qquad\quad (1/\sqrt{2})\,\Delta r_1 - (1/\sqrt{2})\,\Delta r_2 = A'R_{2a}' + B'R_{2b}' \end{aligned}$$

where A, B, A', and B' are constants. Substituting for R_{2a}' and R_{2b}' on the right-hand side of these equations from Eqs. 12.17, and equating coefficients, it is found that

$$A = -1/2, \quad B = \sqrt{3}/2, \quad A' = -\sqrt{3}/2, \quad B' = -1/2$$

That is, the transformation caused by the operation C_3^+ is written

$$C_3^+\begin{bmatrix} R_{2a}' \\ R_{2b}' \end{bmatrix} = \begin{bmatrix} -1/2 & \sqrt{3}/2 \\ -\sqrt{3}/2 & -1/2 \end{bmatrix}\begin{bmatrix} R_{2a}' \\ R_{2b}' \end{bmatrix}$$

Thus the character of the transformation matrix for this operation is -1. In a similar manner, it can be shown that the characters for the operations I and σ_v are 2 and 0, respectively.

The symmetry coordinates for the three angle coordinates are similar to those in Eqs. 12.17. Therefore the complete **U** matrix is written

$$\begin{bmatrix} R_1'(A_1) \\ R_2'(A_1) \\ R_{3a}'(E) \\ R_{4a}'(E) \\ R_{3b}'(E) \\ R_{4b}(E) \end{bmatrix} = \begin{bmatrix} 1/\sqrt{3} & 1/\sqrt{3} & 1/\sqrt{3} & 0 & 0 & 0 \\ 0 & 0 & 0 & 1/\sqrt{3} & 1/\sqrt{3} & 1/\sqrt{3} \\ 2/\sqrt{6} & -1/\sqrt{6} & -1/\sqrt{6} & 0 & 0 & 0 \\ 0 & 0 & 0 & 2/\sqrt{6} & -1/\sqrt{6} & -1/\sqrt{6} \\ 0 & 1/\sqrt{2} & -1/\sqrt{2} & 0 & 0 & 0 \\ 0 & 0 & 0 & 0 & 1/\sqrt{2} & -1/\sqrt{2} \end{bmatrix}\begin{bmatrix} \Delta r_1 \\ \Delta r_2 \\ \Delta r_3 \\ \Delta\alpha_{23} \\ \Delta\alpha_{31} \\ \Delta\alpha_{13} \end{bmatrix} \tag{12.18}$$

The **G** matrix of a pyramidal XY_3 molecule has already been calculated (see Table I-10). By using Eq. 12.13, the new $\mathbf{G}'$ matrix becomes

$$\mathbf{G}' = \left[\begin{array}{cc|cc|cc} A+2B & C+2D & \multicolumn{2}{c|}{0} & \multicolumn{2}{c}{0} \\ C+2D & E+2F & & & & \\ \hline \multicolumn{2}{c|}{0} & A-B & C-D & \multicolumn{2}{c}{0} \\ & & C-D & E-F & & \\ \hline \multicolumn{2}{c|}{0} & \multicolumn{2}{c|}{0} & A-B & C-D \\ & & & & C-D & E-F \end{array}\right] \tag{12.19}$$

Here A, B, and so forth, denote the elements in Table I-10. The **F** matrix transforms similarly. Therefore it is necessary only to solve two quadratic equations for the A_1 and E species.

For the tetrahedral XY_4 molecule shown in Fig. I-9*e*, group theory (Secs. I-7, I-8) predicts one A_1 and one F_2 stretching, and one E and one F_2 bending vibration. The **U** matrix for the four stretching coordinates becomes

$$\begin{bmatrix} R'_1(A_1) \\ R'_{2a}(F_2) \\ R'_{2b}(F_2) \\ R'_{2c}(F_2) \end{bmatrix} = \begin{bmatrix} 1/2 & 1/2 & 1/2 & 1/2 \\ 1/\sqrt{6} & 1/\sqrt{6} & -2/\sqrt{6} & 0 \\ 1/\sqrt{12} & 1/\sqrt{12} & 1/\sqrt{12} & -3/\sqrt{12} \\ -1/\sqrt{2} & 1/\sqrt{2} & 0 & 0 \end{bmatrix} \begin{bmatrix} \Delta r_1 \\ \Delta r_2 \\ \Delta r_3 \\ \Delta r_4 \end{bmatrix} \tag{12.20}$$

whereas the **U** matrix for the six bending coordinates becomes

$$\begin{bmatrix} R'_1(A_1) \\ R'_{2a}(E) \\ R'_{2b}(E) \\ R'_{3a}(F_2) \\ R'_{3b}(F_2) \\ R'_{3c}(F_2) \end{bmatrix} =$$

$$\begin{bmatrix} 1/\sqrt{6} & 1/\sqrt{6} & 1/\sqrt{6} & 1/\sqrt{6} & 1/\sqrt{6} & 1/\sqrt{6} \\ 2/\sqrt{12} & -1/\sqrt{12} & -1/\sqrt{12} & -1/\sqrt{12} & -1/\sqrt{12} & 2/\sqrt{12} \\ 0 & 1/2 & -1/2 & 1/2 & -1/2 & 0 \\ 2/\sqrt{12} & -1/\sqrt{12} & -1/\sqrt{12} & 1/\sqrt{12} & 1/\sqrt{12} & -2/\sqrt{12} \\ 1/\sqrt{6} & 1/\sqrt{6} & 1/\sqrt{6} & -1/\sqrt{6} & -1/\sqrt{6} & -1/\sqrt{6} \\ 0 & 1/2 & -1/2 & -1/2 & 1/2 & 0 \end{bmatrix} \begin{bmatrix} \Delta\alpha_{12} \\ \Delta\alpha_{23} \\ \Delta\alpha_{31} \\ \Delta\alpha_{14} \\ \Delta\alpha_{24} \\ \Delta\alpha_{34} \end{bmatrix} \tag{12.21}$$

The symmetry coordinate $R'_1(A_1)$ in Eq. 12.21 represents a *redundant coordinate* (see Eq. 8.1). In such a case, a coordinate transformation reduces the order of the matrix by one, since all the **G** matrix elements related to this coordinate become zero. Conversely, this result provides a general method of finding redundant coordinates. Suppose the elements of the **G** matrix are calculated in terms of internal coordinates such as those in Table I-10. If a suitable combination of internal coordinates is made so that $\sum_j G_{ij} = 0$ (where j refers to all the equivalent internal coordinates), such a combination is a redundant coordinate. By using the **U** matrices in Eqs. 12.20 and 12.21, the problem of solving a tenth order secular equation for a tetrahedral XY_4 molecule is reduced to that of solving two first order (A_1 and E) and one quadratic (F_2) equation.

I-13. POTENTIAL FIELDS AND FORCE CONSTANTS

Using Eqs. 11.1 and 12.1, the potential energy of a bent XY_2 molecule is written

$$2V = f_{11}(\Delta r_1)^2 + f_{11}(\Delta r_2)^2 + f_{33}r^2(\Delta\alpha)^2 + 2f_{12}(\Delta r_1)(\Delta r_2) + 2f_{13}r(\Delta r_1)(\Delta\alpha) + 2f_{13}r(\Delta r_2)(\Delta\alpha) \quad (13.1)$$

This type of potential field is called a *generalized valence force* (GVF) field.* It consists of stretching and bending force constants as well as the interaction force constants between them. When using such a potential field, four force constants are needed to describe the potential energy of a bent XY_2 molecule. Since only three vibrations are observed in practice, it is impossible to determine all four force constants simultaneously.† One method used to circumvent this difficulty is to calculate the vibrational frequencies of isotopic molecules (e.g., D_2O and HDO for H_2O), assuming the same set of force constants. This method is satisfactory, however, only for simple molecules. As molecules become more complex, the number of interaction force constants in the generalized valence force field becomes too large to allow any reliable evaluation.

In another approach, Shimanouchi[37] introduced the *Urey-Bradley force* (UBF) *field*[38] which consists of stretching and bending force constants, as well as repulsive force constants between nonbonded atoms. The general form of the potential field is given by

$$V = \sum_i [\tfrac{1}{2}K_i(\Delta r_i)^2 + K_i' r_i(\Delta r_i)] + \sum_i [\tfrac{1}{2}H_i r_{i\alpha}{}^2(\Delta\alpha_i)^2 + H_i' r_{i\alpha}{}^2(\Delta\alpha_i)] + \sum_i [\tfrac{1}{2}F_i(\Delta q_i)^2 + F_i' q_i(\Delta q_i)] \quad (13.2)$$

* A potential field consisting of stretching and bending force constants only is called a simple valence force field.

† However, the range of these force constants can be determined mathematically.[36]

Here Δr_i, $\Delta\alpha_i$, and Δq_i are the changes in the bond lengths, bond angles, and distances between nonbonded atoms, respectively. The symbols K_i, K'_i, H_i, H'_i, F_i, and F'_i represent the stretching, bending, and repulsive force constants, respectively. Furthermore, r_i, $r_{i\alpha}$, and q_i are the values of the distances at the equilibrium positions and are inserted to make the force constants dimensionally similar.

Using the relation

$$q_{ij}^2 = r_i^2 + r_j^2 - 2r_i r_j \cos\alpha_{ij} \tag{13.3}$$

and considering that the first derivatives can be equated to zero in the equilibrium case, the final form of the potential field becomes

$$\begin{aligned} V = {} & \tfrac{1}{2}\sum_i \left[K_i + \sum_{j(\neq i)} (t_{ij}^2 F'_{ij} + s_{ij}^2 F_{ij})\right](\Delta r_i)^2 \\ & + \tfrac{1}{2}\sum_{i<j} [H_{ij} - s_{ij}s_{ji}F'_{ij} + t_{ij}t_{ji}F_{ij}](r_{ij}\,\Delta\alpha_{ij})^2 \\ & + \sum_{i<j} [-t_{ij}t_{ji}F'_{ij} + s_{ij}s_{ji}F_{ij}](\Delta r_i)(\Delta r_j) \\ & + \sum_{i<j} [t_{ij}s_{ji}F'_{ij} + t_{ji}s_{ij}F_{ij}](r_j/r_i)^{1/2}(\Delta r_i)(r_{ij}\,\Delta\alpha_{ij}) \end{aligned} \tag{13.4}$$

Here

$$\begin{aligned} s_{ij} &= (r_i - r_j\cos\alpha_{ij})/q_{ij} \\ s_{ji} &= (r_j - r_i\cos\alpha_{ij})/q_{ij} \\ t_{ij} &= (r_j\sin\alpha_{ij})/q_{ij} \\ t_{ji} &= (r_i\sin\alpha_{ij})/q_{ij} \end{aligned} \tag{13.5}$$

In a bent XY_2 molecule, Eq. 13.4 becomes

$$\begin{aligned} V = {} & \tfrac{1}{2}(K + t^2F' + s^2F)[(\Delta r_1)^2 + (\Delta r_2)^2] \\ & + \tfrac{1}{2}(H - s^2F' + t^2F)(r\,\Delta\alpha)^2 \\ & + (-t^2F' + s^2F)(\Delta r_1)(\Delta r_2) \\ & + ts(F' + F)(\Delta r_1)(r\,\Delta\alpha) \\ & + ts(F' + F)(\Delta r_2)(r\,\Delta\alpha) \end{aligned} \tag{13.6}$$

where

$$\begin{aligned} s &= r(1 - \cos\alpha)/q \\ t &= (r\sin\alpha)/q \end{aligned}$$

Comparing Eqs. 13.6 and 13.1, the following relations are obtained between the force constants of the generalized valence force field and those of the Urey-Bradley force field.

$$\begin{aligned} f_{11} &= K + t^2F' + s^2F \\ r^2 f_{33} &= (H - s^2F' + t^2F)r^2 \\ f_{12} &= -t^2F' + s^2F \\ r f_{13} &= ts(F' + F)r \end{aligned} \tag{13.7}$$

Although the Urey-Bradley field has four force constants, F' is usually taken as $-\frac{1}{10}F$, with the assumption that the repulsive energy between nonbonded atoms is proportional to $1/r$.[9]* Thus only three force constants, K, H, and F, are needed to construct the **F** matrix.

The number of the force constants in the Urey-Bradley field is, in general, much smaller than that in the generalized valence force field. In addition, the former field has the advantages that (1) the force constants have a clearer physical meaning than those of the generalized valence force field, and (2) they are often transferable from molecule to molecule. For example, the force constants obtained for $SiCl_4$ and $SiBr_4$ can be used for $SiCl_3Br$, $SiCl_2Br_2$, and $SiClBr_3$. Mizushima, Shimanouchi, and their co-workers[13] and Overend and Scherer[39] have given many examples that demonstrate transferability of the force constants in the Urey-Bradley force field. This property of the Urey-Bradley force constants is highly useful in calculations for complex molecules. It should be mentioned, however, that ignorance of the interactions between non-neighboring stretching vibrations and between bending vibrations in the Urey-Bradley field sometimes causes difficulties in adjusting the force constants to fit the observed frequencies. In such a case, it is possible to improve the results by introducing more force constants.[39, 40]

The normal coordinate analysis developed in Secs. I-11 to I-13 has already been applied to a number of molecules of various structures. In Parts II and III, references are cited for each type. Appendix IV lists the **G** and **F** matrix elements for typical molecules.

I-14. SOLUTION OF THE SECULAR EQUATION

Once the **G** and **F** matrices are obtained, the next step is to solve the matrix secular equation:

$$|\mathbf{GF} - \mathbf{E}\lambda| = 0 \tag{11.6}$$

In diatomic molecules, $\mathbf{G} = G_{11} = 1/\mu$ and $\mathbf{F} = F_{11} = K$. Then $\lambda = G_{11}F_{11}$ and $\tilde{\nu} = \sqrt{\lambda}/2\pi c = \sqrt{K/\mu}/2\pi c$ (Eq. 2.6). If the units of mass and force constant are atomic weight and mdyn/A (or 10^5 dynes/cm), respectively,† λ is related to $\tilde{\nu}$ (cm^{-1}) by

$$\tilde{\nu} = 1302.83\sqrt{\lambda}$$

or

$$\lambda = 0.58915\left(\frac{\tilde{\nu}}{1000}\right)^2 \tag{14.1}$$

*This assumption does not cause serious error in final results, since F' is small in most cases.

†Although the bond distance is involved in both the **G** and the **F** matrices, it is canceled during multiplication of the **G** and **F** matrix elements. Therefore any unit can be used for the bond distance.

As an example, for the HF molecule $\mu = 0.9573$ and $K = 9.65$ in these units. Then, from Eqs. 2.6 and 14.1, $\tilde{\nu}$ is 4139 cm^{-1}.

The **F** and **G** matrix elements of a bent XY_2 molecule are given in Eqs. 12.15 and 12.16, respectively. The secular equation for the A_1 species is quadratic:

$$|\mathbf{GF} - \mathbf{E}\lambda| = \begin{vmatrix} G_{11}F_{11} + G_{12}F_{21} - \lambda & G_{11}F_{12} + G_{12}F_{22} \\ G_{21}F_{11} + G_{22}F_{21} & G_{21}F_{12} + G_{22}F_{22} - \lambda \end{vmatrix} = 0 \qquad (14.2)$$

If this is expanded into an algebraic equation, the following result is obtained:

$$\lambda^2 - (G_{11}F_{11} + G_{22}F_{22} + 2G_{12}F_{12})\lambda + (G_{11}G_{22} - G_{12}^2)(F_{11}F_{22} - F_{12}^2) = 0 \qquad (14.3)$$

For the H_2O molecule,

$$\mu_1 = \mu_H = 1/1.008 = 0.99206$$
$$\mu_3 = \mu_O = 1/15.995 = 0.06252$$
$$r = 0.96(\text{A}), \quad \alpha = 105°$$
$$\sin\alpha = \sin 105° = 0.96593$$
$$\cos\alpha = \cos 105° = -0.25882$$

Then the **G** matrix elements of Eq. 12.16 are

$$G_{11} = \mu_1 + \mu_3(1 + \cos\alpha) = 1.03840$$

$$G_{12} = -\frac{\sqrt{2}}{r}\mu_3 \sin\alpha = -0.08896$$

$$G_{22} = \frac{1}{r^2}[2\mu_1 + 2\mu_3(1 - \cos\alpha)] = 2.32370$$

If the force constants in terms of the generalized valence force field are selected as

$$f_{11} = 8.428, \quad f_{12} = -0.105$$
$$f_{13} = 0.252, \quad f_{33} = 0.768$$

the **F** matrix elements of Eq. 12.15 are

$$F_{11} = f_{11} + f_{12} = 8.32300$$
$$F_{12} = \sqrt{2}\,rf_{13} = 0.35638$$
$$F_{22} = r^2 f_{33} = 0.70779$$

Using these values, we find that Eq. 14.3 becomes

$$\lambda^2 - 10.22389\lambda + 13.86234 = 0$$

The solution of this equation gives

$$\lambda_1 = 8.61475 \qquad \lambda_2 = 1.60914$$

If these values are converted to $\tilde{\nu}$ through Eq. 14.1, we obtain

$$\tilde{\nu}_1 = 3824 \text{ cm}^{-1} \qquad \tilde{\nu}_2 = 1653 \text{ cm}^{-1}$$

With the same set of force constants, the frequency of the B_2 vibration is calculated as

$$\begin{aligned} \lambda_3 &= G_{33}F_{33} = [\mu_1 + \mu_3(1 - \cos\alpha)](f_{11} - f_{12}) \\ &= 9.13681 \\ \tilde{\nu}_3 &= 3938 \text{ cm}^{-1} \end{aligned}$$

The observed frequencies corrected for anharmonicity are: $\omega_1 = 3825$ cm^{-1}, $\omega_2 = 1654$ cm^{-1}, and $\omega_3 = 3936$ cm^{-1}.

If the secular equation is third order, it gives rise to a cubic equation:

$$\begin{aligned} &\lambda^3 - (G_{11}F_{11} + G_{22}F_{22} + G_{33}F_{33} + 2G_{12}F_{12} + 2G_{13}F_{13} + 2G_{23}F_{23})\lambda^2 \\ &+ \left\{ \begin{vmatrix} G_{11} & G_{12} \\ G_{21} & G_{22} \end{vmatrix} \begin{vmatrix} F_{11} & F_{12} \\ F_{21} & F_{22} \end{vmatrix} + \begin{vmatrix} G_{12} & G_{13} \\ G_{22} & G_{23} \end{vmatrix} \begin{vmatrix} F_{12} & F_{13} \\ F_{22} & F_{23} \end{vmatrix} \right. \\ &+ \begin{vmatrix} G_{11} & G_{13} \\ G_{21} & G_{23} \end{vmatrix} \begin{vmatrix} F_{11} & F_{13} \\ F_{21} & F_{23} \end{vmatrix} + \begin{vmatrix} G_{11} & G_{12} \\ G_{31} & G_{32} \end{vmatrix} \begin{vmatrix} F_{11} & F_{12} \\ F_{31} & F_{32} \end{vmatrix} \\ &+ \begin{vmatrix} G_{12} & G_{13} \\ G_{32} & G_{33} \end{vmatrix} \begin{vmatrix} F_{12} & F_{13} \\ F_{32} & F_{33} \end{vmatrix} + \begin{vmatrix} G_{11} & G_{13} \\ G_{31} & G_{33} \end{vmatrix} \begin{vmatrix} F_{11} & F_{13} \\ F_{31} & F_{33} \end{vmatrix} \\ &+ \begin{vmatrix} G_{21} & G_{22} \\ G_{31} & G_{32} \end{vmatrix} \begin{vmatrix} F_{21} & F_{22} \\ F_{31} & F_{32} \end{vmatrix} + \begin{vmatrix} G_{22} & G_{23} \\ G_{32} & G_{33} \end{vmatrix} \begin{vmatrix} F_{22} & F_{23} \\ F_{32} & F_{33} \end{vmatrix} \\ &\left. + \begin{vmatrix} G_{21} & G_{23} \\ G_{31} & G_{33} \end{vmatrix} \begin{vmatrix} F_{21} & F_{23} \\ F_{31} & F_{33} \end{vmatrix} \right\} \lambda - \begin{vmatrix} G_{11} & G_{12} & G_{13} \\ G_{21} & G_{22} & G_{23} \\ G_{31} & G_{32} & G_{33} \end{vmatrix} \begin{vmatrix} F_{11} & F_{12} & F_{13} \\ F_{21} & F_{22} & F_{23} \\ F_{31} & F_{32} & F_{33} \end{vmatrix} = 0 \end{aligned} \tag{14.4}$$

Thus it is possible to solve the secular equation by expanding it into an algebraic equation. If the order of the secular equation is higher than three, direct expansion such as that just shown becomes too cumbersome. There are several methods of calculating the coefficients of an algebraic equation using indirect expansion.[12] The use of the electronic computer greatly reduces the burden of calculation.[39, 41]

I-15. THE SUM RULE AND THE PRODUCT RULE

Let $\lambda_1, \lambda_2, \cdots, \lambda_n$ be the roots of the secular equation, $|\mathbf{GF} - \mathbf{E}\lambda| = 0$. Then

$$\lambda_1\lambda_2\cdots\lambda_n = |\mathbf{G}|\,|\mathbf{F}| \tag{15.1}$$

holds for a given molecule. Suppose that another molecule has exactly the same $|\mathbf{F}|$ as that in Eq. 15.1. Then a similar relation

$$\lambda_1'\lambda_2'\cdots\lambda_n' = |\mathbf{G}'|\,|\mathbf{F}|$$

holds for the second molecule. It follows that

$$\frac{\lambda_1\lambda_2\cdots\lambda_n}{\lambda_1'\lambda_2'\cdots\lambda_n'} = \frac{|\mathbf{G}|}{|\mathbf{G}'|} \tag{15.2}$$

Since

$$\tilde{\nu} = \frac{1}{2\pi c}\sqrt{\lambda}$$

Eq. 15.2 can be written

$$\frac{\tilde{\nu}_1\tilde{\nu}_2\cdots\tilde{\nu}_n}{\tilde{\nu}_1'\tilde{\nu}_2'\cdots\tilde{\nu}_n'} = \sqrt{\frac{|\mathbf{G}|}{|\mathbf{G}'|}} \tag{15.3}$$

This is called the *product rule*, and its validity has been confirmed by using pairs of molecules such as H_2O and D_2O, CH_4 and CD_4. This rule is also applicable to the product of vibrational frequencies belonging to a single symmetry species.

It is also seen from Eqs. 14.3 and 14.4 that

$$\lambda_1 + \lambda_2 + \cdots + \lambda_n = \sum_n \lambda = \sum_{i,j} G_{ij}F_{ij} \tag{15.4}$$

Let σ_k denote $\sum_{ij} G_{ij}F_{ij}$ for k different molecules, all of which have the same $\mathbf{F}$ matrix. If a suitable combination of molecules is taken so that

$$\begin{aligned}\sigma_1 + \sigma_2 + \cdots + \sigma_k &= (\textstyle\sum G_{ij}F_{ij})_1 + (\sum G_{ij}F_{ij})_2 + \cdots + (\sum G_{ij}F_{ij})_k \\ &= [(\textstyle\sum G_{ij})_1 + (\sum G_{ij})_2 + \cdots + (\sum G_{ij})_k](\sum F_{ij}) \\ &= 0\end{aligned}$$

then it follows that

$$(\textstyle\sum\lambda)_1 + (\sum\lambda)_2 + \cdots + (\sum\lambda)_k = 0 \tag{15.5}$$

This is called the *sum rule*, and it has been verified for such combination as H_2O, D_2O and HDO where

$$2\sigma(\mathrm{HDO}) - \sigma(\mathrm{H_2O}) - \sigma(\mathrm{D_2O}) = 0$$

Such relations between the frequencies of isotopic molecules are highly useful in making band assignments.

I-16. VIBRATIONAL SPECTRA OF CRYSTALS

Because of intermolecular interactions, the symmetry of a molecule is generally lower in the crystalline state than in the gaseous (isolated) state.* This change in symmetry may split the degenerate vibrations and activate infrared (or Raman) inactive vibrations. In addition, the spectra obtained in the crystalline state exhibit *lattice modes*—vibrations due to translatory and rotatory motions of a molecule in the crystalline lattice. Although their frequencies are usually lower than 300 cm^{-1}, they may appear in the high frequency region as the combination bands with internal modes (see Fig. II-12, for example). Thus the vibrational spectra of crystals must be interpreted with caution, especially in the low frequency region.

In order to analyze the spectra of crystals, it is necessary to carry out the site group or factor group analysis described in the following subsection.

(A) Site Group Analysis

According to Halford,[42] the vibrations of a molecule in the crystalline state are governed by a new selection rule derived from *site symmetry*—a local symmetry around the center of gravity of a molecule in a unit cell. The site symmetry can be found by using the following two conditions: (1) the site group must be a subgroup of both the space group of the crystal and the molecular point group of the isolated molecule, and (2) the number of equivalent sites must be equal to the number of molecules in the unit cell. Halford[42] derived a complete table that lists possible site symmetries and the number of equivalent sites for 230 space groups. Suppose that the space group of the crystal, the number of molecules in the unit cell (Z), and the point group of the isolated molecule are known. Then the site symmetry can be found from Halford's table. In general, the site symmetry is lower than the molecular symmetry in an isolated state.

The vibrational spectra of calcite and aragonite crystals are markedly different, although both have the same composition (Sec. II-5). This result can be explained if we consider the difference in site symmetry of the CO_3^{2-} ion between these crystals. According to x-ray analysis, the space group of calcite is $\mathbf{D}_{3d}^{6}$ and Z is two. Halford's table gives

$$\mathbf{D}_3(2), \quad \mathbf{C}_{3i}(2), \quad \infty\mathbf{C}_3(4), \quad \mathbf{C}_i(6), \quad \infty\mathbf{C}_2(6)$$

as possible site symmetries for the space group, $\mathbf{D}_{3d}^{6}$ (the number in front of point group notation indicates the number of distinct sets of sites and that in the brackets denotes the number of equivalent sites for each set). Rule 1

* The symmetry of a molecule may be the lowest in solution (or liquid) because it interacts with randomly oriented molecules.

eliminates all but $\mathbf{D}_3(2)$ and $\mathbf{C}_{3i}(2)$. Rule 2 eliminates the latter since $\mathbf{C}_{3i}$ is not a subgroup of $\mathbf{D}_{3h}$. Thus the site symmetry of the CO_3^{2-} ion in calcite must be $\mathbf{D}_3$. On the other hand, the space group of aragonite is $\mathbf{D}_{2h}^{16}$ and Z is four. Halford table gives

$$2\mathbf{C}_i(4), \qquad \infty\mathbf{C}_s(4)$$

Since $\mathbf{C}_i$ is not a subgroup of $\mathbf{D}_{3h}$, the site symmetry of the CO_3^{2-} ion in aragonite must be $\mathbf{C}_s$. Thus the $\mathbf{D}_{3h}$ symmetry of the CO_3^{2-} ion in an isolated state is lowered to $\mathbf{D}_3$ in calcite and to $\mathbf{C}_s$ in aragonite. Then the selection rules are changed as shown in Table I-11.

TABLE I-11. CORRELATION TABLE FOR $\mathbf{D}_{3h}$, $\mathbf{D}_3$, $\mathbf{C}_{2v}$, and $\mathbf{C}_s$

Point Group	ν_1	ν_2	ν_3	ν_4
$\mathbf{D}_{3h}$	A_1'(R)	A_2''(I)	E'(I, R)	E'(I, R)
$\mathbf{D}_3$	A_1(R)	A_2(I)	E(I, R)	E(I, R)
$\mathbf{C}_{2v}$	A_1(I, R)	B_1(I, R)	A_1(I, R) + B_2(I, R)	A_1(I, R) + B_2(I, R)
$\mathbf{C}_s$	A'(I, R)	A''(I, R)	A'(I, R) + A'(I, R)	A'(I, R) + A'(I, R)

There is no change in the selection rule in going from the free CO_3^{2-} ion to calcite. In aragonite, however, ν_1 becomes infrared active, and ν_3 and ν_4 each split into two bands. The observed spectra of calcite and aragonite are in good agreement with these predictions (see Table II-15).

(B) Factor Group Analysis

A more complete analysis including lattice modes can be made by the method of factor group analysis developed by Bhagavantam and Venkatarayudu.[43] In this method, we consider all the normal vibrations for an entire Bravais unit cell.* Figure I-12 illustrates the Bravais cell of calcite. It consists of the following symmetry elements: I, $2S_6$, $2S_6^2 \equiv 2C_3$, $S_6^3 \equiv i$, $3C_2$ and $3\sigma_v$ (glide plane). These elements are exactly the same as those of the point group, $\mathbf{D}_{3d}$, although the last element is a glide plane rather than a plane of symmetry in a single molecule.

It is possible to derive the 230 space groups by combining operations possessed by the 32 crystallographic point groups† with operations such as

* Every molecule (or ion) in a Bravais unit cell can be superimposed on that of the neighboring unit cell by simple translation. In the case of calcite, the Bravais unit cell is the same as the crystallographic unit cell. However, this is not always the case.

† In crystals, the number of point groups is limited to 32 since only C_1, C_2, C_3, C_4, and C_6 are possible.

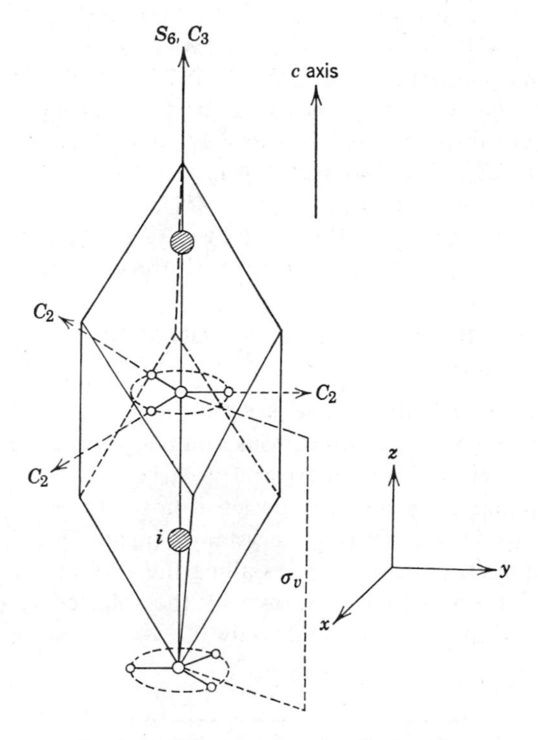

Fig. I-12. *The Bravais unit cell of calcite.*

pure translation, screw rotation (translation + rotation) and glide plane reflection (translation + reflection). If we regard the translations that carry a point in a unit cell into the equivalent point in another cell as identity, we define the 230 factor groups that are the subgroups of the corresponding space groups. In the case of calcite, the factor group consists of the symmetry elements described above, and is denoted by the same notation as that used for the space group ($\mathbf{D}_{3d}^6$). The site group discussed previously is a subgroup of a factor group.

Since the Bravais cell contains 10 atoms, it has $3 \times 10 - 3 = 27$ normal vibrations excluding three translational motions of the cell as a whole.* These 27 vibrations can be classified into various symmetry species of the factor group, $\mathbf{D}_{3d}^6$ using the procedure similar to that described previously for internal vibrations (Sec. I-7). At first, we calculate the characters of representations corresponding to the entire freedom possessed by the Bravais unit cell

* These three motions give "acoustic modes" that propagate sound waves through the crystal.

($\chi_R(N)$), translational motions of the whole unit cell ($\chi_R(T)$), translatory lattice modes ($\chi_R(T')$), rotatory lattice modes ($\chi_R(R')$), and internal modes ($\chi_R(n)$) using the equations given in Table I-12. Then each of these characters is resolved into the symmetry species of the point group, $\mathbf{D}_{3d}$. The final results show that three internal modes (A_{2u} and $2E_u$), three translatory modes (A_{2u} and $2E_u$), and two rotatory modes (A_{2u} and E_u) are infrared active; and three internal modes (A_{1g} and $2E_g$), one translatory mode (E_g), and one rotatory mode (E_g) are Raman active. As will be shown in Sec. I-17, these predictions are in perfect agreement with the observed spectra.

I-17. INFRARED DICHROISM AND POLARIZATION OF RAMAN LINES

As stated in Secs. I-7 and I-8, it is possible, by using group theory, to classify the normal vibrations into various symmetry species. Experimentally, measurements of infrared dichroism and polarization of Raman lines are useful in determining the symmetry species of normal vibrations.

Suppose that we irradiate a single crystal of calcite by polarized infrared radiation whose electric vector vibrates along the c axis (z direction) in Fig. I-12. Then the infrared spectrum shown by the solid curve of Fig. I-13 is obtained.[44] According to Table I-12, only the A_{2u} vibrations are activated

TABLE I-12. FACTOR GROUP ANALYSIS OF CALCITE CRYSTAL

$\mathbf{D}_{3d}^6$	I	$2S_6$	$2S_6^2$ ($\equiv 2C_3$)	S_6^3 ($\equiv i$)	$3C_2$	$3\sigma_v$	N	T	T'	R'	n		
A_{1g}	1	1	1	1	1	1	1	0	0	0	1		$\alpha_{xx}+\alpha_{yy}, \alpha_{zz}$
A_{1u}	1	−1	1	−1	1	−1	2	0	1	0	1		
A_{2g}	1	1	1	1	−1	−1	3	0	1	1	1		
A_{2u}	1	−1	1	−1	−1	1	4	1	1	1	1	T_z	
E_g	2	−1	−1	2	0	0	4	0	1	1	2		$(\alpha_{xx}-\alpha_{yy}, \alpha_{xy}), (\alpha_{xz}\ \alpha_{yz})$
E_u	2	1	−1	−2	0	0	6	1	2	1	2	(T_x, T_y)	
$N_R(p)$	10	2	4	2	4	0							
$N_R(s)$	4	2	4	2	2	0							
$N_R(s-v)$	2	0	2	0	2	0							
$\chi_R(N)$	30	0	0	−6	−4	0							
$\chi_R(T)$	3	0	0	−3	−1	1							
$\chi_R(T')$	9	0	0	−3	−1	−1							
$\chi_R(R')$	6	0	0	0	−2	0							
$\chi_R(n)$	12	0	0	0	0	0							

Number of Vibrations (N, T, T', R', n) obtained from $\chi_R(N)$, $\chi_R(T)$, $\chi_R(T')$, $\chi_R(R')$, $\chi_R(n)$, respectively.

p, total number of atoms in the Bravais unit cell
s, total number of molecules (ions) in the unit cell
v, total number of monoatomic molecules (ions) in the unit cell
$N_R(p)$, number of atoms unchanged by symmetry operation, R
$N_R(s)$, number of molecules (ions) whose center of gravity is unchanged by symmetry operation, R
$N_R(s-v)$, $N_R(s)$ minus number of monoatomic molecules (ions) unchanged by symmetry operation, R
$\chi_R(N) = N_P(p)\{\pm(1+2\cos\theta)\}$, character of representation for entire freedom possessed by the unit cell
$\chi_R(T) = \{\pm(1+2\cos\theta)\}$, character of representation for translational motions of the whole unit cell
$\chi_R(T') = \{N_R(s)-1\}\{\pm(1+2\cos\theta)\}$, character of representation for translatory lattice modes
$\chi_R(R') = N_R(s-v)\{\pm(1+2\cos\theta)\}$, character of representation for rotatory lattice modes
$\chi_R(n) = \chi_R(N) - \chi_R(T) - \chi_R(T') - \chi_R(R')$, character of representation for internal modes
Note that + and − signs are for proper and improper rotations, respectively. The symbol θ should be taken as defined in Sec. I-7.

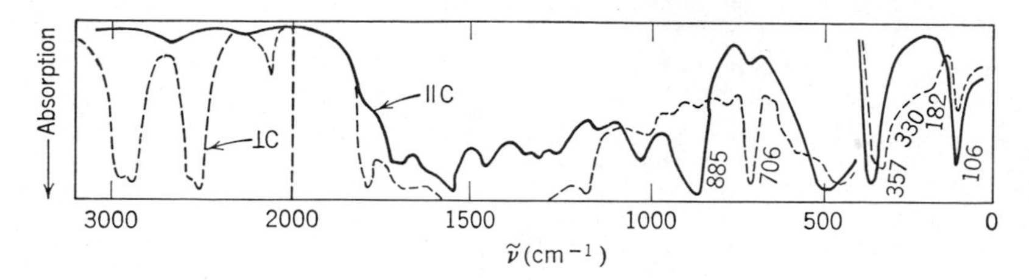

Fig. I-13. *Infrared dichroism of calcite.*[44]

under such a condition. Thus the three bands observed at 885(v), 357(t) and 106(r) cm^{-1} are assigned to the A_{2u} species. The spectrum shown by the dotted curve is obtained if the direction of polarization is perpendicular to the c axis (x, y plane). In this case, only the E_u vibrations should be infrared active. Therefore the five bands observed at 1484(v), 706(v), 330(t), 182(t), and 106(r) cm^{-1} are assigned to the E_u species. As shown above, polarized infrared studies of single crystals provide valuable information about the symmetry properties of normal vibrations if the crystal structure is known from other sources.

The ratio of the absorption intensities in the directions parallel and perpendicular to a crystal axis is called the *dichroic ratio*, and defined by

$$R = \frac{\int_{\text{band}} \varepsilon_{\parallel}(\tilde{\nu})\, d\tilde{\nu}}{\int_{\text{band}} \varepsilon_{\perp}(\tilde{\nu})\, d\tilde{\nu}} \tag{17.1}$$

Here $\varepsilon_{\parallel}$ and $\varepsilon_{\perp}$ denote the absorption coefficients (Sec. I-19) for radiation whose electric vector vibrates in the direction parallel and perpendicular, respectively, to the crystal axis. In the case of calcite, the maximum dichroic ratio is expected, since the carbonate ions are oriented parallel to each other. The dichroic ratio will be smaller if the molecules are not parallel to each other in a crystal lattice.

Polarized Raman spectra provide more information about the symmetry properties of normal vibrations than polarized infrared spectra. Again consider a single crystal of calcite. According to Table I-12, the A_{1g} vibrations become Raman active if either one of the polarizability components, α_{xx}, α_{yy}, and α_{zz} is changed. Suppose that we irradiate a calcite crystal from the y direction using the polarized radiation whose electric vector vibrates parallel to the z axis (see Fig. I-14), and observe the Raman scattering in the x direction with its polarization in the z direction. This condition is abbreviated as $y(zz)x$. In this case, Eq. 5.7 of Sec. I-5 is simplified to $P_z = \alpha_{zz} E_z$ because $E_x = E_y = 0$ and $P_x = P_y = 0$. Since α_{zz} belongs to the A_{1g} species, only the

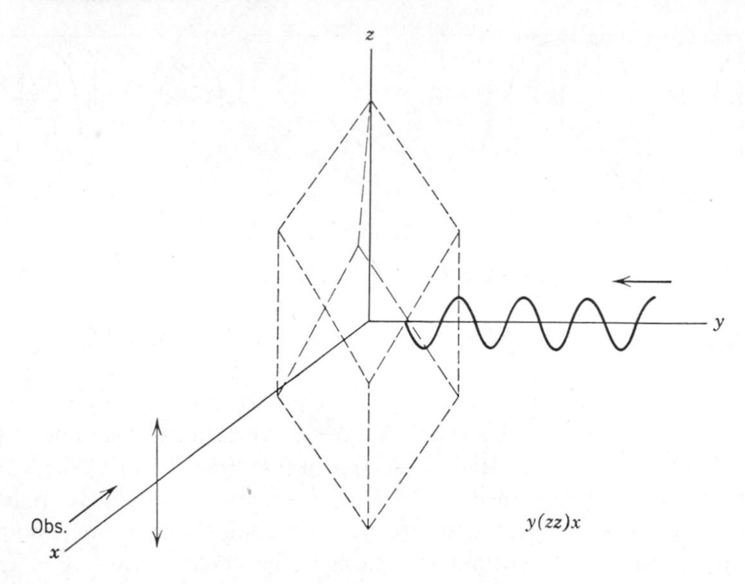

Fig. I-14

A_{1g} vibrations are observed under this condition. Figure I-15(*c*) illustrates the Raman spectrum obtained with this condition. Thus the strong Raman line at 1088 cm^{-1}(v) is assigned to the A_{1g} species. Both the A_{1g} and E_g vibrations are observed if the $z(xx)y$ condition is used. The Raman spectrum (Fig. I-15 (*a*)) shows that five Raman lines (1088(v), 714(v), 283(r), 156(t), and 1434(v) cm^{-1} (not shown)) are observed under this condition. Since the 1088 cm^{-1} line belongs to the A_{1g} species, the remaining four must belong to the E_g species. These assignments can also be confirmed by measuring Raman spectra using the $y(xy)x$ and $x(zx)y$ conditions (Fig. I-15(*b*) and (*d*)). These experiments were originally performed by Bhagavantum[46] using a mercury line as a Raman source. However, recently developed gas lasers are ideal for such experiments, since they provide strong and completely polarized radiation.

Evidently, the method described above is not applicable to crystalline powders, liquids, and solutions. In these cases, it is possible to measure the degree of depolarization defined below. Suppose that the incident radiation (natural light) is traveling in the y direction and the Raman spectrum is observed in the x direction. If the spectrum is resolved into $y(\|)$ and $z(\perp)$ components by using an analyzer, the ratio of the intensities in these two directions, ρ_n,

$$\rho_n = \frac{I_{\|}(y)}{I_{\perp}(z)} \tag{17.2}$$

is called the *degree of depolarization*.

It can be shown theoretically[12, 13, 32] that ρ_n of the normal vibration associated with normal coordinate Q_a is expressed by

$$\rho_n = \frac{I_{\parallel}}{I_{\perp}} = \frac{6\beta^2}{45\alpha^2 + 7\beta^2} \tag{17.3}$$

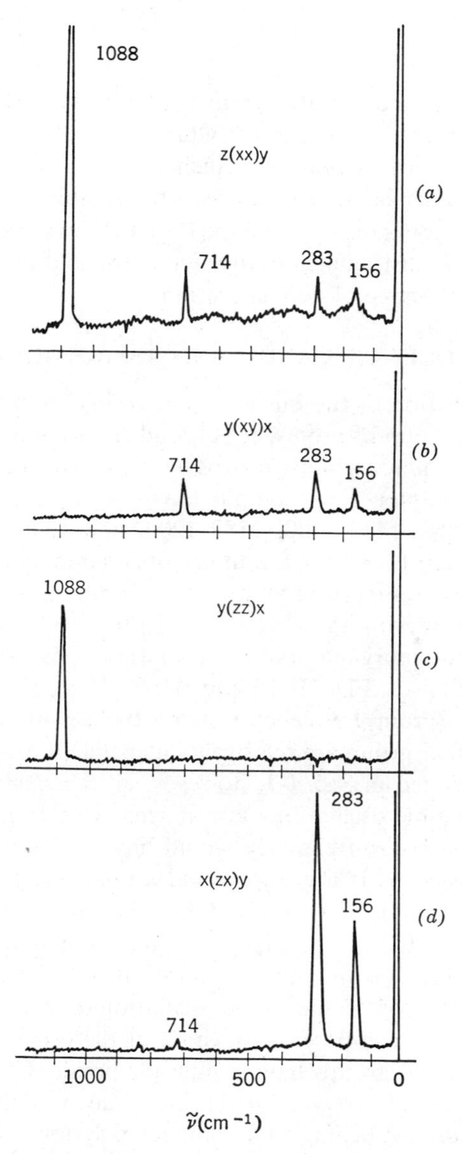

Fig. I-15. *Polarized Raman spectra of calcite.*[45]

where

$$\alpha = \frac{1}{3}\left[\left(\frac{\partial \alpha_{xx}}{\partial Q_a}\right)_0 + \left(\frac{\partial \alpha_{yy}}{\partial Q_a}\right)_0 + \left(\frac{\partial \alpha_{zz}}{\partial Q_a}\right)_0\right]$$

$$\beta^2 = \frac{1}{2}\left[\left(\frac{\partial \alpha_{xx}}{\partial Q_a} - \frac{\partial \alpha_{yy}}{\partial Q_a}\right)_0^2 + \left(\frac{\partial \alpha_{yy}}{\partial Q_a} - \frac{\partial \alpha_{zz}}{\partial Q_a}\right)_0^2 + \left(\frac{\partial \alpha_{zz}}{\partial Q_a} - \frac{\partial \alpha_{xx}}{\partial Q_a}\right)_0^2 + 6\left\{\left(\frac{\partial \alpha_{xy}}{\partial Q_a}\right)_0^2 + \left(\frac{\partial \alpha_{yz}}{\partial Q_a}\right)_0^2 + \left(\frac{\partial \alpha_{xz}}{\partial Q_a}\right)_0^2\right\}\right]$$

α become zero for all nontotally symmetric vibrations. This rule holds for molecules with and without degenerate vibrations.[8] Therefore $\rho_n = \frac{6}{7} = 0.857$ for nontotally symmetric vibrations. Such Raman lines are said to be *depolarized.* On the other hand, α is not zero for totally symmetric vibrations. Then ρ_n is in the range $0 < \rho_n < \frac{6}{7}$, and the Raman lines are said to be *polarized.* Thus it is possible to distinguish totally symmetric and nontotally symmetric vibrations from the degree of depolarization.

I-18. GROUP FREQUENCIES AND BAND ASSIGNMENTS

From the observation of the infrared spectra of a number of compounds having a common group of atoms, it is found that, regardless of the rest of the molecule, this common group absorbs over a narrow range of frequencies, called the *group frequency*. For example, the group frequencies of the methyl group are 3000–2860, 1470–1400, 1380–1200, and 1200–800 cm^{-1}. Group frequencies have been found for a number of organic and inorganic groups, and they have been summarized as *group frequency charts,* [3,4,27] which are highly useful in identifying the atomic groups from infrared spectra. Group frequency charts for inorganic and coordination compounds are shown in Appendix V as well as in Figs. II-15 and II-16 of Part II.

The concept of a group frequency rests on the assumption that the vibrations of the particular group are relatively independent of those of the rest of the molecule. As stated in Sec. I-3, however, all the nuclei of the molecule perform their harmonic oscillations in a normal vibration. Thus an *isolated vibration,* which the group frequency would have to be, cannot be expected in polyatomic molecules. If a group includes relatively light atoms such as hydrogen (OH, NH, NH_2, CH, CH_2, CH_3, etc.) or relatively heavy atoms such as the halogens (CCl, CBr, CI, etc.), when compared to other atoms in the molecule, the idea of an isolated vibration may be justified, since the amplitude (or velocity) of the harmonic oscillation of these atoms is relatively larger or smaller when compared with those of the other atoms in the same molecule. Vibrations of groups having multiple bonds (C≡C, C≡N, C=C, C=N, C=O, etc.) may also be relatively independent of the rest of the molecule if the groups do not belong to a conjugated system.

If atoms of similar mass are connected by bonds of similar strength (force constant), the amplitude of oscillation is similar for each atom of the whole system. Therfore it is not possible to isolate the group frequencies in a system like

$$-\mathrm{O}-\overset{|}{\underset{|}{\mathrm{C}}}-\overset{|}{\underset{|}{\mathrm{C}}}-\mathrm{N}\langle$$

A similar situation may occur in a system in which resonance effects average out the single and multiple bonds by conjugation. Examples of this effect are seen in the metal chelate compounds of β-diketones, α-diimines and oxalic acid (discussed in Part III). Where the group frequency approximation is permissible, the mode of vibration corresponding to this frequency can be inferred empirically from the band assignments obtained theoretically for simple molecules. If *coupling* between various group vibrations is serious, it is necessary to make a theoretical analysis for each individual compound, using a method like the following one.

As stated in Sec. I-3, the generalized coordinates are related to the normal coordinates by

$$q_k = \sum_i B_{ki} Q_i \tag{3.10}$$

In matrix form, this is written

$$\mathbf{q} = \mathbf{B}_q \mathbf{Q} \tag{18.1}$$

It can be shown[12] that the internal coordinates are also related to the normal coordinates by

$$\mathbf{R} = \mathbf{L}\mathbf{Q} \tag{18.2}$$

This is written more explicitly as

$$\begin{aligned} R_1 &= l_{11}Q_1 + l_{12}Q_2 + \cdots + l_{1N}Q_N \\ R_2 &= l_{21}Q_1 + l_{22}Q_2 + \cdots + l_{2N}Q_N \\ &\cdots\cdots\cdots\cdots\cdots\cdots\cdots\cdots \\ R_i &= l_{i1}Q_1 + l_{i2}Q_2 + \cdots + l_{iN}Q_N \end{aligned} \tag{18.3}$$

In a normal vibration in which the normal coordinate Q_N changes with frequency ν_N, all the internal coordinates, $R_1, R_2 \cdots R_i$, change with the same frequency. The amplitude of oscillation is, however, different for each internal coordinate. The relative ratio of the amplitudes of the internal coordinates in a normal vibration associated with Q_N is given by

$$l_{1N} : l_{2N} : \cdots : l_{iN} \tag{18.4}$$

If one of these elements is relatively large compared to the others, the normal vibration is said to be predominantly due to the vibration caused by the change of this coordinate.

The ratio of l's given by Eq. 18.4 can be obtained as a column matrix (or eigenvector) $\boldsymbol{l}_N$, which satisfies the relation:[12, 13]

$$\mathbf{GF}\boldsymbol{l}_N = \boldsymbol{l}_N \lambda_N \tag{18.5}$$

It consists of i elements, $l_{1N}, l_{2N}, \cdots l_{iN}$, i being the number of internal coordinates, and can be calculated if the **G** and **F** matrices are known. An assembly by columns of the l elements obtained for each λ gives the relation:

$$\mathbf{GFL} = \mathbf{L\Lambda} \tag{18.6}$$

where Λ is a diagonal matrix whose elements consist of λ values.

As an example, calculate the **L** matrix of the H_2O molecule using the results obtained in Sec. I-14. The **G** and **F** matrices for the A_1 species are:

$$\mathbf{G} = \begin{bmatrix} 1.03840 & -0.08896 \\ -0.08896 & 2.32370 \end{bmatrix} \qquad \mathbf{F} = \begin{bmatrix} 8.32300 & 0.35638 \\ 0.35638 & 0.70779 \end{bmatrix}$$

with $\lambda_1 = 8.61475$ and $\lambda_2 = 1.60914$. The **GF** product becomes

$$\mathbf{GF} = \begin{bmatrix} 8.61090 & 0.30710 \\ 0.08771 & 1.61299 \end{bmatrix}$$

The **L** matrix can be calculated from Eq. 18.6.

$$\begin{bmatrix} 8.61090 & 0.30710 \\ 0.08771 & 1.61299 \end{bmatrix} \begin{bmatrix} l_{11} & l_{12} \\ l_{21} & l_{22} \end{bmatrix} = \begin{bmatrix} l_{11} & l_{12} \\ l_{21} & l_{22} \end{bmatrix} \begin{bmatrix} 8.61475 & 0 \\ 0 & 1.60914 \end{bmatrix}$$

However, this equation gives only the ratios of $l_{11} : l_{21}$ and $l_{12} : l_{22}$. In order to determine their values, it is necessary to use the following normalization condition:

$$\mathbf{L\tilde{L}} = \mathbf{G} \tag{18.7}^*$$

Then the final result is

$$\begin{bmatrix} l_{11} & l_{12} \\ l_{21} & l_{22} \end{bmatrix} = \begin{bmatrix} 1.01683 & -0.06686 \\ 0.01274 & 1.52432 \end{bmatrix}$$

This result indicates that, in the normal vibration Q_1, the relative ratio of amplitudes of two internal coordinates, R_1 (symmetric OH stretching) and R_2 (HOH bending) is 1.0168 : 0.0127. Therefore, this vibration (3824 cm^{-1}) is assigned to an almost pure OH stretching mode. The relative ratio of amplitudes for the Q_2 vibration is -0.0669 : 1.5243. Thus this vibration is assigned to an almost pure HOH bending mode.

* This equation can be derived as follows. According to Eq. 11.3, $2T = \tilde{\dot{\mathbf{R}}}\mathbf{G}^{-1}\dot{\mathbf{R}}$. On the other hand, Eq. 18.2 gives $\dot{\mathbf{R}} = \mathbf{L}\dot{\mathbf{Q}}$ and $\tilde{\dot{\mathbf{R}}} = \tilde{\dot{\mathbf{Q}}}\tilde{\mathbf{L}}$. Thus $2T = \tilde{\dot{\mathbf{Q}}}\tilde{\mathbf{L}}\mathbf{G}^{-1}\mathbf{L}\dot{\mathbf{Q}}$. Comparing this with $2T = \tilde{\dot{\mathbf{Q}}}\mathbf{E}\dot{\mathbf{Q}}$ (matrix form of Eq. 3.11), we obtain $\tilde{\mathbf{L}}\mathbf{G}^{-1}\mathbf{L} = \mathbf{E}$ or $\mathbf{L}\tilde{\mathbf{L}} = \mathbf{G}$.

In other cases, the l values do not provide band assignments that are expected empirically.[47] This occurs because the dimension of l for a stretching coordinate is different from that for a bending coordinate. Morino and Kuchitsu[47] have proposed that the potential energy distribution of a normal vibration Q_N defined by

$$V(Q_N) = \tfrac{1}{2}Q_N^2 \sum_{ij} F_{ij} l_{iN} l_{jN} \qquad (18.8)^*$$

gives a better measure for making band assignments. In general, the value of $F_{ij} l_{iN} l_{jN}$ is large when $i = j$. Therefore, the $F_{ii} l_{iN}^2$ terms are most important in determining the distribution of the potential energy. Thus the ratios of the $F_{ii} l_{iN}^2$ terms provide a measure of the relative contribution of each internal coordinate R_i to the normal coordinate Q_N. If any $F_{ii} l_{iN}^2$ term is exceedingly large compared with others, the vibration is assigned to the mode associated with R_i. If $F_{ii} l_{iN}^2$ and $F_{jj} l_{jN}^2$ are relatively large compared with others the vibration is assigned to a mode associated with both R_i and R_j (coupled vibration).

As an example, let us calculate the potential energy distribution for the H_2O molecule. Using the **F** and **L** matrices obtained previously, we find that the $\tilde{\mathbf{L}}\mathbf{F}\mathbf{L}$ matrix is calculated to be

$$\begin{bmatrix} \begin{pmatrix} l_{11}^2F_{11} + l_{21}^2F_{22} + 2l_{21}l_{11}F_{12} \\ 8.60551 \quad 0.00011 \quad 0.00923 \end{pmatrix} & \text{zero} \\ \text{zero} & \begin{pmatrix} l_{12}^2F_{11} + l_{22}^2F_{22} + 2l_{12}l_{22}F_{12} \\ 0.03721 \quad 1.64459 \quad -0.07264 \end{pmatrix} \end{bmatrix}$$

Then the potential energy distribution in each normal vibration $(F_{ii}l_{iN}^2)$ is given by

$$\begin{array}{c} \\ R_1 \\ R_2 \end{array} \begin{array}{cc} \lambda_1 & \lambda_2 \\ \left[\begin{array}{c} 8.60551 \\ 0.00011 \end{array}\right. & \left.\begin{array}{c} 0.03721 \\ 1.64459 \end{array}\right] \end{array}$$

More conveniently, the result is expressed by calculating $(F_{ii} l_{iN}^2 / \sum F_{ii} l_{iN}^2) \times 100$ for each coordinate:

$$\begin{array}{c} \\ R_1 \\ R_2 \end{array} \begin{array}{cc} \lambda_1 & \lambda_2 \\ \left[\begin{array}{c} 99.99 \\ 0.01 \end{array}\right. & \left.\begin{array}{c} 2.21 \\ 97.79 \end{array}\right] \end{array}$$

* According to Eq. 11.1, the potential energy is written $2V = \tilde{\mathbf{R}}\mathbf{F}\mathbf{R}$. Using Eq. 18.2, this is written $2V = \tilde{\mathbf{Q}}\tilde{\mathbf{L}}\mathbf{F}\mathbf{L}\mathbf{Q}$. On the other hand, Eq. 3.12 can be written $2V = \tilde{\mathbf{Q}}\mathbf{\Lambda}\mathbf{Q}$. A comparison of these two expression gives $\mathbf{\Lambda} = \tilde{\mathbf{L}}\mathbf{F}\mathbf{L}$. If this is written for one normal vibration whose frequency is λ_N, we have

$$\lambda_N = \sum_{ij} \tilde{l}_{Ni} F_{ij} l_{jN} = \sum_{ij} F_{ij} l_{iN} l_{jN}$$

Then the potential energy due to this vibration is expressed by Eq. 18.8.

In this case, the final results are the same whether the band assignments are made based on the **L** matrix or the potential energy distribution: Q_1 is the symmetric OH stretching and Q_2 is the HOH bending. In other cases, different results may be obtained depending upon which criterion is used for band assignments.

A more rigorous method of determining the vibrational mode is to draw the displacements of individual atoms in terms of rectangular coordinates. Similar to Eq. 18.2, the relationship between the rectangular and normal coordinates is given by

$$\mathbf{X} = \mathbf{L}_x\mathbf{Q} \tag{18.9}$$

The $\mathbf{L}_x$ matrix can be obtained from the relationship:[48]

$$\mathbf{L}_x = \mathbf{M}^{-1}\tilde{\mathbf{B}}\mathbf{G}^{-1}\mathbf{L} \tag{18.10}$$

The matrices on the right have already been defined.

I-19. INTENSITY OF INFRARED ABSORPTION[49,50]

The absorption of strictly monochromatic light (ν) is expressed by the Lambert-Beer law:

$$I_\nu = I_{0,\nu}\, e^{-\alpha_\nu pl} \tag{19.1}$$

where I_ν is the intensity of the light transmitted by a cell of length l containing a gas at pressure p, $I_{0,\nu}$ is the intensity of the incident light, and α_ν is the absorption coefficient for unit pressure. The true integrated absorption coefficient A is defined by

$$A = \int_{\text{band}} \alpha_\nu \, d\nu = \frac{1}{pl}\int_{\text{band}} ln\left(\frac{I_{0,\nu}}{I_\nu}\right) d\nu \tag{19.2}$$

where the integration is carried over the entire frequency region of a band.

In practice, I_ν and $I_{0,\nu}$ cannot be measured accurately, since no spectrophotometers have an infinite resolving power. Therefore, we measure instead the apparent intensity T_ν:

$$T_\nu = \int_{\text{slit}} I(\nu)g(\nu, \nu')\, d\nu \tag{19.3}$$

where $g(\nu, \nu')$ is a function indicating the amount of light of the frequency ν when the spectrophotometer reading is set at ν'. Then the apparent integrated absorption coefficient B is defined by

$$B = \frac{1}{pl}\int_{\text{band}} ln \frac{\int_{\text{slit}} I_0(\nu)g(\nu, \nu')\, d\nu}{\int_{\text{slit}} I(\nu)g(\nu, \nu')\, d\nu}\, d\nu' \tag{19.4}$$

It can be shown that

$$\lim_{pl \to 0} (A - B) = 0 \tag{19.5}$$

if I_0 and α_ν are constant within the slit width used. (This condition is approximated by using a narrow slit.) In practice, we plot B/pl against pl, and extrapolate the curve to $pl \to 0$. In order to apply this method to gaseous molecules, it is necessary to broaden the vibrational-rotational bands by adding a high pressure inert gas (pressure broadening).

For liquids and solutions, p and α in previous equations are replaced by M (molar concentration) and ε (molar absorption coefficient), respectively. However, the extrapolation method just described is not applicable, since experimental errors in determining B values become too large at low concentration or at a small cell length. The true integrated absorption coefficient of a liquid can be calculated if we assume that the shape of an absorption band is represented by the Lorentz equation and that the slit function is triangular.[51]

Theoretically, the true integrated absorption coefficient A_N of the Nth normal vibration is given by [12]

$$A_N = \frac{n\pi}{3c}\left[\left(\frac{\partial \mu_x}{\partial Q_N}\right)_0^2 + \left(\frac{\partial \mu_y}{\partial Q_N}\right)_0^2 + \left(\frac{\partial \mu_z}{\partial Q_N}\right)_0^2\right] \tag{19.6}$$

where n is the number of molecules per cc, and c is the velocity of light. As is shown by Eq. 18.2, an internal coordinate R_i is related to a set of normal coordinates by

$$R_i = \sum_N L_{iN} Q_N \tag{19.7}$$

or

$$Q_N = \sum_i (L^{-1})_{Ni} R_i \tag{19.8}$$

If the additivity of the bond dipole moment is assumed, it is possible to write

$$\frac{\partial \mu}{\partial Q_N} = \sum_i \left(\frac{\partial \mu}{\partial R_i}\right) L_{iN} \tag{19.9}$$

Then, Eq. 19.6 is written

$$\begin{aligned} A_N &= \frac{n\pi}{3c}\left[\left(\sum_i \frac{\partial \mu_x}{\partial R_i} L_{iN}\right)_0^2 + \left(\sum_i \frac{\partial \mu_y}{\partial R_i} L_{iN}\right)_0^2 + \left(\sum \frac{\partial \mu_z}{\partial R_i} L_{iN}\right)_0^2\right] \\ &= \frac{n\pi}{3c}\sum_i \left[\left(\frac{\partial \mu_x}{\partial R_i}\right)_0^2 + \left(\frac{\partial \mu_y}{\partial R_i}\right)_0^2 + \left(\frac{\partial \mu_z}{\partial R_i}\right)_0^2\right](L_{iN})^2 \end{aligned} \tag{19.10}$$

This equation shows that the intensity of an infrared band depends upon the values of $\partial \mu/\partial R$ terms as well as of the L matrix elements.

In the case of Raman spectra, the equation corresponding to Eq. 19.6 is shown to be [32]

$$I_N = \frac{KM(\nu_0 - \Delta\nu)^4}{\Delta\nu[1 - \exp(-h\,\Delta\nu/kT)]}\,45\left(\frac{\partial\bar{\alpha}}{\partial Q_N}\right)^2\left(\frac{6}{6 - 7\rho_n}\right) \tag{19.11}$$

where K is a constant, M the molar concentration of a solution, $\bar{\alpha}$ the mean molecular polarizability ($\frac{1}{3}(\alpha_{xx} + \alpha_{yy} + \alpha_{zz})$), ν_0 the frequency of incident light, $\Delta\nu$ the frequency shift, and ρ_n the depolarization ratio defined by Eq. 17.2.

References

Introduction to Molecular Spectroscopy

1. W. Brügel, *Einfuhrung in die Ultrarotspektroskopie*, Verlag, Weinheim, 1954.
2. G. M. Barrow, *Introduction to Molecular Spectroscopy*, McGraw-Hill, 1962.
3. N. B. Colthup, L. H. Daly, and S. E. Wiberley, *Introduction to Infrared and Raman Spectroscopy*, Academic Press, 1964.
4. H. A. Szymanski, *IR: Theory and Practice on Infrared Spectroscopy*, Plenum Press, 1964.
5. J. J. Charette, *An Introduction to the Theory of Molecular Structure*, Reinhold, 1966.
6. C. N. Banwell, *Fundamentals of Molecular Spectroscopy*, McGraw-Hill, 1966.

Theory of Molecular Vibrations

7. H. Eyring, J. Walter, and G. E. Kimball, *Quantum Chemistry*, Wiley, 1944.
8. G. Herzberg, *Molecular Spectra and Molecular Structure, II: Infrared and Raman Spectra of Polyatomic Molecules*, Van Nostrand, 1945.
9. G. Herzberg, *Molecular Spectra and Molecular Structure, I: Spectra of Diatomic Molecules*, Van Nostrand, 1950.
10. A. B. F. Duncan, "Theory of Infrared and Raman Spectra," in W. West, *Chemical Applications of Spectroscopy*, Interscience, 1956.
11. G. W. King, *Spectroscopy and Molecular Structure*, Holt, Rinehart and Winston, 1964.
12. E. B. Wilson, J. C. Decius, and P. C. Cross, *Molecular Vibrations*, McGraw-Hill, 1955.
13. S. Mizushima and T. Shimanouchi, *Infrared Absorption and the Raman Effect*, Kyoritsu, Tokyo, 1958.

Symmetry, Group Theory and Matrix Theory

14. H. H. Jaffé and M. Orchin, *Symmetry in Chemistry*, Wiley 1965.
15. D. Schonland, *Molecular Symmetry*, Van Nostrand, 1965.
16. F. A. Cotton, *Chemical Applications of Group Theory*, Interscience, 1963.
17. V. Heine, *Group Theory in Quantum Mechanics*, Pergamon, London, 1960.
18. A. C. Aitken, *Determinants and Matrices*, Interscience, 1951.
19. P. S. Dwyer, *Linear Computations*, Wiley, 1951.
20. R. A. Frazer, W. J. Duncan, and A. R. Collar, *Elementary Matrices*, Cambridge, 1960.

20*a*. J. R. Ferraro and J. S. Ziomek, *Introductory Group Theory*, Plenum Press, 1969.

Review of Infrared Spectra

21. F. A. Cotton, "The Infrared Spectra of Transition Metal Complexes," in J. Lewis and R. G. Wilkins, *Modern Coordination Chemistry*, Interscience, 1960.
22. K. E. Lawson, *Infrared Absorption of Inorganic Substances*, Reinhold, 1961.
23. H. Siebert, *Anwendungen der Schwingungs-spektroskopie in der Anorganischen Chemie*, Springer-Verlag, 1966.
24. D. A. Adams, *Metal-Ligand and Related Vibrations*, Edward Arnold, 1967.
25. R. N. Jones and C. Sandorfy, "The Application of Infrared and Raman Spectrometry to the Elucidation of Molecular Structure," in W. West, *Chemical Applications of Spectroscopy*, Interscience, 1956.
26. J. Lecomte, "Spectroscopie dans l'Infrarouge," *Handbuch der Physik*, Vol. XXVI, Springer-Verlag, 1957.
27. L. J. Bellamy, *Infrared Spectra of Complex Molecules*, Wiley, 1958.
28. K. Nakanishi, *Infrared Absorption Spectroscopy*, Holden-Day, 1962.

Instrumentation

29. G. R. Harrison, R. C. Lord, and J. R. Loofbourow, *Practical Spectroscopy*, Prentice-Hall, 1948.
30. W. J. Potts, Jr., *Chemical Infrared Spectroscopy, Vol.* I. *Techniques*, Wiley, 1963.

Raman Spectra

31. S. Mizushima, "Raman Effect," *Handbuch der Physik*, Vol. XXVI, Springer-Verlag, 1957.
32. H. A. Szymanski (ed.), *Raman Spectroscopy, Theory and Practice*, Plenum Press, 1967.

General References

33. E. B. Wilson, *J. Chem. Phys.*, **7**, 1047 (1939); **9**, 76 (1941).
34. J. C. Decius, *J. Chem. Phys.*, **16**, 1025(1948).
35. T. Shimanouchi, *J. Chem. Phys.*, **25**, 660(1956).
36. A. Fadini, *Fortschritte Raman-Spektroskopie*, Weimer, 1966.
37. T. Shimanouchi, *J. Chem. Phys.*, **17**, 245, 734, and 848(1949).
38. H. C. Urey and C. A. Bradley, *Phys. Rev.*, **38**, 1969(1931).
39. J. Overend and J. R. Scherer, *J. Chem. Phys.*, **32**, 1289, 1296, and 1720. (1960); **33**, 446(1960); **34**, 547(1961); **36**, 3308(1962).
40. T. Shimanouchi, *Pure Appl. Chem.*, **7**, 131(1963).
41. J. H. Schachtschneider and R. G. Snyder, *Spectrochim. Acta*, **19**, 117 (1963).
42. R. S. Halford, *J. Chem. Phys.*, **14**, 8(1946).
43. S. Bhagavantam and T. Venkatarayudu, *Proc. Ind. Acad. Sci.*, **9A**, 224(1939); "Theory of Groups and Its Application to Physical Problems," Andhra Univ., Waltair, 1951.
44. M. Tsuboi, "Infrared Absorption Spectra," Vol. 6, Nankodo, Tokyo, 1958, p. 41.
45. S. P. S. Porto, J. A. Giordmaine and T. C. Damen, *Phys. Rev.*, **147**, 608 (1966).
46. S. Bhagavantam, *Proc. Ind. Acad. Sci.*, **11A**, 62(1940).
47. Y. Morino and K. Kuchitsu, *J. Chem. Phys.*, **20**, 1809(1952).
48. B. L. Crawford and W. H. Fletcher, *J. Chem. Phys.*, **19**, 141(1951).
49. E. B. Wilson and A. J. Wells, *J. Chem. Phys.*, **14**, 578(1946).
50. A. M. Thorndike, E. B. Wilson and A. J. Wells, *J. Chem. Phys.*, **15**, 157(1947).
51. D. A. Ramsay, *J. Am. Chem. Soc.*, **74**, 72(1952).

Inorganic Compounds

Part II

II-1. DIATOMIC MOLECULES

As shown in Sec. I-2, diatomic molecules have only one vibration along the chemical bond; its frequency is given by

$$\tilde{\nu} = \frac{1}{2\pi c}\sqrt{\frac{K}{\mu}}$$

where K is the force constant, μ the reduced mass, and c the velocity of light. In homopolar X—X molecules ($\mathbf{D}_{\infty h}$), the vibration is not infrared active but is Raman active, whereas in heteropolar X—Y molecules ($\mathbf{C}_{\infty v}$) it is both infrared and Raman active. Table II-1*a* lists the observed fundamental frequencies, together with ω_e values corrected for anharmonicity (see Sec. I-2). Except for the ions, the values were obtained in the gas phase. A large amount of information about these diatomic molecules is available, but only the fundamental frequencies of chemically interesting compounds are reviewed briefly here.

TABLE II-1*a*. VIBRATIONAL FREQUENCIES OF DIATOMIC MOLECULES (CM^{-1})

Molecule	Observed Frequency	ω_e	Molecule	Observed Frequency	ω_e
H_2	4161.13	4395.24	$N^{14}O^{16}$	1876.11	1904.03
HD	3632.06	3817.09	$N^{15}O^{16}$	1843.04	–
				1842.76	–
D_2	2993.55	3118.46	$[NO]^+$	2220	–
HF^{19}	3961.64	4138.52	$C^{12}O^{16}$	2143.16	2170.21
DF	–	2998.25	$C^{13}O^{16}$	2096.07	2121.41
TF	2443.8	2508.54	$C^{12}N^{14}$	–	2068.71
HCl^{35}	2886.01	2989.74	F_2	892.1	–
HCl^{37}	2883.89	–	$Cl_2{}^{35}$	557	564.9
DCl^{35}	2091.05	2144.77	Br_2	316.8	323.2
DCl^{37}	2088.05	2141.82	I_2	213.3*	214.57
TCl^{35}	1739.10	1775.86	FCl^{35}	773.88	–
TCl^{37}	1735.51	1772.11	FCl^{37}	766.61	793.2
HBr^{79}	2558.76	2649.67	FBr	665*	671
HBr^{81}	2558.40	2648.60	FI	604*	610
TBr	1519.26	1550.06	$Cl^{35}Br$	439.5	–
HI	2229.60	2309.53	$Cl^{35}I$	381.5	384.18
HO	–	3735.21	BrI	266.8*	268.4
DO	–	2720.9	K[CN]	2080	–
$N_2{}^{14}$	2331	2359.61	Na[OH]	3637.4	–
$O_2{}^{16}$	1555	1580.36	Na[OD]	2681.1	–

For alkali halide vapor, see Ref. 1.
T = tritium (H^3).
* From band spectra.

In 1954, Pimentel and his co-workers[2] developed the matrix isolation method to study the vibrational spectra of free radicals and other unstable molecules. The principle of this method is to trap unstable molecules in a solid matrix of inert gases such as Ar and Kr, and to observe their spectra at low temperatures. Using this method, infrared spectra of a number of free radicals have been observed, and Table II-1*b* lists the observed frequencies of diatomic free radicals thus obtained.

TABLE II-1*b*. INFRARED FREQUENCIES OF DIATOMIC FREE RADICALS IN RARE GAS MATRICES (cm^{-1})

Radical	Frequency	References
$N^{14}H$	3133	3, 4
$N^{14}D$	2323	3, 4
NF	1115	5
NCl	818.5, 825	5
NBr	691	5
OH	3596	6
OD	2680	6
OF	1028.5	7

Hydrogen halides polymerize in the condensed phases. Hydrogen fluoride polymerizes even in the gaseous phase. Infrared spectra of monomeric and polymeric HCl and HBr in solid rare gas matrices have been reported.[8,9] Hydrogen halides also form molecular compounds with organic solvents. Table II-2 indicates the effects on the frequency of polymerization and association with the solvent. In mixed crystals of HCl and HBr at low temperatures, Hiebert and Hornig[14] have found that the H—Br stretching frequency in the mixed crystals is higher than in pure HBr crystals, whereas the H—Cl stretching frequency in the mixed crystals is lower than in pure HCl crystals.

TABLE II-2. VIBRATIONAL FREQUENCIES OF HYDROGEN HALIDES IN VARIOUS PHASES (cm^{-1})

State	HF	HCl	HBr	HI
Gas (monomer)	3962	2886	2558	2230
Gas (polymer)	3500–3400, 3380–3330[10,11]	–	–	–
Liquid	3375[12]	–	–	–
Solid	3420, 3060[13]	2746, 2704[14]	2438, 2404[14]	2120[14]
Organic solvent (mesitylene)	–	2712[15] 2393 (ether)	2416[15]	2132[15]

Alkali halides dimerize in the gaseous state; a normal coordinate analysis has been carried out by Berkowitz[16] using this electrostatic model:

$$\begin{array}{ccc} & \overset{+}{M} & \\ {}^{-}X & & X^{-} \\ & \overset{+}{M} & \end{array}$$

Linevsky[17] developed a technique to condense alkali halide vapor produced at high temperature into rare gas matrices. This technique has been used widely to study the structure of metal halide vapors. For example, Redington[18] and Snelson[19] observed the infrared spectra of LiF, Li_2F_2 and Li_3F_3.

Nitric oxide (NO) exists as a dimer in the condensed phases at low temperatures.[20] Fateley et al.[21] have shown that the NO stretching frequencies are 1883 (monomer), 1862 and 1768 (*cis* dimer) and 1740 (*trans* dimer) cm^{-1}. According to Millen and Watson,[22] the nitrosonium ion ($[NO]^+$) in nitric acid absorbs at 2220 cm^{-1}.

Halogens form molecular compounds with organic solvents. For example, the band at 213 cm^{-1} (Raman active) of gaseous I_2 is shifted to 201 cm^{-1} in benzene solution,[23] and the band at 381.5 cm^{-1} of gaseous ICl is shifted to 275 cm^{-1} in pyridine solution.[24]

The cyanide ion ($[CN]^-$) is easily characterized by a relatively sharp and weak band at 2250–2050 cm^{-1}. Table II-3 lists the CN stretching frequencies of simple cyanides. It is interesting to note that phosphorus tricyanide ($P(CN)_3$) exhibits a CN stretching band at 2204 cm^{-1},[28] whereas boron tricyanide ($B(CN)_3$) has no absorption in this region because of a marked decrease in the CN bond order.[29] Miller and his co-workers[30] assigned the infrared and Raman spectra of $P(CN)_3$, $As(CN)_3$ ($\mathbf{C}_{3v}$ symmetry), and the $[C(CN)_3]^-$ ion ($\mathbf{D}_{3h}$ symmetry). Vibrational spectra of $S(CN)_2$ have been studied by Long and Steele.[31] Langseth and Møller[32] have done a normal

TABLE II-3. THE C≡N STRETCHING FREQUENCIES OF SIMPLE CYANIDES (CM^{-1})

Compound	Infrared	Raman	References
NaCN	2080	2085	25
KCN	2080	2081	25
$Ba(CN)_2$	2080	–	25
AgCN[a]	2178	–	25
CuCN[a]	2172	–	26
AuCN[a]	2239	–	27

[a] In these compounds, two metal atoms are bridged by the CN group through essentially covalent bonds.

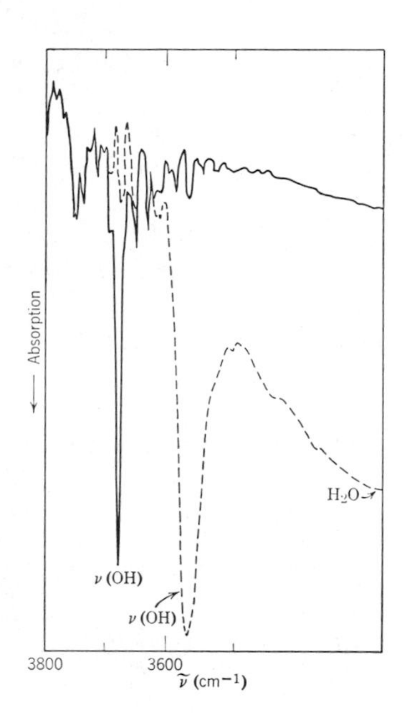

Fig. II-1*a*. Infrared spectra of LiOH (solid line) and LiOH · H_2O (broken line).[33]

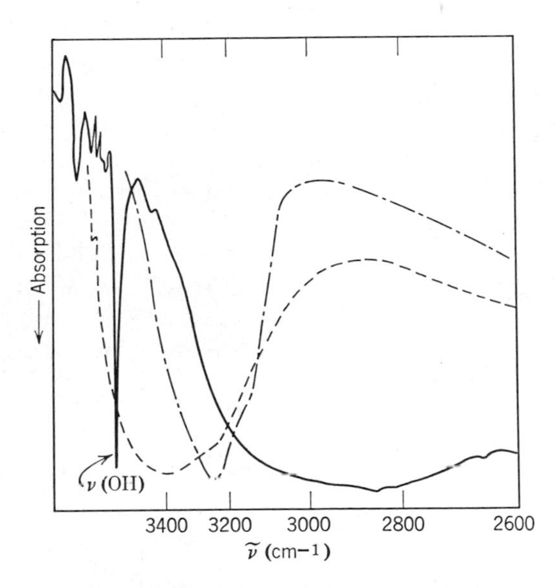

Fig. II-1*b*. Infrared spectra of LiOH · H_2O (295°K) (solid line); H_2O liquid (295°K) (broken line); H_2O solid (180°K) (dot-dash line).[33]

coordinate analysis for the linear cyanogen molecule, N≡C—C≡N. For cyano complexes, see Sec. III-5.

The hydroxyl ion ($[OH]^-$) is characterized by a sharp band at 3700-3500 cm^{-1}. In general, the hydroxyl OH stretching band is sharper and at a higher frequency than the OH stretching bands of water. Figures II-1*a* and II-1*b* compare the infrared spectra of LiOH, $LiOH \cdot H_2O$, and H_2O.[33] Table II-4 gives the OH stretching frequencies in various metal hydroxides. The relation between the OH stretching frequency and the electronegativity of the metal in a series of compounds of the type $(C_6H_5)_3M$—OH has been discussed by West and Baney.[50]

TABLE II-4. THE O—H STRETCHING FREQUENCIES IN METAL HYDROXIDES (CM^{-1})*

Compound	Infrared	Raman[44]	References
LiOH	3678	3664	33–37
$LiOH \cdot H_2O$	3574	3563	33, 38
NaOH	3637	3633	39
KOH	3600	–	40, 41
$Ca(OH)_2$	3644	3618	42, 43
$Mg(OH)_2$	3698	–	44, 45

* For other metal hydroxides, see Refs. 46–49.

Some metal hydroxides exhibit M—O—H bending bands at 1200–600 cm^{-1}, corresponding to the C—O—H bending bands at 1350–600 cm^{-1} in organic alcohols.[51,52] The presence of such absorption bands has been reported in various metal hydroxides[53–56] and in weak acids such as H_3BO_3[57] and H_5IO_6.[58] The frequency of this vibration may depend on the strength of the M—O bond as well as on that of hydrogen bonds. Hartert and Glemser[59] have found a relation between the O—H distance and the O—H stretching and M—O—H bending frequencies. Williams and Page[60] have shown that $Na_2SnO_3 \cdot 3H_2O$ should be written $Na_2[Sn(OH)_6]$, since this compound exhibits an M—O—H bending band at 898 cm^{-1} instead of H—O—H bending band near 1630 cm^{-1}.

II-2. LINEAR TRIATOMIC MOLECULES

(1) X_3 and YXY Molecules ($D_{\infty h}$)

The three normal modes of vibration of linear X_3 and YXY molecules have already been shown in Fig. I-5. The frequency ν_1 is not infrared active but is Raman active, whereas ν_2 and ν_3 are infrared active but not Raman active (mutual exclusion rule). Table II-5a lists the fundamental frequencies for compounds of these types.

TABLE II-5*a*. VIBRATIONAL FREQUENCIES OF LINEAR X_3 AND YXY MOLECULES

Molecule[a]	State	ν_1	ν_2	ν_3	References
$C^{12}O_2$	gas	(1337)[b]	667	2349	61
	solid (−190°C)	–	660	2344	62, 63
			653		
	aqueous soln	–	—	2342	64
$C^{13}O_2$	gas	–	(649)	2284	65
	solid (−190°C)	–	637	2280	62, 63
$C^{14}O_2$	gas	–	632	2226	65
CS_2	gas	658	397	1533	66
	liquid	657	397	1510	66
CSe_2	gas	(368)	(308)	1303	66
	liquid	368	300	1267	66
XeF_2	gas	513	213	564	67
				550	
KrF_2	gas	449	233	596	68
				580	
$HgCl_2$	gas	360	70	413	69, 70
$HgBr_2$	gas	225	41	293	69, 70
HgI_2	gas	156	33	237	69, 70
$K[N_3]$	solid	1344	645	2041	71, 72
$NH_4[N_3]$	solid (phase I)	1345	625	2030	73, 74
$K[HF_2]$	solid	(620)	1233	1473	75, 76
	aqueous soln	–	1206	1536	75
$K[DF_2]$	solid	–	885	1045	75
	aqueous soln	–	873	1102	75
$Na[HF_2]$	solid	675	1200	1550	76
$Na[DF_2]$	solid	–	860	1140	76
$((CH_3)_4N)[HCl_2]$	solid	210	1185	1575	77, 78
			1080		
$((CH_3)_4N)[DCl_2]$	solid	210	880	1200	77
			830		
$((C_3H_7)_4N)[ClHBr]$	solid	172	1145	1650	79
				1550	
$((C_3H_7)_4N)[ClDBr]$	solid	165	838	1250	79
$((C_2H_5)_4N)[ClHF]$	solid	275	863	2710	80
			823		
$((C_2H_5)_4N)[BrHF]$	solid	220	740	2900	80
$((C_4H_9)_4N)[IHF]$	solid	180	635	3145	80
$[NO_2](ClO_4)$	solid	1396	570	2360	81, 82
$[NO_2]^+$	conc. HNO_3	1400	–	2360	83
$Na_2[CN_2]$	solid	1234	598	2120	84
$[UO_2]^{2+}$	solid	856	–	931	105, 85

[a] Frequencies listed are for ions in square brackets.
[b] Average of two frequencies split by the Fermi resonance. (See Sec. I-10).

The infrared and Raman spectra of gaseous XeF_2 and KrF_2 clearly indicate that they are linear and symmetric. Infrared spectra of NiF_2 and $NiCl_2$ have been observed in an Ar matrix; ν_3 is at 780 cm^{-1} for NiF_2 and at 520 cm^{-1} for $NiCl_2$.[86] For other gaseous transition metal halides, see Ref. 87. Snelson[88] has obtained the infrared spectra of monomeric alkaline earth dihalides in rare gas matrices; he has shown that BeF_2, $BeCl_2$, and MgF_2 are linear whereas CaF_2, SrF_2, and BaF_2 are nonlinear. BeF_2 and $BeCl_2$ are dimeric in the gaseous state,[89] although they are polymeric in the crystalline state.[90] Infrared spectra of LiO, Li_2O(linear), and Li_2O_2(rhombic)

X—Be<(X)(X)>Be—X   X = F, Cl

have also been obtained in rare gas matrices.[91]

Infrared spectra of triatomic anions of the type, X-H-Y^- (X and Y are halogens), have been obtained as tetraalkylammonium salts. In addition to those listed in Table II-5*a*, the infrared spectra of the DI_2^-, DBr_2^-, $DClI^-$, $DClNO_3^-$, and $HClNO_3^-$ anions are reported.[92] Recently, vibrational spectra of a number of trihalide anions have been obtained (Table II-5*b*). Most of these anions are linear or almost linear.

TABLE II-5*b*. VIBRATIONAL FREQUENCIES OF TRIHALIDE ANIONS (CM^{-1})

Compound	State	ν_1	ν_2	ν_3	References
$((C_3H_7)_4N)[Cl_3]$	solid	268	(165)	242	93
$[Br_3]^-$	aqueous soln	170	–	210	94
$Cs[I_3]$	solid	103	69	149	95, 96
Cs[BrII]	solid	117	84	168	95
Cs[IBrBr]	solid	143	98	178	95, 96
Cs[ClICl]	solid	268	129	218	95
Cs[BrICl]	solid	198	145–125	180	95
Cs[FClF]	solid	510–478	–	636	97

Long and co-workers[98] have done a normal coordinate analysis for carbon suboxide (C_3O_2) assuming a linear model ($\mathbf{D}_{\infty h}$), whereas Rix[99] proposed a bent model ($\mathbf{C}_{2h}$). This controversy still remains unsettled even after a recent infrared study by Miller and Fateley.[100] The infrared and Raman spectra of C_3S_2 have been obtained and interpreted on the basis of a linear structure by Smith and Leroi.[101]

The presence of the nitronium ion ($[NO_2]^+$) in mixtures of concentrated HNO_3 and H_2SO_4 has been demonstrated by Ingold and co-workers.[102]

$$HNO_3 + 2H_2SO_4 \rightleftharpoons [NO_2]^+ + 2[HSO_4]^- + [H_3O]^+$$

Marcus and Fresco[83] made an extensive infrared study of mixtures of strong acids.

The configuration of the uranyl ion ($[UO_2]^{2+}$) in simple salts has been a subject of considerable interest in past years. If it is linear, only ν_2 and ν_3 are infrared active but neither is Raman active. If it is bent, all vibrations are both infrared and Raman active. Thus it is theoretically possible to distinguish these two structures by the activity of the three fundamentals in both spectra. Although the linear structure is favored by many investigators,[85,103] the ν_2 and ν_3 vibrations have been observed in the Raman spectrum. In order to explain this anomaly, Sutton[104] postulated that ν_2 and ν_3 appear in the Raman spectrum, both because of polarization of the two U=O bonds and because of the asymmetrical field produced by a complexing anion. Jones and Penneman[105] have studied the infrared spectra of a series of compounds having the structures $XO_2(ClO_4)_2$ and $[XO_2(CH_3COO)_3]^-$ where X is Np, U, Pu, or Am; they have found that the X=O stretching force constant decreases in going from Np to Am. Their finding contradicts the results of an x-ray study on $Na[XO_2(CH_3COO)_3]$ crystals which suggests that the X=O distance decreases in the order U > Np > Pu > Am. In other words, the shortest X=O bond, which occurs in the Am compound, has the smallest X=O stretching force constant in this series. They explained this apparent contradiction by postulating that the bond, though shortened by contraction of the electron shells of the metal, is weakened by interaction with the extra valence shell electrons. On the basis of the x-ray and infrared data of a number of compounds containing $[UO_2]^{2+}$, Jones[106] derived a general formula relating the U=O distance and the U=O stretching force constant. For the infrared spectra and structures of simple $[UO_2]^{2+}$ salts, see Refs. 107 and 108. McGlynn et al.[109] have found that, in a series of complexes of the type $K_xUO_2L_y(NO_3)_2$, both the ν_3 and ν_1 frequencies of the UO_2 group decrease as L is changed along the spectrochemical series,

$$[CN]^- > en^* > NH_3 > [NCS]^- > [ONO]^- > py^* > H_2O > F^- > [NO_3]^-$$

which Tsuchida[110] obtained from the ultraviolet spectral study of cobaltic complexes.

(2) XYZ Molecules ($C_{\infty v}$)

In linear XYZ molecules, the three normal modes of vibration shown in Fig. II-2 are both infrared and Raman active. Table II-6*a* gives the fundamental frequencies for compounds of this type. Table II-6*b* lists the infrared frequencies of linear triatomic free radicals observed in rare gas matrices.

* In this equation, en = ethylenediamine and py = pyridine.

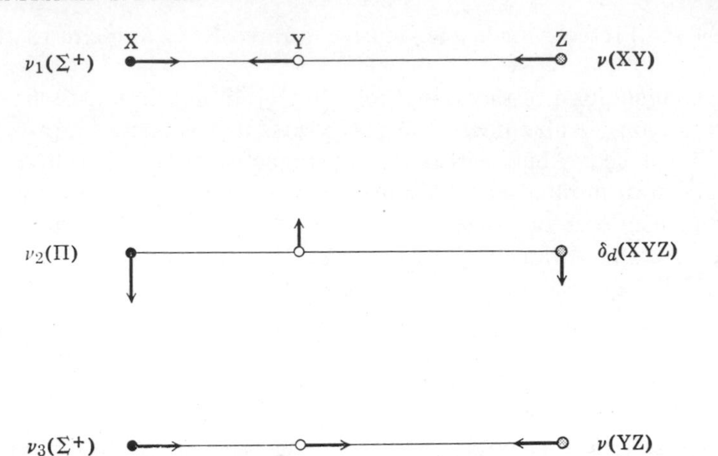

Fig. II-2. Normal modes of vibration of linear XYZ molecules.

For the hydrogen cyanide tetramer, Webb and colleagues[135] proposed structure I, whereas Wadsten and Andersson[136] preferred the dimeric structure II on the basis of the results of x-ray and infrared studies.

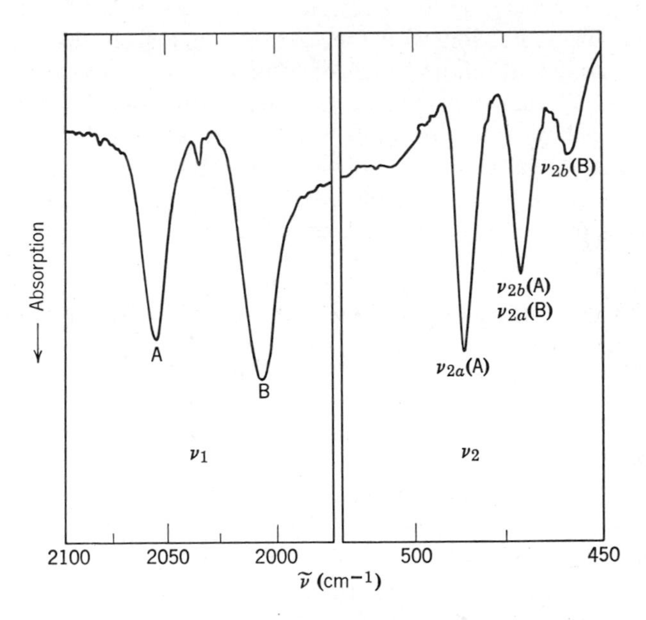

Fig. II-3. Infrared spectra of mixture of $KN^{14}C^{12}S^{32}$ (A) and $KN^{14}C^{13}S^{32}$ (B) in the solid state.[127]

Table II-6a. Vibrational Frequencies of Linear XYZ Molecules (cm^{-1})

Molecule [XYZ]	State	ν_1	ν_2	ν_3	References
$HC^{12}N$	gas	3311	712	2097	111
$HC^{13}N$	gas	3295	706	–	112
$DC^{12}N$	gas	2630	569	1925	111
$DC^{13}N$	gas	2585	–	1916	112
TCN	gas	2460	513	1724	113
FCN	gas	1077	449	2290	114, 115
ClCN	gas	714	380	2219	116–118
	liquid	730	394	2206	119
	solid (−180°C)	734	398	2212	116
BrCN	gas	574	342.5	2200	116, 120
	liquid	568	357	2191	121
	solid (−180°C)	572.5	363.5	2194	116
ICN	liquid	470	321	2158	122
	solid (−180°C)	452	329	2716	116
$N^{14}N^{14}O$	gas	2224	589	1286	123
	solid (80°K)	2238	591	1293	124
$N^{14}N^{15}O$	gas	2178	576	1281	123
	solid (80°K)	–	–	1196	124
$N^{115}N^{14}O$	gas	2203	586	1271	123
	solid (80°K)	2220	–	1280	124
$N^{15}N^{15}O$	gas	2156	572	1266	123
SCO	gas	859	524	2064	125, 120
SCSe	gas	1435	(355)	506	117, 66
	liquid	(1408)	(350)	502	117, 66
SCTe	CS_2 soln	1347	(337)	423	117, 66
$K[N^{14}C^{12}S^{32}]$	solid (mull)	2041	–	747	126
	solid (KBr disk)	2053	486, 471	748	126
	aqueous soln	2066	470	743	127
$K[N^{14}C^{13}S^{32}]$	solid	2006	470, 458	743	127
$K[N^{14}C^{12}S^{34}]$	solid	2052	487, 470	739	127
$K[N^{14}C^{13}S^{34}]$	solid	2005	470, 458	733	127
K[NCSe]	solid	2070	424, 416	558	128
$K[NC^{12}O]$	solid	2165	637, 628	1207	129, 130
$K[NC^{13}O]$	solid	2112	620, 612	1195	129, 30

TABLE II-6*b*. INFRARED FREQUENCIES OF LINEAR TRIATOMIC FREE RADICALS (cm^{-1})

X-Y-Z	$\nu(XY)$	$\delta(XYZ)$	$\nu(YZ)$	References
$H—N^{14}—C^{12}$	3583	535	2032	131
$D—N^{14}—C^{12}$	2733	413	1940	131
$H—C^{13}—O$	2488	1084	1821	132
$H—C^{12}—O$	2488	1090	1861	132
C—C—O	1074	381	1978	133
$C^{12}—N^{14}—N^{14}$	2847	393	1241	134

$$H_2N-C-C\equiv N$$
$$\|$$
$$H_2N-C-C\equiv N$$

I

$$N\equiv C-C(H)=N-H$$

II

However, Penfold and Lipscomb[137] have shown from x-ray analysis that structure I is correct for this compound.

Figure II-3 shows the effect of isotopic substitution on the infrared spectra of KNCS crystals as obtained by Jones.[127] As Table II-6*a* indicates, the doubly degenerate ν_2 vibration splits into two bands in the NCS and NCO salts because the crystal field removes the degeneracy. Raman spectra of thiocyanato salts of alkali metals have been studied by several investigators.[138,139] The structure of thiocyanogen trichloride cannot be Cl_3-SCN, but it is either

$$Cl-S-C(Cl)=N-Cl \quad \text{or} \quad Cl_2C=N-S-Cl$$

since a strong band is observed at 1600 cm^{-1} instead of a C≡N stretching band near 2100 cm^{-1}.[140,141] Nelson and Pullin[142] have obtained the infrared spectra of thiocyanogen, $(SCN)_2$, and thiocyanogen halides, X-SCN(X: halogen).

Vibrational spectra of $P(NCO)_3$ and $PO(NCO)_3$ have been studied by Miller and Baer,[143] and those of $P(NCS)_3$ and $PO(NCS)_3$ by Oba et al.[144] For $Si(NCO)_4$ and $Ge(NCO)_4$, see Refs. 145 and 146, respectively. Normal coordinate analyses of linear triatomic molecules have been carried out by many investigators.[120,147–149]

II-3. BENT TRIATOMIC MOLECULES

Bent triatomic molecules have the three normal modes of vibration that are shown in Fig. I-.5 The vibrations are both infrared and Raman active,

TABLE II-7. VIBRATIONAL FREQUENCIES OF BENT XY_2 AND X_3 MOLECULES (cm^{-1})

Molecule	State	ν_1	ν_2	ν_3	References
H_2O^{16}	gas	3657	1595	3756	150
	liquid	3219	1627	3445	151
	solid (−78°C)	3400	1620	3220	152
H_2O^{18}	gas	(3647)	(1586)	(3744)	153
HDO^{16}	gas	2727	1402	3707	150
	liquid	2520	1455	3405	154
	solid (−190°C)	2416	1490	3275	152
D_2O^{16}	gas	2671	1178	2788	150
	solid (−190°C)	(2520)	1210	2432	152
D_2O^{18}	gas	2657	1169	2764	155
THO	gas	–	1324	3720	156
TDO	gas	–	–	2735	156
T_2O	gas	–	996	2370	156
H_2S	gas	2615	1183	(2627)	157
	solid (66°K)	2532, 2523	1186, 1171	2544	158
HDS	gas	–	1090	–	159
	solid (66°K)	--	1026	2535	158
D_2S	gas	1892	934	2000	159
	solid (66°K)	1843, 1835	857	1854	158
H_2Se	gas	2345	1034	2358	160
D_2Se	gas	1687	741	1697	160
O_3	gas	1110	705	1043	161
$NH_4[O_3]$	solid	1260	800	1140	162
F_2O	gas	929	461	826	163
Cl_2O	solid	631	296	671	164
ClO_2	gas	943	445	1111	165
$[ClF_2]^+$	solid	810	558, 520	813	116
$Na[ClO_2]$	aqueous soln	790	400	(840)	167
$N^{14}O_2$	gas	1318	750	1618	168
	solid	–	750	1624	21
$N^{15}O_2$	gas	1306	740	1580	168
$Na[N^{14}O_2)$	solid (mull)	1328	828.2	1261	169
$Na[N^{15}O_2]$	solid (mull)	1303	824	–	169
SO_2^{16}	gas	1151	518	1362	170, 171
	solid (−180°C)	1147	521	1330, 1308	172
	aqueous soln	1157	–	1332	64
$SO^{16}O^{18}$	gas	1122	507	1341	173

TABLE II-7 *continued*

Molecule	State	ν_1	ν_2	ν_3	References
SO_2^{18}	gas	–	–	1316	173
SeO_2	gas	910	(400)	967	174
SCl_2	liquid	514	208	535	175
$N^{14}H_2$	CO matrix	–	1499	3220	176
CF_2	Ar matrix	1222	668	1102	177
SiF_2	gas	855	345	872	178
GeF_2	gas	692	263	663	179
CCl_2	Ar matrix	721[a]	–	748	180
NO_2	Ar matrix	–	749	1610	181
Al_2O	Ar matrix	715	(238)	994	182

[a] Tentative band assignment.

TABLE II-8. VIBRATIONAL FREQUENCIES OF BENT XYZ MOLECULES (CM^{-1})

Molecule (XYZ)	ν_1 ν(XY)	ν_2 δ(XYZ)	ν_3 ν(YZ)	References
HOCl	3626	1242	739	188
DOCl	2674	911	739	188
ONF	1844	521	766	189
$ON^{14}Cl^{35}$	1800	332	605	190
$ON^{15}Cl^{35}$	1769	331	590	190
$ON^{14}Cl^{37}$	–	325	–	191
ONBr	1801	(265)	542	192
NSCl	1325	273	414	193
NSF	1372	640	366	194
OOH[a]	1101	1389	3414	195
OOF[a]	1500	376	586	196, 197
$FC^{12}O^{16}$,[a]	1018	626	1855	198
ClCO[a]	570	281	1880	199

[a] Radical.

whether the molecule is symmetrical (XY_2 and X_3, $\mathbf{C}_{2v}$) or asymmetrical (XYZ and XXY, $\mathbf{C}_s$). Tables II-7 and II-8 list the fundamental frequencies for a number of bent triatomic molecules. The last parts of both tables give the frequencies of free radicals observed in various matrices. Table II-7 indicates that, in most compounds, the antisymmetric stretching frequency (ν_3) is higher than the symmetric one (ν_1). However, this is not true for O_3, F_2O, $[NO_2]^-$, and H_2O (ice).

Vibrational spectra of ices in various crystalline phases have been studied extensively by Whalley and his coworkers.[183] Vibrational frequencies of water in various organic solvents have been studied by Greinacher et al.[184] For example, dioxane solutions of water exhibit three bands at 3518, 1638, and 3584 cm^{-1}. Apparently, hydrogen bonding between water and dioxane is responsible for the shifts of the stretching modes to lower frequencies and for the shift of the bending mode to a higher frequency. The vibrational spectra of lattice water and coordinated water will be discussed in Sec. III-3.

Solutions of hydrogen sulfide (H_2S) in organic solvents have been studied by Josien and Saumagne.[185] Their results show, for example, that the band at 2627 cm^{-1} in the gaseous state is shifted to 2482 cm^{-1} in pyridine solution.

Anbar and colleagues[186] have compared the infrared frequencies of $AgNO_2^{16}$ and $AgNO_2^{18}$ in Nujol mulls. For the frequencies of other nitrites, see Ref. 187. The selenium dioxide (SeO_2) polymer in the crystalline state has been studied by Giguère and Falk.[174]

II-4. PYRAMIDAL FOUR-ATOM MOLECULES

(1) XY_3 Molecules (C_{3v})

The four normal modes of vibration of a pyramidal XY_3 molecule are shown in Fig. II-4. These four vibrations are both infrared and Raman active. Table II-9 lists the fundamental frequencies of molecules of the XH_3 type. Several bands marked by an asterisk (*) in Table II-9 are split in two because of *inversion doubling*. This arises in pyramidal XY_3 molecules, where the two configurations shown in the sketch are equally probable. If the potential barrier between these two configurations is small, the molecule may resonate between the two structures. As a result, each vibrational level splits into two (positive and negative). Transitions between levels of different sign are allowed in the infrared spectrum, whereas those between levels of the same sign are allowed in the Raman spectrum. The transition between the two levels at $v = 0$ is also observed in the microwave region

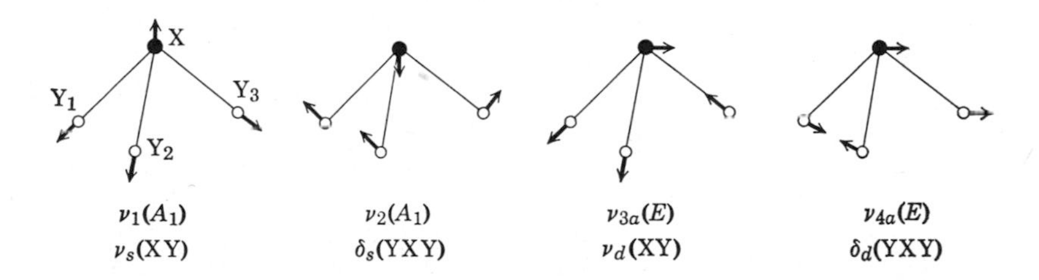

Fig. II-4. Normal modes of vibration of pyramidal XY_3 molecules.

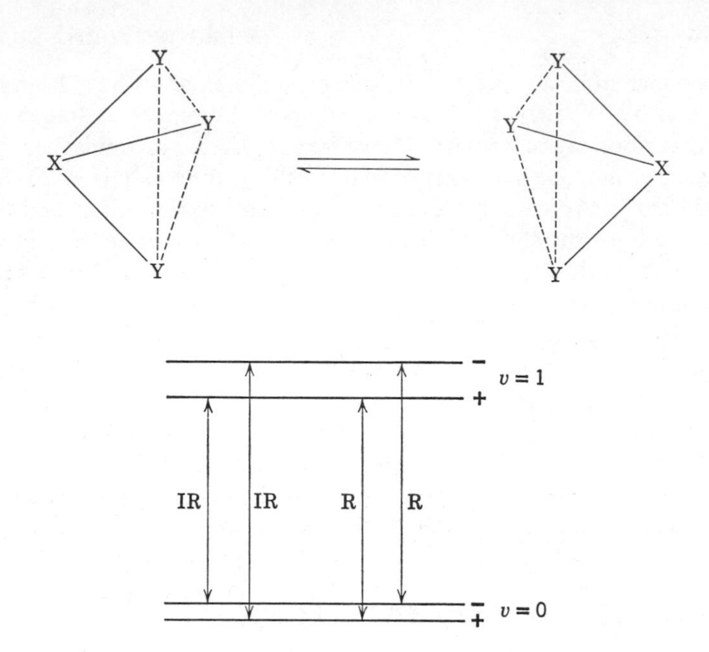

TABLE II-9. VIBRATIONAL FREQUENCIES OF PYRAMIDAL XY_3 MOLECULES (cm^{-1})

Molecule	State	ν_1	ν_2	ν_3	ν_4	References
NH_3	gas	3336 3338 }*	932 968 }*	3414	1628	200, 201
	solid (−190°C)	3223	1060	3378	1646	202
ND_3	gas	2420	746 749 }*	2556	1191	203
	solid (−190°C)	2318	815	2500	1196	202
$N^{15}H_3$	gas	3335	926 961 }*	–	1625	204
NT_3	gas	2014	657	2185	996	205
PH_3	gas	2327	990 992 }*	2421	1121	206
PD_3	gas	1694	730	(1698)	806	206
AsH_3	gas	2122	906	2185	1005	206
AsD_3	gas	1534	660	–	714	206
SbH_3	gas	1891	782	1894	831	207
SbD_3	gas	1359	561	1362	593	207
$[OH_3]ClO_4$	solid	3285	1175	3100	1577	208
$[OH_3]NO_3$	solid	2780	1135	2780	1680	209
$[OH_3]HSO_4$	solid	2840	1160	2840	1620	209

($\tilde{\nu}$ = 0.79 cm^{-1}). If the potential barrier is sufficiently high and if the three Y groups are not identical, optical isomers may be anticipated. Weston[210] has shown through calculation of the potential barrier that phosphorus compounds may have optical isomers at low temperatures, and compounds of arsenic and antimony may be optically active even at room temperature.

Water in acid hydrates such as $HClO_4 \cdot H_2O$ exists as the hydronium ion ($[OH_3]^+$). The infrared spectrum of the hydronium ion has been studied by several investigators.[211–214] Gillard and Wilkinson[215] have shown from infrared spectra that $H_2PtCl_6 \cdot 2H_2O$ should be formulated as $(H_3O)_2(PtCl_6)$. For infrared and Raman studies of related compounds, see the following references: liquid ammonia (216, 217), $NH_3 \cdot H_2O$ crystal (218), partially deuterated ammonia (219, 220) and partially deuterated phosphine (221). For a normal coordinate analysis of XH_3 compounds, see Ref. 222–224.

Table II-10 lists the fundamental frequencies of halogen compounds of the XY_3 type. It is interesting that the hydrogen compounds in Table II-9 have symmetric stretching and bending frequencies (ν_1 and ν_2) which are lower than the antisymmetric stretching and bending frequencies (ν_3 and ν_4), respectively, while the opposite prevails in the halogen compounds in

TABLE II-10. VIBRATIONAL FREQUENCIES OF PYRAMIDAL XY_3 MOLECULES (CM^{-1})

Molecule	ν_1	ν_2	ν_3	ν_4	References
NF_3	1032	647	905	493	225, 226
PF_3	892	487	860	344	227, 228
AsF_3	707	341	644	274	229
PCl_3	507	260	494	189	230
$AsCl_3$	412	194	387	155	230
$SbCl_3$	377	164	356	128	230, 231
$BiCl_3$	288	130	242	100	232
$[GeCl_3]^-$	320	162	253	139	233
$[SnCl_3]^-$	297	128	256	103	234
$[SeCl_3]^+$	416	234	395	186	239
$[TeCl_3]^+$	391	185	367	139	239
PBr_3	392	161	392	116	230
$AsBr_3$	(284)	128	275	98	235
$SbBr_3$	254	101	245	81	236
$BiBr_3$	196	104	169	90	237
$[SnBr_3]^-$	211	83	181	65	234
PI_3	303	111	325	79	238
AsI_3	226	102	201	74	237
SbI_3	177	89	147	71	237
BiI_3	145	90	115	71	237

Table II-10. The vibrational spectra of chloroammonia (NH_xCl_{3-x}) and nitrogen trichloride (NCl_3) have been studied by Moore.[240] Schatz[241] has done a normal coordinate analysis for NF_3.

Table II-11 lists the fundamental frequencies of various ions of the XY_3 type. The frequencies ν_1 and ν_3 are near each other for all these ions. Thus the stretching vibrations are usually observed as one strong, broad band. The infrared spectra of a number of metal chlorates ($[ClO_3]^-$) have been

TABLE II-11. VIBRATIONAL FREQUENCIES OF PYRAMIDAL XY_3 IONS AND MOLECULES (CM^{-1})

XY_3	State	ν_1	ν_2	ν_3	ν_4	References
$[ClO_3]^-$	solid (R)	930	620	975	486	242, 243
	solid (IR)	910	617	960	493	242, 244
	soln (R)	930	610	982	479	245
$[BrO_3]^-$	soln (R)	806	421	836	356	245, 246
$[IO_3]^-$	soln (R)	754	373	774	355, 332	247, 248
$[SO_3]^{2-}$	soln (R)	967	620	933	469	249, 250
	solid (IR)	1010	633	961	496	249, 251
$[SeO_3]^{2-}$	soln (R)	807	432	737	374	252, 253
$[TeO_3]^{2-}$	soln (R)	758	364	703	326	252
XeO_3	soln (R)	780	344	833	317	254

obtained by Rocchiccioli.[243] Dasent and Waddington[247] studied the infrared spectra of metal iodates ($[IO_3]^-$) and related compounds. It was suggested that extra bands observed at 480–420 cm^{-1} may be due to the metal-oxygen vibration. Figure II-5 shows the infrared spectra of $KClO_3$ and KIO_3 in the solid state. Rocchiccioli[251] has measured the infrared spectra of a number of metal sulfites ($[SO_3]^{2-}$). According to Depaigne-Delay et al.,[255] the stannate ion ($[SnO_3]^{2-}$) is planar in the Pb(II) and Ba(II) salts, and slightly

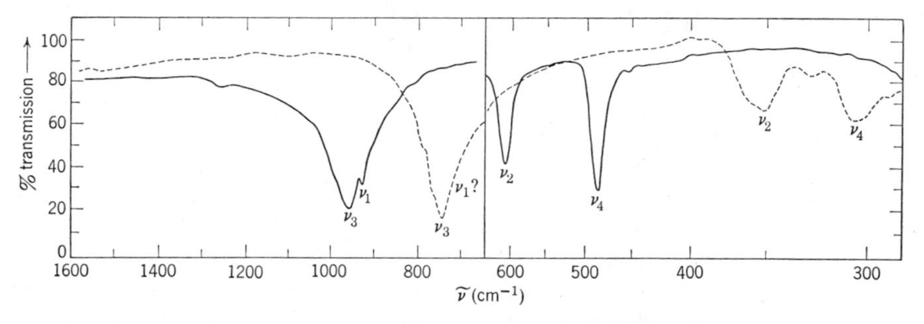

Fig. II-5. Infrared spectra of $KClO_3$ (solid line) and KIO_3 (broken line).

pyramidal in the Cu(II) salt. Interpretations of the vibrational spectra of metal titanates ($[TiO_3]^{2-}$) are based on the distorted octahedral TiO_6 unit ($\mathbf{D}_{4h}$ symmetry) in the crystalline state.[256]

(2) ZXY_2 Molecules ($\mathbf{C}_s$)

Substitution of a Z atom for one Y atom in the XY_3 molecule lowers the symmetry from $\mathbf{C}_{3v}$ to $\mathbf{C}_s$. As a result, the degenerate vibrations split into two bands. Thus six vibrations are observed, all of which are infrared and Raman active. The relation between $\mathbf{C}_{3v}$ and $\mathbf{C}_s$ is shown in Table II-12. Table II-13 lists the fundamental frequencies of pyramidal ZXY_2 molecules. The second compound from the bottom is selenious acid, the band assignments for which

TABLE II-12. RELATIONSHIP BETWEEN $\mathbf{C}_{3v}$ AND $\mathbf{C}_s$

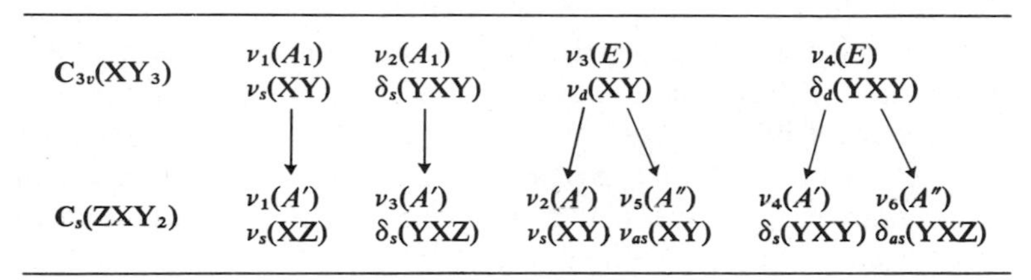

$\mathbf{C}_{3v}(XY_3)$	$\nu_1(A_1)$ $\nu_s(XY)$	$\nu_2(A_1)$ $\delta_s(YXY)$	$\nu_3(E)$ $\nu_d(XY)$		$\nu_4(E)$ $\delta_d(YXY)$	
	↓	↓	↙	↘	↙	↘
$\mathbf{C}_s(ZXY_2)$	$\nu_1(A')$ $\nu_s(XZ)$	$\nu_3(A')$ $\delta_s(YXZ)$	$\nu_2(A')$ $\nu_s(XY)$	$\nu_5(A'')$ $\nu_{as}(XY)$	$\nu_4(A')$ $\delta_s(YXY)$	$\nu_6(A'')$ $\delta_{as}(YXZ)$

TABLE II-13. VIBRATIONAL FREQUENCIES OF PYRAMIDAL ZXY_2 MOLECULES (CM^{-1})

Molecule (ZXY_2)	$\nu_1(A')$	$\nu_2(A')$	$\nu_3(A')$	$\nu_4(A')$	$\nu_5(A'')$	$\nu_6(A'')$	References
HNF_2	3193	972	500	1307	888	1424	257
$HNCl_2$	3279	666	–	1002	687	1295	258
$ClNH_2$	686	–	1553	1032	3380	–	258
$ClNF_2$	692	918	552	366	842	382	259, 260
$ClPF_2$	545	864	411	(308)	852	259	261
$BrPF_2$	459	858	233	391	849	112	261
$FPCl_2$	838	512	328	(200)	521	268	261
$FPBr_2$	824	398	258	(126)	423	(220)	261
OSF_2	1308	801	526	326	721	393	262–264
$OSCl_2$	1229	490	194	344	443	284	262, 265, 266
$OSBr_2$	1121	405	120	267	379	223	267
$[FSO_2]^-$	496	1105	–	595	1182	–	268
$[FSeO_2]^-$	440	882	415	282	909	348	268
$OSeF_2$	1012	664	373	278	605	308	269
$OSeCl_2$	995	388	161	279	347	255	269
$[OSeO_2^*]^{2-}$	831	702	430	336	690	364	270, 271
$FClO_2$	630	1106	547	402	1271	367	272, 273

were made on the assumption that each OH group is a single atom (O*) having the same mass as the OH group. Simon and Paetzold[274] made an extensive study of the vibrational spectra of selenium compounds. Cotton and Horrocks[275] have carried out a normal coordinate analysis of thionyl halides (X_2SO; X = F, Cl, or Br).

II-5. PLANAR FOUR-ATOM MOLECULES

(1) XY_3 Molecules (D_{3h})

The four normal modes of vibration of planar XY_3 molecules are shown in Fig. II-6. The vibrations ν_2, ν_3, and ν_4 are infrared active, and ν_1, ν_3, and ν_4 are Raman active. Table II-14 lists the observed fundamental frequencies of molecules of this type. The infrared spectrum of monomeric AlF_3 was obtained in an Ar matrix.[282] The spectrum of dimeric Al_2F_6 was interpreted on the basis of a planar-bridge structure (D_{2h} symmetry).[282] Infrared spectra of boron mixed halides, such as BF_2Cl and $BFCl_2$,[289,290] BHF_2,[291,292] $BHCl_2$,[293] and $BHBr_2$,[294] have been measured. Normal coordinate analyses on these planar XY_3- and XYZ_2-type molecules have been carried out by many investigators.[295–301]

Gerding and Nijveld[302] have found from Raman spectra that liquid sulfur trioxide (SO_3) exists as a cyclic trimer. Gillespie and Robinson[303]

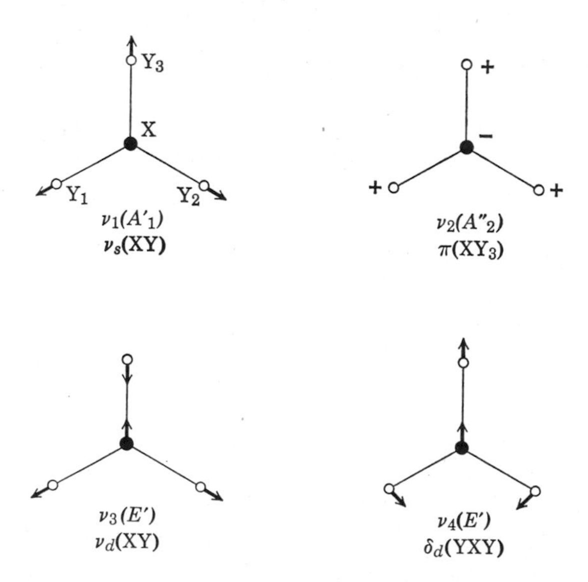

Fig. II-6. Normal modes of vibration of planar XY_3 molecules.

made detailed band assignments for this compound. Duval and Lecomte[304] have measured the infrared spectra of a number of metal orthoborates and metaborates. Infrared spectra of rare earth orthoborates have been measured.[305] Hisatsune and Suarez[306] obtained the infrared spectra of the BO_2^- and $(BO_2)_3^{3-}$ (cyclic) ions in KBr pellets. Although orthoboric acid (H_3BO_3) does not belong to this group, the observed frequencies are listed in Table II-14, with the OH group regarded as a single atom. For more detailed assignments for boric acid, see Refs. 57 and 307–309. The infrared spectra of metaboric acid ($[HBO_2]_3$, cyclic trimer) and its deuterated compound have been studied by Parsons.[310]

TABLE II-14. VIBRATIONAL FREQUENCIES OF PLANAR XY_3 MOLECULES (CM^{-1})

Molecule	ν_1	ν_2	ν_3	ν_4	References
$B^{10}F_3$	888	718	1505	482	276, 277
$B^{11}F_3$	888	691	1454	480	276, 278
$B^{10}Cl_3$	471	480	995	244	279, 280
$B^{11}Cl_3$	471	460	956	243	279, 280
$B^{10}Br_3$	278	395	856	150	281
$B^{11}Br_3$	278	375	820	150	281
$B^{10}I_3$	190	352	737	100	281
$B^{11}I_3$	190	336	704	100	281
AlF_3	–	300	965	270	282
SO_3	1068	496	1391	529	283, 284
$La[BO_3]$	939	718 790	1275	595 614	285
$In[BO_3]$	–	740 765	1260	672	285
$H_3[BO_3]$	1060	668 648	1490- 1428	545	57
$[CS_3]^{2-}$	520	505	905	320	279, 286, 287
$[CSe_3]^{2-}$	316	420	802	185	288

Table II-15 lists the observed infrared and Raman frequencies of some carbonates and nitrates. Calcite and aragonite exhibit different spectra, although their chemical compositions are the same. As discussed in Sec. I-16, this is due to the difference in crystal structure. A similar difference is also seen in the comparison of the Raman spectra of magnesite and dolomite (calcite type) and strontianite (aragonite type). Louisfert[318] and Buijs and Schutte[319] have measured the infrared spectra of a number of metal carbonates and nitrates. For complex carbonates, see Sec. III-3(a). Decius[320] has noted that the ν_2 vibration of KNO_3 (aragonite type) splits slightly because of coupling between neighboring ions. In the $NaNO_3$ (calcite type) crystal, this coupling is small because the distance between two neighboring ions is much greater than that in the KNO_3 crystal. Ferraro[321] has measured the infrared spectra of a number of metal nitrates and has found that the $\mathbf{D}_{3h}$ symmetry of the free ion is progressively lowered to $\mathbf{C}_{2v}$ in going from a monovalent to a tetravalent metal. Similar observations were made by Addison and Gatehouse[322] for transition metal nitrates. Raman spectra of metal nitrates in a molten state have been observed.[323,324]

TABLE II-15. VIBRATIONAL FREQUENCIES OF CARBONATES AND NITRATES (CM^{-1})

Compound		ν_1	ν_2	ν_3	ν_4	References
$CaCO_3$ (calcite)	IR	–	879	1429–1492	706	311
	R	1087	–	1432	714	311
$CaCO_3$ (aragonite)	IR	180	866	1492, 1504	706, 711	311
	R	1084	852	1460	704	311
$MgCO_3$ (magnesite)	R	1096	–	1460	735	312
$CaMg[CO_3]_2$ (dolomite)	R	1099	–	1444	724	312
$SrCO_3$(strontianite)	R	1076	850	1449, 1438, 1406	703	313
$KNO_3{}^{16}$	IR	1049	828	1768, 974 *	716	314
$KNO_3{}^{18}$	IR	1028	817	1755, 969 *	705	314
$NaNO_3$	IR	–	831	1405	692	311
	R	1068	–	1385	726	311, 315
NH_4NO_3	IR	1050	830	1350	715	316, 317
$TlNO_3$	IR	1044	822	1410–1280	697, 713	316
$Pb[NO_3]_2$	IR	1018	807,831	1400–1310	723	316

* According to Anbar et al.,[314] ν_3(1390 cm^{-1}) splits into two bands in decalin mull. It seems more reasonable, however, to assign the strong and broad band observed at ca. 1400 cm^{-1} to the ν_3 mode.

(2) $ZXY_2(C_{2v})$ and $ZXYW(C_s)$ Molecules

If one of the Y atoms of a planar XY_3 molecule is replaced by a Z atom, the symmetry is lowered to $\mathbf{C}_{2v}$. If two of the Y atoms are replaced by two different atoms, W and Z, the symmetry is lowered further to $\mathbf{C}_s$. As a result, the selection rules are changed as already shown in Table I-11. For both cases, all six vibrations become active in the infrared and Raman spectra. Table II-16 lists the observed fundamental frequencies of molecules of these two types. Band assignments shown for HNO_3 and DNO_3 assume the OH group to be a single atom having the same mass as the OH group. The O—H stretching, N—O—H bending and O—H twisting vibrations are observed at 3560, 1335, and 465 cm^{-1}, respectively, in the infrared spectrum of HNO_3 vapor.[325] It is interesting to note that ClF_3 and BrF_3 are planar T-shaped molecules having $\mathbf{C}_{2v}$ symmetry. For normal coordinate analyses of the planar XY_3 molecules, see Refs. 337–339. For those of the planar ZXY_2 molecules, see Refs. 329, 340, 341, and 342.

II-6. OTHER FOUR-ATOM MOLECULES

(1) X_2Y_2 Molecules

Although several different models have been suggested for molecules like H_2O_2, the nonplanar ($\mathbf{C}_2$) (twisted about the X_1—X_2 axis) and planar *cis* ($\mathbf{C}_{2v}$) models are most probable. Figure II-7 shows the six normal modes of vibration and the band assignments for these two models. It is not easy to distinguish between the two models from the vibrational spectra, since the only difference between the two occurs in the ν_6 vibration. This is infrared inactive and Raman depolarized in the planar model but infrared active and

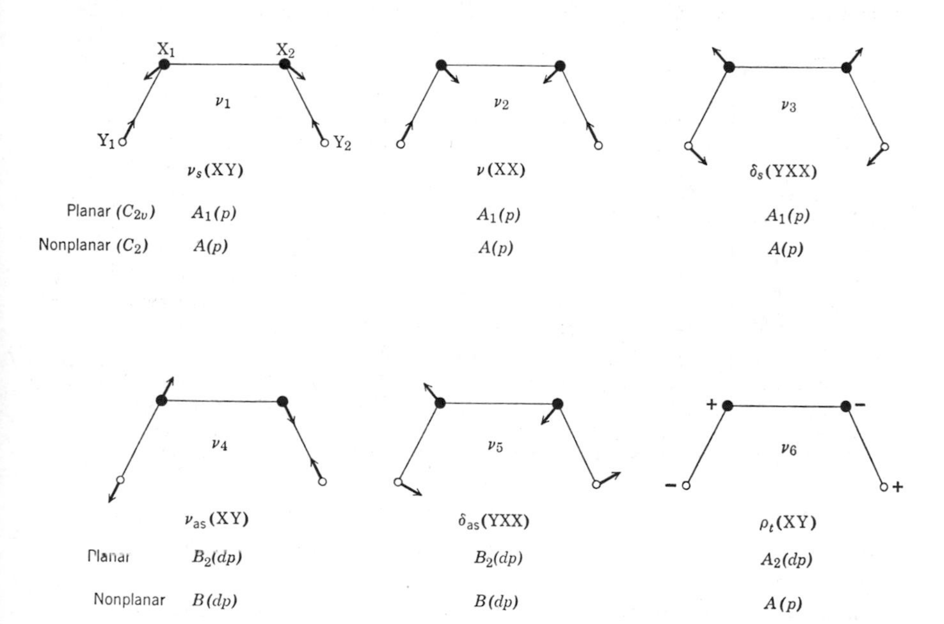

Fig. II-7. Normal modes of vibration of nonlinear X_2Y_2 molecules (p: polarized; dp; depolarized).[358]

TABLE II-16. VIBRATIONAL FREQUENCIES OF ZXY_2 AND ZXYW MOLECULES (cm^{-1})*

Y–X(–Y)(–Y) (D_{3h})	$\nu_1(A_1')$ $\nu_s(XY)$	$\nu_2(A_2'')$ $\pi(XY_3)$	$\nu_3(E')$ $\nu_d(XY)$		$\nu_4(E')$ $\delta_d(YXY)$		
Z–X(–Y)(–Y) (C_{2v})	$\nu_1(A_1)$ $\nu(XZ)$	$\nu_6(B_1)$ $\pi(ZXY_2)$	$\nu_2(A_1)$ $\nu_s(XY)$	$\nu_4(B_2)$ $\nu_{as}(XY)$	$\nu_3(A_1)$ $\delta_s(ZXY)$	$\nu_5(B_2)$ $\delta_{as}(ZXY)$	
Z–X(–Y)(–W) (C_s)	$\nu_1(A')$ $\nu(XZ)$	$\nu_6(A'')$ $\pi(ZXYW)$	$\nu_2(A')$ $\nu(XY)$	$\nu_4(A')$ $\nu(XW)$	$\nu_3(A')$ $\delta(ZXY)$	$\nu_5(A')$ $\delta(ZXW)$	References
(HO)–NO_2	886	765	1320	1710	–	583	325
(DO)–NO_2	888	764	1313	1685	–	543	325
F–NO_2	822	742	1312	1793	460	570	326
Cl–NO_2	794	651	1293	1685	411	367	327, 328
O=CF_2	1928	774	965	1249	584	626	329, 330
O=CCl_2	1827	580	569	849	285	440	329–331
O=CBr_2	1828	512	425	757	181	350	329
O=CClF	1868	667	776	1095	501	415	329–332
O=CBrCl	1828	547	517	806	240	372	329
O=CBrF	1874	620	721	1068	398	335	329, 333
O=CHF	1837	–	2981	1065	1343	663	334
O=CDF	1797	857	2262	1073	968	658	334
S=CF_2	1368	622	787	1189	526	417	335
S=CCl_2	1137	437	505	816	–	–	335
F–ClF_2	528	364	752	703	326	434	336
F–BrF_2	(528)	(289)	674	613	(300)	(384)	336

* The band assignments for some of these compounds differ among investigators.

Raman polarized in the nonplanar model. A recent neutron diffraction study[343] on solid H_2O_2 indicates that the dihedral angle between two OOH planes is about 90°.

Table II-17 lists the fundamental frequencies of some molecules belonging to this group. Infrared spectra of α-GaOOH and its deutero analog has been measured.[365] $NH_3 \cdot H_2O_2$ exists as ammonium hydroperoxide ($[NH_4]^+[OOH]^-$) in the crystalline state. [366,367] Fehér and co-workers[368,354] have studied the Raman spectra of hydrogen polysulfides

(H_2S_x, $x = 2$–8). In alkaline and alkaline earth peroxides such as sodium peroxide (Na_2O_2), the absence of the O—O stretching vibration near 880 cm^{-1} in the infrared spectra may demonstrate the presence of the O—O ion, since the O—O stretching vibration of the ion should be infrared inactive.[369] Ketelaar and colleagues[356,358] have obtained the frequencies of S_2Cl_2 and S_2Br_2 from simultaneous transitions* observed in the infrared spectra of mixtures of each of these compounds with CS_2. As Table II-17 shows, these values are in good agreement with those of the Raman spectra. Normal coordinate analyses have been done for H_2O_2[345] and X_2S_2, where X is a halogen.[358] For the infrared spectrum of HNSO, see Ref. 370.

TABLE II-17. VIBRATIONAL FREQUENCIES OF X_2Y_2 MOLECULES (CM^{-1})

Molecule	State		ν_1	ν_2	ν_3	ν_4	ν_5	ν_6	References
H_2O_2	gas	(IR)	–	(890)	–	3610	1260	465, 575	344–346
	liquid	(IR)	–	878	–	3360	1350	635	344–347
	liquid	(R)	3364	880	1402	–	–	–	348, 349
	solid	(IR)	–	878	1430	3320	1380	472, 660, 792	344, 347, 349–351
D_2O_2	gas	(IR)	–	–	(1007)	2661	923, 947	–	344, 347
	liquid	(IR)	–	878	–	2482	1004	538	41, 347
	liquid	(R)	2472	880	1013	–	–	–	348
	solid	(IR)	–	880	–	2470	1000	480	344, 351
H_2S_2	gas	(IR)	–	–	–	2557	897	–	352, 353
	liquid	(IR)	(2509)	509	–	–	882	–	352
	liquid	(R)	2509	509	883	–	–	–	354
	solid	(IR)	(2495)	501	(868)	2480	890	–	352
F_2S_2	liquid	(IR)	745	526	–	807	–	–	355
Cl_2S_2	liquid	(IR)	438	448	203	538	242	–	356,357
	liquid	(R)	437	451	207	541	241	104	358, 359
Br_2S_2	liquid	(IR)	–	354	176	531	196	–	356
	liquid	(R)	302	355	172	529	200	66	360
	liquid	(IR)	302	531	175	355	198	66	361
Cl_2Se_2	liquid	(R)	288	367	130	418	146	87	360
Br_2Se_2	liquid	(R)	204	265	94	292	106	50	360
cis-N_2F_2	liquid	(R)	896	1525	341	952	737	(550)	362
cis-N_2F_2	liquid	(R)	896	1524	552	952	737	–	363
trans-N_2F_2	gas	(IR)	(1010)	(1636)	(592)	989	421	360	363
trans-N_2F_2	liquid	(R)	1010	1522	600	–	–	–	364
	gas	(IR)	–	–	–	900	423	364	364

*They are similar to the combination bands in one molecule, but here vibrations of two different molecules are combined in the mixture because of transient molecular collisions.

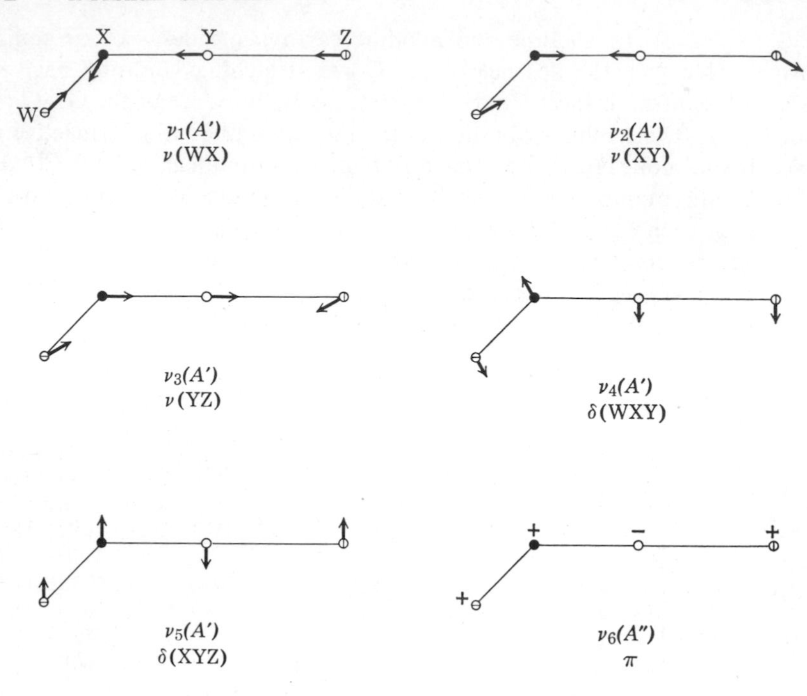

Fig. II-8. Normal modes of vibration of nonlinear WXYZ molecules.

(2) Other Planar Molecules (C_s)

Planar four-atom molecules of the XY_3, XYZY, and WXYZ types have six modes of vibration such as shown in Fig. II-8. All these vibrations are both infrared and Raman active. Table II-18 lists the fundamental frequencies of some molecules of these types. For normal coordinate analysis, see Refs. 385 and 386. Of the two isomers of nitrous acid (HONO), the *trans* form is more stable than the *cis* form.[382] Palm[387] has done a normal coordinate analysis of these two isomers. For the infrared spectra of X—SCN molecules (X: a halogen), see Ref. 142.

II-7. NITROGEN COMPOUNDS OF VARIOUS STRUCTURES

So far the normal vibrations have been discussed individually for each general type of molecular structure. It is more convenient, however,to group nitrogen compounds of various structures here and to discuss their group frequencies, since most of the structures rarely appear in the compounds of other elements.

TABLE II-18. VIBRATIONAL FREQUENCIES OF PLANAR FOUR-ATOM MOLECULES (CM^{-1})

Molecule	State	ν_1	ν_2	ν_3	ν_4	ν_5	ν_6	References
HN_3	gas	3336	2140	1274	1150	522	672	371
	solid(80°K)	3090	2162	1299	1180	–	–	371
DN_3	gas	2480	2141	1183	955	498	638	371
	solid (80°K)	2308	2155	1230	977	–	–	371
HNCO	gas	3531	2274	1327	797	572	(670)	372, 373
	solid	3133	2246	1326	–	–	–	
HNCS	gas	3536	1963	995	817	469	600	374, 375
	Ar matrix	3505	1979	988	577	461	–	376
DNCS	Ar matrix	2623	1938	–	548	366	–	376
HOCN	N_2 matrix	3530	1098	2294	1241	460	–	377
DOCN	N_2 matrix	2590	1093	2292	957	437	–	377
HCNO	solid	3335	1251	2190	–	538	–	378
HNCO	gas	3531	2274	1527	777	660	578	379, 380
DNCO	gas	2635	2235	1310	460	767	603	381
cis-HONO	gas	3426	1640	1292	856	–	637	382-384
trans-HONO	gas	3590	1696	1260	794	598	543	
cis-DONO	gas	2530	1616	–	816	–	508	
trans-DONO	gas	2650	1690	1018	739	591	416	

(1) Oxides and Acids

The oxides of nitrogen are NO, N_2O, N_2O_2, $[N_2O_2]^{2-}$, N_2O_3, N_2O_4, N_2O_5, and so forth. We have already described NO and N_2O in Secs. II-1 and II-2, respectively. The structures of the other oxides are reported to be

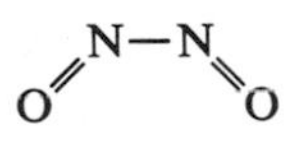

C_{2v} (*cis*)
C_{2h} (*trans*

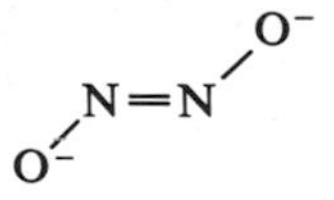

C_{2h} (*trans*)

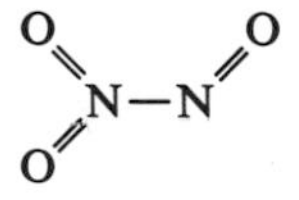

C_s (planar)

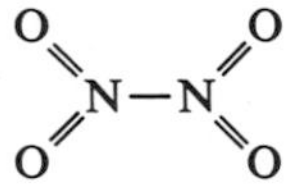

V_h (planar)
V_d (nonplanar)

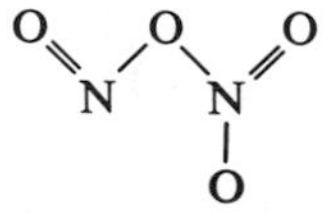

C_s (planar)

$N_2O_5 = [NO_2]^+[NO_3]^-$

Ionic crystal

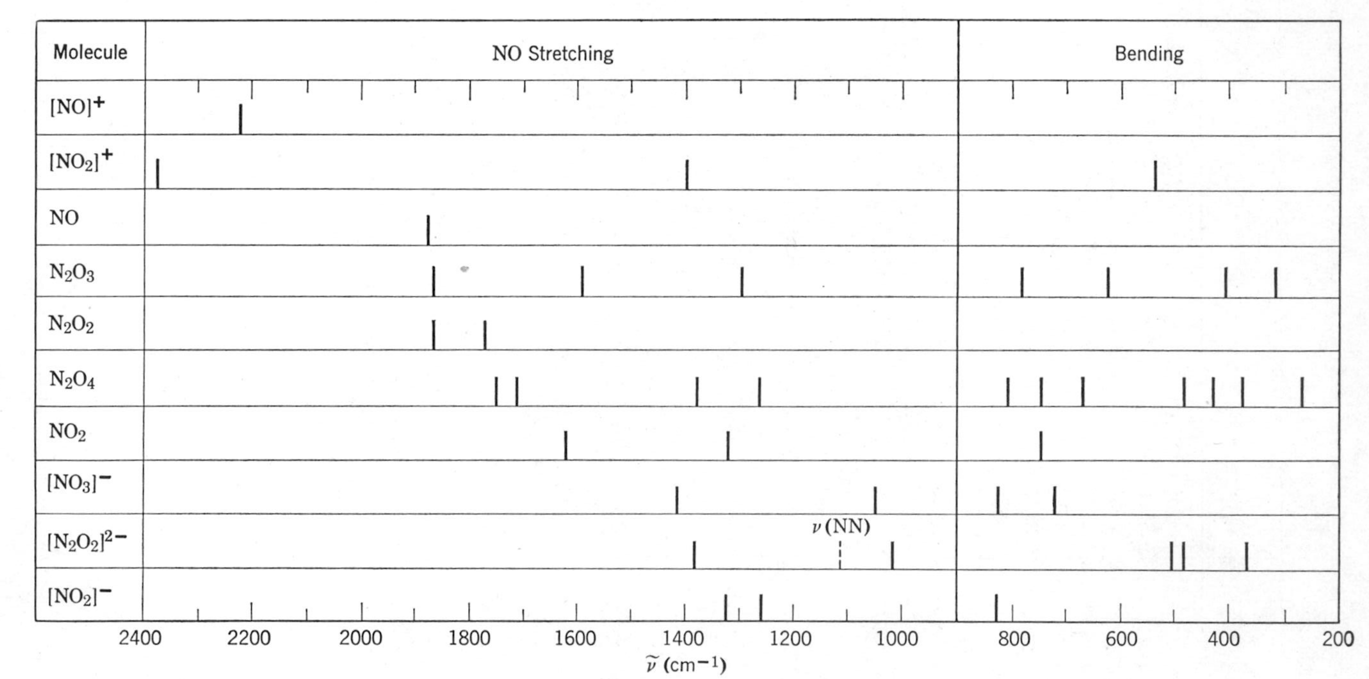

Fig. II-9. Distribution of stretching and bending frequencies in various nitrogen oxides.

The observed fundamental frequencies and band assignments are given in Table II-19. Using the matrix isolation method, Fateley and colleagues[21] have found that (1) NO exists as a monomer and as two dimers (*cis* and *trans*), (2) NO_2 exists as a monomer and as three dimers (planar, nonplanar, and asymmetrical ONO—NO_2), and (3) N_2O_3 exists as ON—NO_2 and ONONO. Teranishi and Decius[395] have shown that the spectrum of crystalline N_2O_5 can be interpreted on the basis of the ionic structure, $[NO_2]^+[NO_3]^-$, previously suggested by x-ray analysis. Hisatsune et al.[398] obtained the infrared spectrum of the unstable covalent form of N_2O_5 at liquid nitrogen temperature. The vibrational spectra of $H_2N_2O_2$[370] and HNO_3[399] have also been analyzed.

TABLE II-19. VIBRATIONAL FREQUENCIES OF VARIOUS NITROGEN OXIDES (CM^{-1})

Molecule	State	Frequencies and Band Assignments	References
N_2O_2	−190°C in CO_2 (IR)	*cis*: ν_{as}(NO), 1768; ν_s(NO), 1862 *trans*: ν_{as}(NO), 1740	21
N_2O_3	−150°C (IR)	1863, 1589, 1297, 783, 627, 407, 313, 253(R)	388
$N_2{}^{14}O_4$ (V_h)	(IR) (R)	$\nu_1(A_g)$–ν(NO), 1380; $\nu_2(A_g)$–$\delta(NO_2)$, 808; $\nu_3(A_g)$–ν(NN), 266; $\nu_4(A_u)$–$\rho_t(NO_2)$?; $\nu_5(B_{1g})$–(NO), 1712; $\nu_6(B_{1g})$–$\rho_r(NO_2)$, 482; $\nu_7(B_{1u})$–$\rho_w(NO_2)$, 429; $\nu_8(B_{2g})$–$\rho_w(NO_2)$, 672; $\nu_9(B_{2u})$–ν(NO), 1748; $\nu_{10}(B_{2u})$–$\rho_r(NO_2)$, 381; $\nu_{11}(B_{3u})$–ν(NO), 1262; $\nu_{12}(B_{3u})$–$\delta(NO_2)$, 750	389–394
$N_2{}^{14}O_5$	−190°C (IR)	NO_2^+ ion: ν_1(1400), ν_2(538), ν_3(2375) NO_3^- ion: ν_1(1050), ν_2(824), ν_3(1413), ν_4(722)	395
$[N_2{}^{14}O_2]^{2-}$ (C_{2h})	(R) (IR)	$\nu_1(A_g)$–ν(NO), 1419; $\nu_2(A_g)$–ν(NN), 1121; $\nu_3(A_g)$–δ(NNO), (696); $\nu_4(B_u)$–δ(NNO), (371); $\nu_5(B_u)$ ν(NO), 1031 $\nu_6(A_u)$–π, 492	396, 397, 370

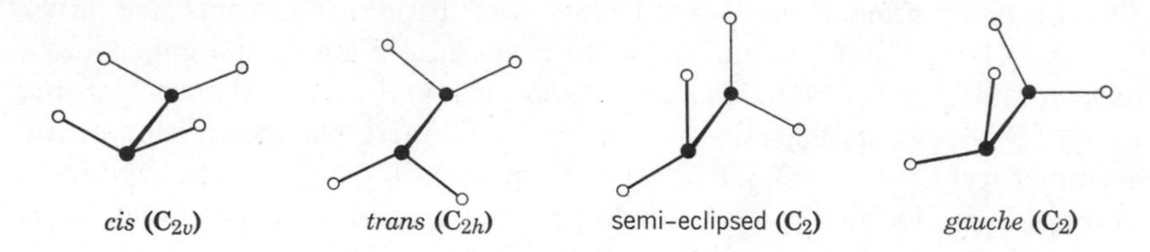

Fig. II-10. Various conformations of hydrazine.

Figure II-9 illustrates the frequency distribution of the NO stretching and ONO bending vibrations in various oxides. It is interesting that the NO stretching frequency varies over a wide range (24000–1000 cm^{-1}) depending on the bond order. As a result, a narrow frequency range cannot be specified for the NO group.

(2) Hydrazine and Hydroxylamine

Owing to internal rotation, the hydrazine molecule may have conformation isomers such as those in Fig. II-10. The equilibria between these isomers may depend on the temperature at which the measurement is made. Table II-20 gives band assignments and fundamental frequencies of hydrazine and hydroxylamine based on C_2 and C_s symmetry, respectively. Infrared spectrum of N_2F_4 has been assigned on the basis of C_2 symmetry.[406] The infrared spectra of hydrazine dihalides ($[N_2H_6]X_2$, X = F or Cl) have been studied by Snyder and Decius,[407] and of hydroxylamine halides ($[NH_3OH]X$, X = Cl, Br or I) by Frasco and Wagner.[408]

II-8. TETRAHEDRAL AND SQUARE-PLANAR, FIVE-ATOM MOLECULES

(1) Tetrahedral XY_4 Molecules (T_d)

Figure II-11 illustrates the four normal modes of vibration of a tetrahedral XY_4 molecule. All the four vibrations are Raman active, whereas only ν_3

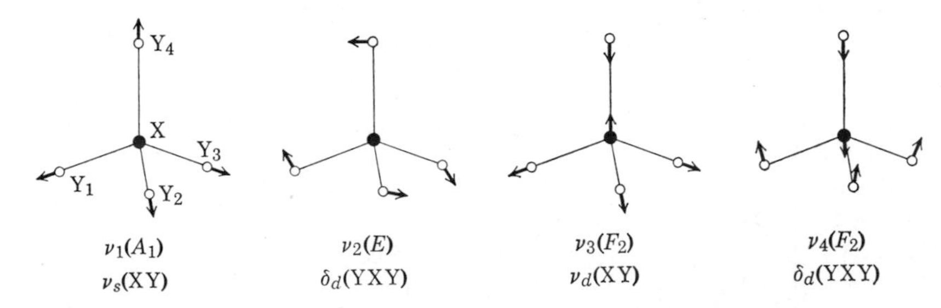

Fig. II-11. Normal modes of vibration of tetrahedral XY_4 molecules.

TABLE II-20. VIBRATIONAL FREQUENCIES AND BAND ASSIGNMENTS FOR HYDRAZINE AND HYDROXYLAMINE (cm^{-1})

$H_2N—NH_2$ [400–403]			$H_2N—OH$ [404, 405]		
Gas (IR)	Assignments		Film (IR)	Assignments	
3325	$\nu_1(A)$	$\nu(NH)$	3245, 3173	$\nu_1(A')$	$\nu(NH)$
–	$\nu_2(A)$	$\nu(NH)$	2867	$\nu_2(A')$	$\nu(OH)$
1493	$\nu_3(A)$	$\delta(HNH)$	1515	$\nu_3(A')$	$\delta(HNH)$
1098	$\nu_4(A)$	$\rho_r(NH_2)$	1191	$\nu_4(A')$	$\delta(NOH)$
–	$\nu_5(A)$	$\nu(NN)$	912	$\nu_5(A')$	$\nu(NO)$
780	$\nu_6(A)$	$\rho_w(NH_2)$	950	$\nu_6(A')$	$\rho_r(NH_2)$
–	$\nu_7(A)$	$\rho_t(NH_2)$	3302	$\nu_7(A'')$	$\nu(NH)$
3350	$\nu_8(B)$	$\nu(NH)$	867	$\nu_8(A'')$	$\rho_w(NH_2)$
3280	$\nu_9(B)$	$\nu(NH)$	535	$\nu_9(A'')$	$\rho_t(NH_2)$
1587, 1628	$\nu_{10}(B)$	$\delta(HNH)$			
1275	$\nu_{11}(B)$	$\rho_r(NH_2)$			
966, 933	$\nu_{12}(B)$	$\rho_w(NH_2)$			

and ν_4 are infrared active. The fundamental frequencies of XH_4 molecules are listed in Table II-21. Woodward and Roberts[421] have shown that the relationship $kr^3 = $ a constant ($k =$ stretching force constant; $r =$ bond distance) holds for a series of isoelectronic XH_4 molecules. Longuet-Higgins and Brown[422] have shown that the frequency, ν_2, is given by $4\pi^2\nu_2{}^2 = 9\sqrt{6}\,e^2/32mr^3$, where r is the bond distance and m and e are the mass and charge of the proton. Normal coordinate analyses on tetrahedral XY_4-type molecules have been made by a number of investigators using the GVF and UBF fields.[423–426]

Among the many XH_4 molecules in Table II-21, the NH_4^+ ion is chemically the most important. The infrared spectrum of NH_4Cl is shown in Fig. II-12. Hornig and co-workers[427–430] have made an extensive study of the infrared spectra of ammonium halide crystals. One of their findings is that the combination band between $\nu_4(F_2)$ and ν_6 (rotatory lattice vibration) is observed if the ammonium ion does not rotate freely in the crystal lattice. This holds for NH_4F, NH_4Cl, and NH_4Br. In HN_4I (phase I), however, the band is not observed. Therefore the NH_4^+ ion may rotate more freely in this compound than in other halides. For other work concerning ammonium halides, see Refs. 431–436.

Table II-22 lists the fundamental frequencies of neutral tetrahalogeno molecules. It is interesting to note that molecules like $TeCl_4$,[456] SF_4,[457,458]

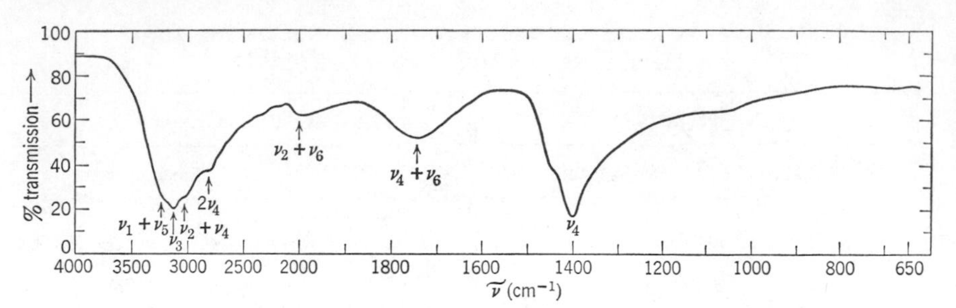

Fig. II-12. The infrared spectrum of NH_4Cl (ν_5, ν_6; lattice modes).

TABLE II-21. VIBRATIONAL FREQUENCIES OF TETRAHEDRAL XH_4 MOLECULES (CM^{-1})

Molecule	ν_1	ν_2	ν_3	ν_4	References
CH_4	2917	1534	3019	1306	409
SiH_4	2180	970	2183	910	409, 410
GeH_4	2106	931	2114	819	411, 409
SnH_4	–	758	1901	677	412
CD_4	2085	1092	2259	996	413, 414
SiD_4	(1545)	(689)	1597	681	415, 410
GeD_4	1504	665	1522	596	411
SnD_4	–	539	1368	487	412
$[N^{14}H_4]^+$	3040	1680	3145	1400	409
$[N^{15}H_4]^+$	–	(1646)	3137	1399	416
$[PH_4]^+$	2304	1040	2370	930	409, 417
$[AsH_4]^+$	2119	1002	2225	845	418
$[ND_4]^+$	2214	1215	2346	1065	409
$[PD_4]^+$	(1625)	777	1740	701	417
			1658	683	
$[NT_4]^+$	–	976	2022	913	416
$[B^{11}H_4]^-$	2264	1210	2244	1080	419, 420
$[B^{10}H_4]^-$	2270	1208	2250	1093	419, 420
$[B^{11}D_4]^-$	(1570)	855	1696	823	419, 420
$[B^{10}D_4]^-$	1604	856	1707	827	419, 420
$[AlH_4]^-$	1790	799	1740	764	421

and SeF_4,[459] are not tetrahedral, but trigonal bipyramidal with one of the three equatorial positions unoccupied ($\mathbf{C}_{2v}$). If an oxygen atom occupies this vacant position, the structure of thionyl tetrafluoride (OSF_4) is obtained. The vibrational spectrum of this compound has been obtained by Goggin et al.[460] In the solid state, however, $TeCl_4$, $TeBr_4$, $SeCl_4$, and $SeBr_4$ consist of the pyramidal XY_3^+ cation and Y^- anion.[461–463]

TABLE II-22. VIBRATIONAL FREQUENCIES OF TETRAHALOGENO MOLECULES (CM^{-1})

Molecule	ν_1	ν_2	ν_3	ν_4	References
$C^{12}F_4$	908	435	1281	628	437, 438
$C^{13}F_4$	908	435	1240	(627)	437
SiF_4	800	268	1010	390	439, 440
GeF_4	738	205	800	260	441, 442
ZrF_4	(600–725)	(200–150)	668	190	443
CCl_4	459	218	790 762	314	444
$SiCl_4$	424	150	608	221	445
$TiCl_4$	388	119	490 506	139	446
VCl_4	383	128	475	128 or 150	447
$GeCl_4$	397	132	451	171	409, 448–450
$ZrCl_4$	388	102	421	112	231
$SnCl_4$	368	106	403	131	445, 444, 451
$PbCl_4$	327	90	348	90	450
CBr_4	267	123	672	183	452
$SiBr_4$	249	90	487	137	452
$GeBr_4$	234	78	328	111	409
$SnBr_4$	220	64	279	88	409
$TiBr_4$	230	74	383	91	453
CI_4	178	90	555	123	454
SiI_4	168	63	405	94	449
GeI_4	159	60	264	80	455
SnI_4	149	47	216	63	455

Table II-23 lists the fundamental frequencies of tetrahedral tetrahalide ions. Except for metal fluorides, the metal-halide stretching and bending vibrations appear in the far-infrared region. Recently, several investigators[473,474,483] have obtained the far-infrared spectra of a number of tetrahedral complex halides. As will be shown in Sec. III-12, the metal-halide stretching bands provide valuable information about the structure of metal complexes. For $[PCl_4]^+$ and $[BCl_4]^-$, see Refs. 484 and 485.

Table II-24 lists the fundamental frequencies of XO_4, XS_4, and $X(OH)_4$ molecules. Since this group includes a number of ions that are chemically important, the infrared spectra of $KClO_4$ and K_2CrO_4 are shown in Fig. II-13. It is seen that ν_3 and ν_4 appear strongly in infrared spectra, and, because of lowering of the symmetry, degenerate vibrations often split and Raman active modes become infrared active in the crystalline state (Sec. II-5 (1)). Recently, Müller and his coworkers have carried out an extensive

TABLE II.23. VIBRATIONAL FREQUENCIES OF TETRAHALIDE ANIONS (cm^{-1})

Ions	ν_1	ν_2	ν_3	ν_4	References
$[B^{11}F_4]^-$	769	353	984	524	464, 465
$[B^{10}F_4]^-$	769	353	1016	529	464, 465
$[AlCl_4]^-$	352	147	490	176	466
$[GaCl_4]^-$	346	114	386	149	467
$[InCl_4]^-$	321	89	337	112	468
$[TlCl_4]^-$	312	60	296	78	469
$[PCl_4]^+$	458	171	653	251	466, 470
$[AsCl_4]^+$	422	156	500	187	471, 470
$[SbCl_4]^+$	353	143	399	153	472, 470
$[MnCl_4]^{2-}$	–	78	284	118	473
$[FeCl_4]^{2-}$	–	77	286	119	473
$[CoCl_4]^{2-}$	–	82	297	130	473
$[NiCl_4]^{2-}$	–	79	289	112	473
$[CuCl_4]^{2-}$	–	77	267	136	473
			248	118	
$[FeCl_4]^-$	330	106	385	133	468
$[ZnCl_4]^{2-}$	282	(100)	292	(100)	474
$[HgCl_4]^{2-}$	267	180	276	192	475
$[GaBr_4]^-$	210	71	278	102	476
$[InBr_4]^-$	197	55	239	79	477
$[TlBr_4]^-$	190	51	209	64	478
$[PBr_4]^+$	227	72	474	140	479, 470
$[MnBr_4]^{2-}$	–	–	221	85	473
$[FeBr_4]^{2-}$	–	–	219	84	473
$[CoBr_4]^{2-}$	–	–	231	91	473
$[NiBr_4]^{2-}$	–	–	231	83	473
			224		
$[CuBr_4]^{2-}$	–	–	216	85	473
			174		
$[ZnBr_4]^{2-}$	172	61	210	82	480
$[CdBr_4]^{2-}$	166	53	183	62	481
$[GaI_4]^-$	145	52	222	73	482
$[InI_4]^-$	139	42	185	58	482
$[TlI_4]^-$	133	–	156	–	469
$[MnI_4]^{2-}$	–	–	185	–	473
$[FeI_4]^{2-}$	–	–	186	–	473
$[CoI_4]^{2-}$	–	–	197	–	473
			192		
$[NiI_4]^{2-}$	–	–	189	–	473
$[ZnI_4]^{2-}$	122	44	170	62	480
$[CdI_4]^{2-}$	117	36	145	44	473

TABLE II-24. VIBRATIONAL FREQUENCIES OF XO_4, XS_4, AND $X(OH)_4$ MOLECULES (cm^{-1})*

Molecule	ν_1	ν_2	ν_3	ν_4	References
$[SiO_4]^{4-}$	819	340	956	527	409, 486
$[PO_4]^{3-}$	938	420	1017	567	487, 488
$[SO_4]^{2-}$	983	450	1105	611	409
$[ClO_4]^-$	928	459	1119	625	489–491
$[VO_4]^{3-}$	824	(305)	790	340	492, 489
$[CrO_4]^{2-}$	830	(330)	765	330	493
$[MnO_4]^-$	845	(355)	910	395	494, 495
$[AsO_4]^{3-}$	837	349	878	463	489
$[SeO_4]^{2-}$	833	335	875	432	409
$[MoO_4]^{2-}$	894	381	833	318	496, 489, 497
$[TeO_4]^{2-}$	647	298	624	357	498, 409
$[TcO_4]^-$	912	(347)	912	325	496, 495
RuO_4	(880)	(293)	913	330	499, 494
$[IO_4]^-$	791	256	853	325	500
$[WO_4]^{2-}$	931	373	833	320	496, 489, 501
$[ReO_4]^-$	971	(350)	918	331	502, 503, 495
OsO_4	971	328	960	328	504, 503
$[AsS_4]^{3-}$	386	171	419	216	489
$[SbS_4]^{3-}$	366	156	380	178	489
$[VS_4]^{3-}$	404	–	470	200	492, 505
$[MoS_4]^{2-}$	485	(175)	466	185	492, 505
$[WS_4]^{2-}$	460	(179)	480	195	492, 505
$[MoSe_4]^{2-}$	255	–	340	120	506
$[WSe_4]^{2-}$	278	–	310	115	506
$[B(OH)_4]^-$	754	379	945	533	507
$[Al(OH)_4]^-$	615	310	(720)	310	508
$[Zn(OH)_4]^{2-}$	470	300	(570)	300	508

* Hydrogenic vibrations are not listed for the last three ions.

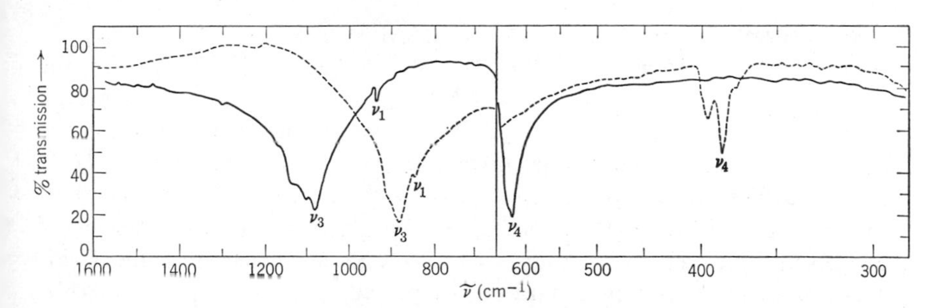

Fig II-13. Infrared spectra of $KClO_4$ (solid line) and K_2CrO_4 (broken line).

study on vibrational spectra of tetrahedral anions of the XO_4 and XS_4 types. Table II-25 gives the Raman frequencies of the $[SO_4]^{2-}$ ion in various crystalline environments.[509] Doubly and triply degenerate vibrations split into two and three components, respectively.

TABLE II-25. RAMAN FREQUENCIES OF THE SO_4^{2-} ION IN VARIOUS COMPOUNDS (CM^{-1})[509]

Compound	ν_1	ν_2	ν_3	ν_4
Na_2SO_4	983	454	1106	622
$BaSO_4$	989	453, 462	1094, 1142, 1167	617, 630, 648
$CaSO_4$	1018	415, 499	1108, 1128, 1160	609, 628, 674
$CaSO_4 \cdot 2H_2O$[a]	1006	415, 499	1115, 1136, 1144	618, 622, 672

[a] For the H_2O bands, see Sec. III-3.

Hass and Sutherland[510] have made a detailed study of the crystal spectrum of gypsum ($CaSO_4 \cdot 2H_2O$). For the spectra of various sulfates in the crystalline state, see Refs. 511–520. The infrared and Raman spectra of sulfuric acid (H_2SO_4) and its deuterated compound have also been reported.[521,522] The spectra of sulfato complexes will be discussed in Sec. III-4(3). References 523–528 describe the spectra of perchloric acid and various perchlorates. For the spectra of $[PO_4]^{3-}$, $[HPO_4]^{2-}$, $[H_2PO_4]^-$, $[HPO_3]^{2-}$, and $[H_2PO_2]^-$, see Refs. 529–537. Duval and Lecomte[537,538] have measured the infrared spectra of a number of metal salts of XO_4^{2-} ions in the low frequency region.

(2) Tetrahedral $ZXY_3(C_{3v})$ and $Z_2XY_2(C_{2v})$ Molecules

If one of the Y atoms of an XY_4 molecule is replaced by a Z atom, the symmetry of the molecule is lowered to $\mathbf{C}_{3v}$. If two of the Y atoms are replaced, the symmetry becomes $\mathbf{C}_{2v}$. This lowering of symmetry splits the degenerate vibrations and activates infrared inactive vibrations, as Table II-26 shows. Thus the number of infrared active vibrations is increased to six in ZXY_3 and to nine in Z_2XY_2 molecules. The fundamental frequencies of hydrogen and halogen compounds of these types are listed in Table II-27.

A good deal of experimental data on compounds of these types is available. A few of these references are given here: $SiH_{4-n}X_n$ (X: a halogen) (544–552);

$SiX_{4-n}Y_n$ (X, Y: halogens) (449, 553); $GeH_{4-n}X_n$ (448, 554–558); $GeX_{4-n}Y_n$ (445, 559–562); $SnX_{4-n}Y_n$ (449, 451); $[BClF_3]^-$ (484); TiX_3Y (563).

TABLE II-26. CORRELATION TABLE FOR T_d, C_{3v} AND C_{2v}

Point Group	ν_1	ν_2	ν_3	ν_4
T_d	A_1(R)	E(R)	F_2(I,R)	F_2(I,R)
C_{3v}	A_1(I,R)	E(I,R)	A_1(I,R) + E(I,R)	A_1(I,R) + E(I,R)
C_{2v}	A_1(I,R)	A_1(I,R) +A_2(R)	A_1(I,R) + B_1(I,R) +B_2(I,R)	A_1(I,R) + B_1(I,R) +B_2(I,R)

TABLE II-27. VIBRATIONAL FREQUENCIES OF TETRAHEDRAL MOLECULES (CM^{-1})*

$T_d(XY_4)$	A_1 ν_s(XY)	E δ_d(YXY)			F_2 ν_d(XY)			F_2 δ_d(YXY)			References
$C_{3v}(ZXY_3)$	A_1 ν(XZ)	E δ(YXY)			A_1 ν(XY)		E ν_d(XY)	A_1 δ(YXY)		E $\rho_r(XY_3)$	
$C_{2v}(Z_2XY_2)$	A_1 ν(XZ)	A_1 δ(YXY)		A_2 $\rho_t(XY_2)$	A_1 ν(XY)	B_1 ν(XY)	B_2 ν(XZ)	A_1 δ(ZXZ)	B_1 $\rho_r(XY_2)$	B_2 $\rho_w(XY_2)$	
CH_4	2914		1526			3020			1306		
DCH_3	2205		1477		2982		3030	1306		1156	
D_2CH_2	2139	1450		1286	2974	3030	2255	1034	1090	1235	539, 540
HCD_3	2992		1046		2141		2269	1299		982	
CD_4	2085		1054			2258			996		
SiH_4	2180		970			2183			910		
D_2SiH_2	1587	944		844	2189	2183	1601	683	743	862	415
$HSiD_3$	2182		683		1573		1598	851		683	
SID_4	1545		689			1597			681		
GeH_4	2106		931			2114			819		
$DGeH_3$	1520		901		2106		2112	820		706	
D_2GeH_2	1512	881		807	2112	2112	1522	620	657	770	411
$HGeD_3$	2112		625		1504		1522	595		792	
GeD_4	1504		665			1522			596		
$FSiH_3$	872		943		2206		2196	990		728	
$ClSiH_3$	551		954		2201		2195	949		664	541
$BrSiH_3$	430		950		2200		2196	930		633	
$ISiH_3$	355		941		2192		2206	903		592	542
$SiCl_4$	424		150			610			221		
$BrSiCl_3$	368		135		545		610	191		205	
Br_2SiCl_2	326	111		122	563	605	508	182	191	174	
$ClSiBr_3$	579		101		288		498	159		173	
$SiBr_4$	249		90			487			137		
$TiCl_4$	389		120			≈ 500			140		
$BrTiCl_3$	326		105		439		{508, 489}	128		136	543
Br_2TiCl_2	294	87		–	462	492	{401, 383}	–	–	125	
$ClTiBr_3$	471		82		263		{398, 388}	110		123	
$TiBr_4$	235		74			393			94		
$GeCl_4$	396		132			453			172		
$BrGeCl_3$	309		116		417		450	160		–	
Br_2GeCl_2	281	94		–	420	444	338	146	155	-	449
$ClGeBr_3$	428		90		257		330	122		137	
$GeBr_4$	235		80			327			112		

* Some of the band assignments given here were made by the author on an empirical basis, and accordingly are subject to change.

Tables II-28*a* and II-28*b* list the fundamental frequencies of compounds containing oxygen and sulfur atoms. The vanadyl group (VO^{2+}), like the uranyl group, forms a number of chelate compounds with organic ligands such as acetylacetone. In all of these complexes, the V=O stretching band is observed between 1050 and 950 cm^{-1}. The stretching vibrations of other metal-oxygen double bonds such as U=O, Mo=O[589,590] Ru=O[591] and Os=O[592] also appear in a similar frequency range. It is interesting to note

TABLE II-28*a*. VIBRATIONAL FREQUENCIES OF ZXY_3 MOLECULES (CM^{-1})

Molecule (ZXY_3)	$\nu_1(A_1)$ $\nu(XY_3)$	$\nu_2(A_1)$ $\nu(XZ)$	$\nu_3(A_1)$ $\delta(XY_3)$	$\nu_4(E)$ $\nu(XY_3)$	$\nu_5(E)$ $\delta(XY_3)$	$\nu_6(E)$ $\rho_r(XY_3)$	References
$[OBF_3]^{2-}$	1048	768	464	1376	530	384	564
$FClO_3$	1061	715	549	1315	589	405	565–567
$[SSO_3]^{2-}$	995	446	669	1123	541	335	489, 568, 569
$[FSO_3]^-$	1082	768	566	1287	592	409	570, 571
$[ClSO_3]^-$	1050	416	585	1195	535	220	569
$[FCrO_3]^-$	911	637	338	955	370	261	572
$[ClCrO_3]^-$	907	438	295	954	365	209	573
$[NOsO_3]^-$	897	1021	309	871	309	372	574, 575
$ClReO_3$	1001	293	435	960	344	196	576
$BrReO_3$	997	195	350	963	332	168	576
$[NReO_3]^{2-}$	878	1022	315	830	273	380	577
$[SReO_3]^-$	938	504	330	891	310	240	578
OVF_3	722	1058	258	806	308	204	579
$OVCl_3$	408	1035	165	504	249	129	580
$OVBr_3$	271	1025	120	400	83	212	235
OPF_3	873	1415	473	990	485	345	581
$OPCl_3$	486	1290	267	581	337	193	582
$OPBr_3$	340	1261	173	488	267	118	583, 582
SPF_3	980	694	445	944	405	276	578, 584, 585
$SPCl_3$	435	753	250	542	250	167	586
$SPBr_3$	299	718	165	438	179	115	585
$[OMoS_3]^{2-}$	465	859	200	475	200	265	587
$[OWS_3]^{2-}$	470	870	200	470	200	260	588

TABLE II-28*b*. VIBRATIONAL FREQUENCIES OF Z_2XY_2 MOLECULES (CM^{-1})

Molecule (Z_2XY_2)	$\nu_1(A_1)$ $\nu(XY)$	$\nu_2(A_1)$ $\nu(XZ)$	$\nu_3(A_1)$ $\delta(YXY)$	$\nu_4(A_1)$ $\delta(ZXZ)$	$\nu_5(A_2)$ $\rho_t(XY_2)$	$\nu_6(B_1)$ $\nu(XY)$	$\nu_7(B_1)$ $\rho_w(XY_2)$	$\nu_8(B_2)$ $\nu(XZ)$	$\nu_9(B_2)$ $\rho_r(XY_2)$	References
O_2SF_2	848	1269	544	545	360	885	545	1502	386	262, 593, 594
O_2SCl_2	405	1182	218	560	388	362	380	1414	282	595, 596
O_2CrF_2	727	1006	(182)	304	(422)	789	(259)	1016	274	597
O_2CrCl_2	465	984	144	357	216	497	263	994	230	598, 599

that the M=O stretching frequency (or force constant) is extremely sensitive to the change in the M=O distance. Thus the latter can be estimated from the former.[590,592]

For other compounds, several references are as follows; $OAsF_3$ (600); $FCrO_3$, $ClCrO_3$ (601); $[SPO_3]^-$ (602); O_2SBrF (603); SPF_2Cl and SPF_2Br (604, 605); $SPFCl_2$ and $SPFBr_2$ (605, 606); $O_2MoS_2{}^{2-}$ and $O_2WS_2{}^{2-}$ (506); Cl_2POF, ClBrPOF, and Br_2POF (607, 608); F_2POCl and F_2POBr (608); Cl_2POBr and Br_2POCl (582); O_2CrFCl (609).

Normal coordinate analyses of tetrahedral XY_4, ZXY_3, and Z_2XY_2 molecules are common. References for some of these theoretical investigations are: XY_4 (610–614); ZXY_3 (615–621).

(3) Square-Planar XY_4 Molecules (D_{4h})

Figure II-14 shows the seven normal modes of vibration of square-planar XY_4 molecules. The vibrations ν_3, ν_6, and ν_7 are infrared active, whereas ν_1, ν_2, and ν_4 are Raman active. Table II-29 gives the observed frequencies for some ions of this type. For normal coordinate analyses, see Refs. 623, 629, and 630.

TABLE II-29. VIBRATIONAL FREQUENCIES OF SQUARE-PLANAR XY_4 MOLECULES (CM^{-1})

Molecule	ν_1	ν_2	ν_3	ν_4	ν_6	ν_7	References
XeF_4	543	235	291	502	586	123	622
$[ICl_4]^-$	288	128	–	261	266	–	623
$[PtCl_4]^{2-}$	333	196	168	306	321	191	628, 624, 623
	–	–	93	–	320	183	625
$[AuCl_4]^-$	347	171	–	324	356	173	623, 624, 626
$[PdCl_4]^{2-}$	310	198	–	275	336	193	628, 627
$[PtBr_4]^{2-}$	205	125	135	190	232	135	628, 625
$[AuBr_4]^-$	212	102	–	196	252	100	624, 623
	–	–	87	–	252	139	625
$[PdBr_4]^{2-}$	192	125	130	165	260	140	628, 627
$[PtI_4]^{2-}$	142	–	105	126	180	127	628

II-9. PHOSPHORUS AND SULFUR COMPOUNDS OF VARIOUS STRUCTURES

In addition to the phosphorus and sulfur compounds mentioned in previous sections, there are many more compounds of these elements and most of them contain tetrahedral phosphorus or sulfur. The band assignments given

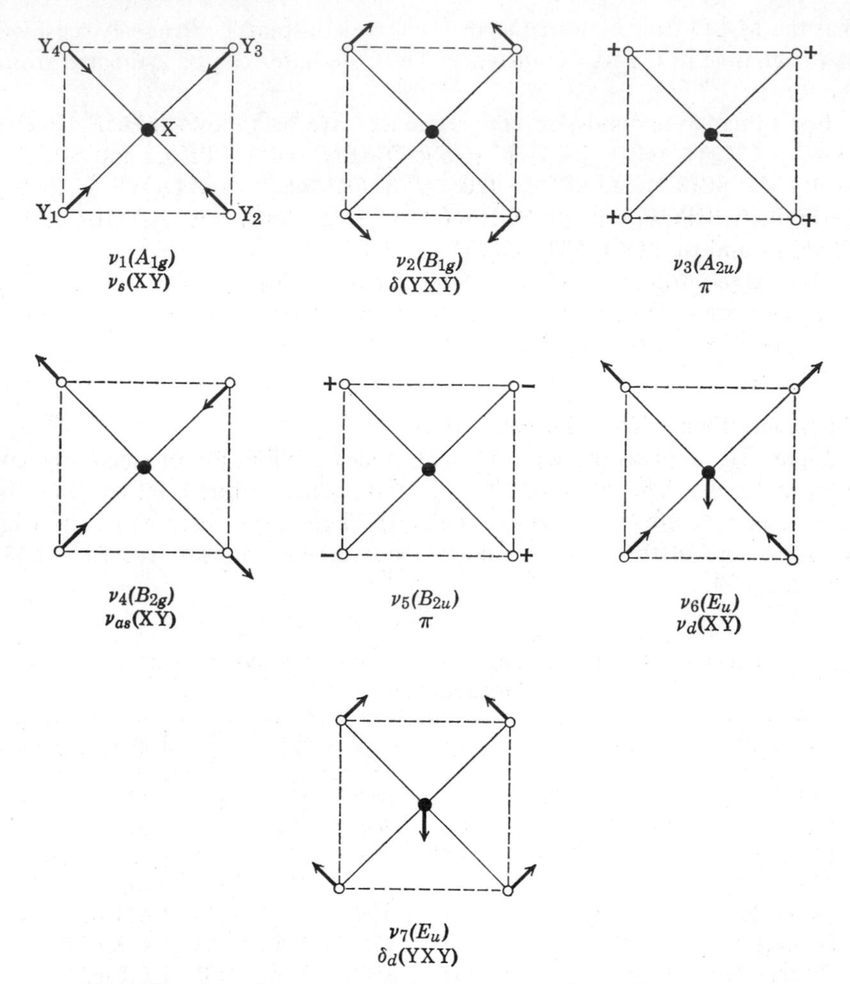

Fig. II-14. Normal modes of vibration of square-planar XY_4 molecules.

here are somewhat unreliable compared with those given in other sections, because most of the references provide only the observed frequencies or empirical band assignments.

(1) Phosphorus Compounds

From measurements of infrared spectra of about 60 salts of phosphorus oxy-acids, Corbridge and Lowe[631] have made a chart of the characteristic absorptions of phosphorus compounds. The group frequency chart in Fig. II-15 is based on references cited in this book. In using a chart of this type, care must be taken because the frequency ranges are affected by many

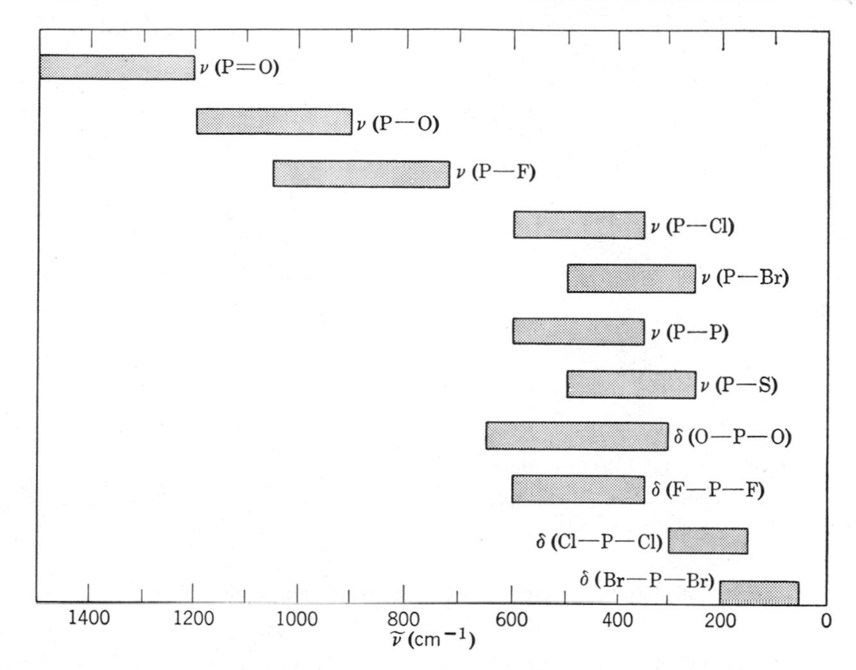

Fig. II-15. Characteristic frequencies of phosphorus compounds.

factors, such as the bond order, the nature of coupling with other modes, the kind of cation, and the effect of hydrogen bonding in the crystalline state.

For individual molecules, only the references are given: P_4 (632–635); P_2H_4 (635, 636); P_2Cl_4 (637); P_2I_4 (638, 639); P_4O_6 (640); P_4O_{10} (641); $[P_2O_6]^{4-}$ and $[H_2P_2O_6]^{2-}$ (642, 643); $[HP_2O_5]^{3-}$ (644); $[P_2O_7]^{4-}$ (643, 645, 646); $[P_2O_8]^{4-}$ (647); $[P_3O_9]^{3-}$ (648, 649); $[P_4O_{12}]^{4-}$ (650, 651); $P_4O_6S_4$ (652); P_4S_3 (653); $(NH_2)_3PO$ (654); HPO_3NH_3 (655); $[PNF_2]_n$ (656, 657); $[PNCl_2]_3$ (658); $[PNBr_2]_3$ (659).

(2) Sulfur Compounds

Simon and Kriegsmann[660] and Siebert [661] have made extensive studies of the vibrational spectra of sulfur compounds. Figure II-16 gives a group frequency chart obtained from the references appearing in this book. For individual molecules, only the references are listed: S_6 (662); S_8 (663–667); $[S_2O_4]^{2-}$ (668); $[S_2O_5]^{2-}$ (669–671); $[SO_5]^{2-}$ (672); $[S_2O_6]^{2-}$ (643); $[S_2O_8]^{2-}$ (673); $[S_3O_6]^{2-}$ (674); $S_2O_5F_2$ and $S_2O_5Cl_2$ (675, 676); $S_2O_6F_2$ (677); $S_3O_8F_2$ and $S_3O_8Cl_2$ (675, 676); S_2O_2 (678); $NH_3^+—SO_3^-$ (sulfamic acid) (679–681); $NH_3^+—OSO_3^-$ (682); N_4S_4 (683, 684); $H_4N_4S_4$ (683); NSF, $S(NF)_2$ and NSF_3 (685–687); S_2F_{10} (688, 689); $[C_2S_7]^{2-}$ (690); $(Cl_2CS)_2$

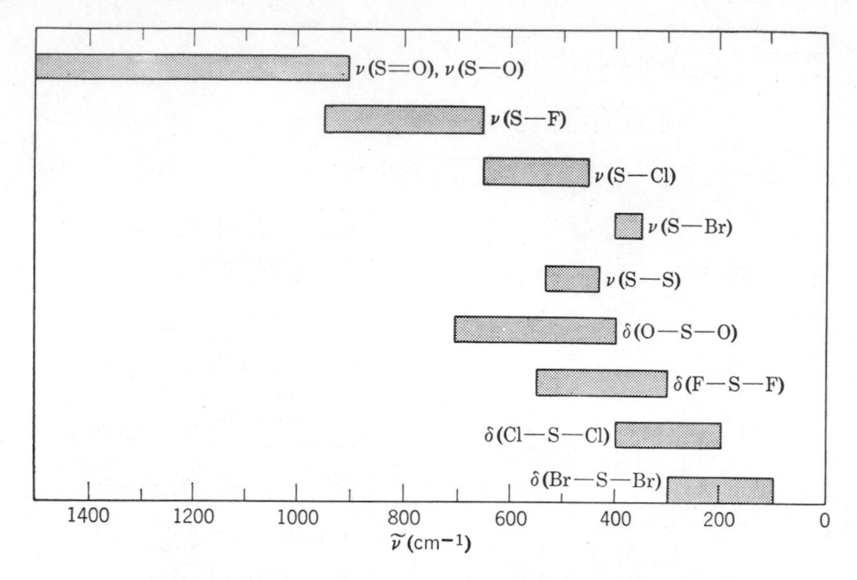

Fig. II-16. Characteristic frequencies of sulfur compounds.

(691). Gillespie and Robinson[692] studied the correlation of stretching frequencies and force constants with bond lengths, bond angles, and bond orders for the SO bond in sulphuryl and thionyl compounds.

II-10. XY_5, XY_6, AND XY_7 MOLECULES

(1) XY_5 Molecules (D_{3h} or C_{4v})

An XY_5 molecule may be a trigonal bipyramid ($\mathbf{D}_{3h}$) or a tetragonal pyramid ($\mathbf{C}_{4v}$). If it is trigonal bipyramidal, six of the eight normal vibrations (A_1', E', and E'') are Raman active, and five (A_2'' and E') are infrared active. If it is tetragonal pyramidal, all nine vibrations are Raman active, but only six (A_1 and E) are infrared active. Thus it is possible to distinguish between these two configurations through the infrared and Raman spectra.

Figure II-17 shows the eight normal modes of vibration of an XY_5 molecule having a trigonal bipyramidal structure. Table II-30 lists the observed fundamental frequencies of this type of molecule. It is interesting to note that in the gaseous and liquid states phosphorus pentachloride (PCl_5) exists as a trigonal bipyramidal molecule, whereas in the crystalline state it has an ionic structure consisting of $[PCl_4]^+[PCl_6]^-$ units. Gerding and Houtgraff[704] have confirmed this structure from the Raman spectrum of crystalline PCl_5. The tetrahedral $[PCl_4]^+$ ion exhibits four Raman lines at 627, 451, 244, and 173 cm^{-1}, whereas the octahedral $[PCl_6]^-$ ion exhibits three Raman lines

TABLE II-30. VIBRATIONAL FREQUENCIES OF TRIGONAL BIPYRAMIDAL XY_5 MOLECULES (CM^{-1})

Molecule	ν_1	ν_2	ν_3	ν_4	ν_5	ν_6	ν_7	ν_8	References
PF_5	817	640	945	576	1026	533	301	514	693, 694
AsF_5	733	642	785	400	809	366	128	388	695
SbF_5	667	264	–	–	716	491	90	228	696
VF_5	719	608	784	331	810	282	200	350	697, 698
PCl_5	395	370	441	301	581	281	100	261	699–701
$SbCl_5$	356	307	371	154	395	172	72	165	700, 702
$NbCl_5$	500	412	420	153	355	170	106	(200)	700, 703
$TaCl_5$	490	410	402	146	365	170	108	194	700

at 358, 285, and 244 cm^{-1}, in agreement with theoretical predictions. Normal coordinate analyses for $\mathbf{D}_{3h}$-type molecules have been made by a number of investigators (Ref. 705–707, 701, 695, and 697). NbF_5 and TaF_5 are tetrameric,[708] whereas $MoCl_5$[709] and $NbCl_5$[710] are dimeric in the crystalline state. Vibrational spectra of $PF_xCl_y(x+y=5)$,[711] PF_3Br_2,[712] PH_2F_3,[713] and SbF_3Cl_2,[714] have also been studied.

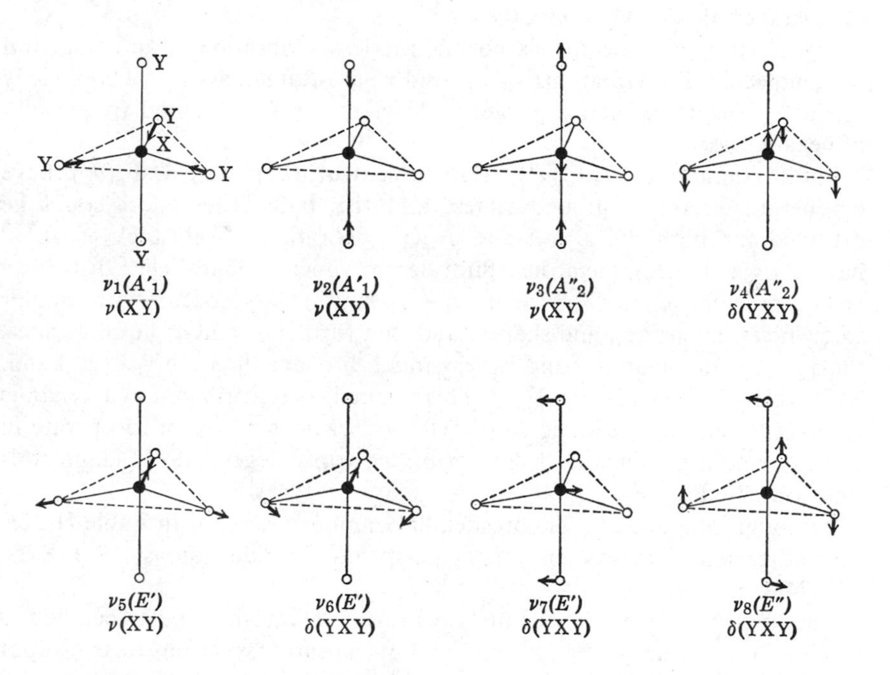

Fig. II-17. Normal modes of vibration of trigonal bipyramidal XY_5 molecules.

Tetragonal pyramidal XY_5 molecules are rare. Table II-31 lists the fundamental frequencies of four compounds belonging to this type. For normal coordinate analysis of this type of compounds, see Refs. 715 and 718. The $XeOF_4$ molecule also takes a tetragonal pyramidal structure, and its vibrational spectrum has been assigned by Begun, Fletcher, and Smith.[715]

TABLE II-31. VIBRATIONAL FREQUENCIES OF TETRAGONAL PYRAMIDAL XY_5 MOLECULES (CM^{-1})

Molecule		ν_1 A_1	ν_2 A_1	ν_3 A_1	ν_4 B_1	ν_5 B_1	ν_6 B_2	ν_7 E	ν_8 E	ν_9 E	References
ClF_5	R	709	538	480	480	346	375	–	–	296	715
	IR	(712)	541	486	(488)	–	–	732	–	302	
BrF_5	R	682	570	365	535	281	312	–	414	237	715
	IR	683	587	369	(547)	–	–	644	415	–	
IF_5	R	698	593	315	575	257	273	–	374	189	715, 716
	IR	710	(595)	318	–	–	–	640	372	–	
$[TeF_5]^-$	R	611	504	282	572	–	231	472	338	–	717
	IR	618	–	283	–	–	–	466	336	164	

(2) Octahedral XY_6 Molecules (O_h)

Figure II-18 indicates the six normal modes of vibration of an octahedral XY_6 molecule. The vibrations ν_1, ν_2, and ν_5 are Raman active, whereas only ν_3 and ν_4 are infrared active. Table II-32*a* gives the fundamental frequencies of hexafluorides.

The hexafluorides $TcF_6(d^1)$, $ReF_6(d^1)$, $RuF_6(d^2)$, and $OsF_6(d^2)$ have degenerate electronic ground states, and the Jahn-Teller effect could be expected for both the $E_g(\nu_2)$ and $F_{2g}(\nu_5)$ vibrations. Weinstock et al.[733] have shown that for metal hexafluorides in which a Jahn-Teller distortion is not possible (e.g., d^0, MoF_6), $\nu_1 + \nu_3$ and $\nu_2 + \nu_3$ occur with similar frequencies, intensities, and shapes; and that for the d^1 and d^2 hexafluorides, the $\nu_2 + \nu_3$ combination band is very much broader than the $\nu_1 + \nu_3$ band. This is clearly shown in Fig. II-19. This anomaly was attributed to a dynamic Jahn-Teller effect. The static Jahn-Teller effect does not seem to operate in these compounds, since no splittings of the triply degenerate fundamentals were observed.

Although only four hexafluorometallate anions are listed in Table II-32*a*, there are many reports on other metal hexafluoride anions. (See Refs. 743–748).

The molecule XeF_6 is definitely distorted from the regular octahedral configuration. This is clearly seen from its infrared spectrum in the vapor phase. If it has O_h symmetry, we should expect one Xe-F stretching (F_{1u})

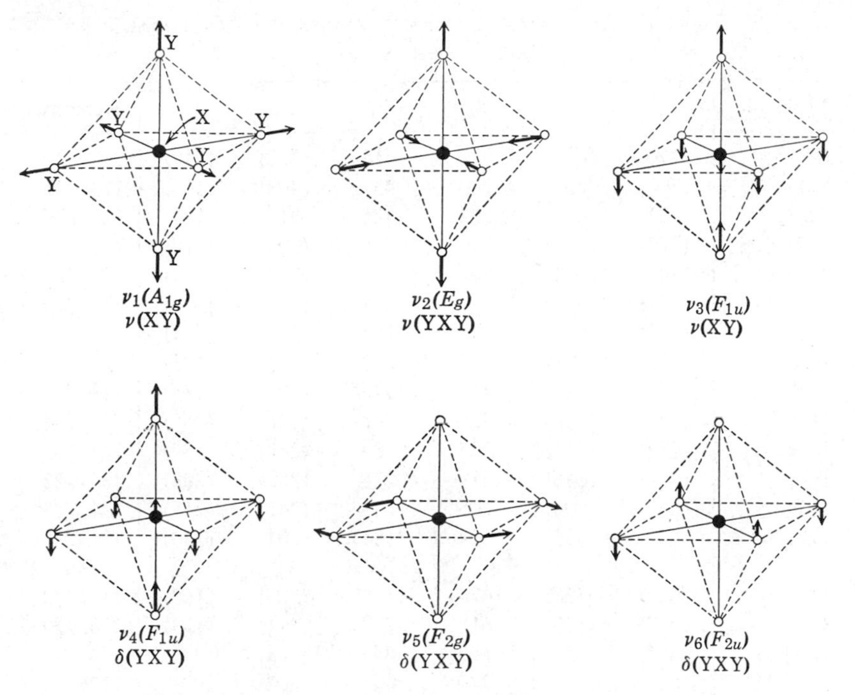

Fig. II-18. Normal modes of vibration of octahedral XY_6 molecules.

band between 650 and 500 cm^{-1}. However, three bands were observed at 612 (strong), 565(shoulder), and 520 (medium) cm^{-1}.[749,750] This result suggests that its symmetry is definitely lower than $\mathbf{O}_h$. A similar result is also obtained for the Raman spectrum; only two Xe—F stretching fundamentals (A_{1g} and E_g) are expected for $\mathbf{O}_h$, while three lines were actually observed: 655 (strong), 635 (medium), and 582 (weak) cm^{-1}.

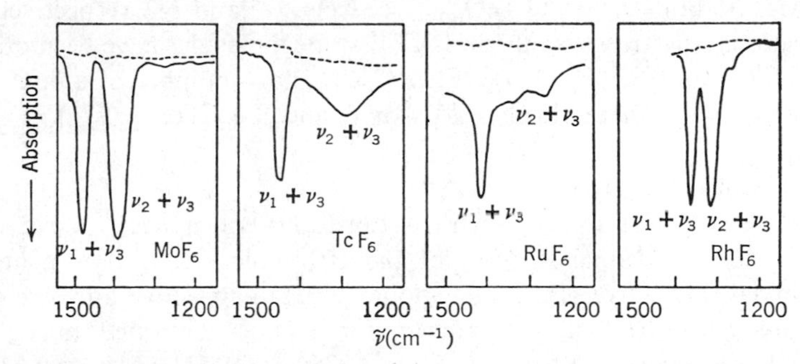

Fig. II-19. Band profiles for ($\nu_1 + \nu_3$) and ($\nu_2 + \nu_3$) for the 4d transition series hexafluorides.[733, 723]

TABLE II-32*a*. VIBRATIONAL FREQUENCIES OF OCTAHEDRAL HEXAFLUORIDES (CM^{-1})

Molecule	ν_1	ν_2	ν_3	ν_4	ν_5	ν_6	References
SF_6	770	(640)	939	614	(522)	(349)	719–723
SeF_6	708	(661)	780	437	(403)	(262)	723–725
TeF_6	701	674	752	325	313	(195)	723–726
CrF_6	(720)	(650)	790	(266)	(309)	(110)	723
MoF_6	741	643	741	262	(312)	(122)	723, 727–730
WF_6	(771)	(673)	711	258	(315)	(134)	723, 729, 730
TcF_6	(712)	(639)	748	265	(297)	(174)	723, 728
ReF_6	755	(671)	715	257	(295)	(193)	723, 731, 732
RuF_6	(675)	(624)	735	275	(283)	(186)	723, 733
OsF_6	(733)	(668)	720	272	(276)	(205)	723, 734, 727
RhF_6	(634)	(592)	724	283	(269)	(189)	723, 733
IrF_6	(701)	(646)	719	276	(258)	(206)	735, 723
PtF_6	(655)	(600)	705	273	(242)	(211)	734, 723
UF_6	667	535	624	(184)	(201)	(140)	724, 730, 736, 723
NpF_6	(648)	(528)	624	(198)	(205)	(165)	737, 723
PuF_6	(628)	(523)	616	(203)	(211)	(173)	738, 739, 723
$[GeF_6]^{2-}$	627	454	598	349	318	–	740
$[SiF_6]^{2-}$	656	466	740	485	403	–	727
$[AsF_6]^-$	679	565	700	400	372	–	741
$[PtF_6]^{2-}$	611	576	571	281	210	(143)	742

Normal coordinate analyses for octahedral XY_6 molecules have been made by many investigators (Refs. 751–757).

Recently, far-infrared spectra of the MX_6 type anions (X = Cl, Br, and I) have been studied extensively. Table II-32*b* lists the fundamental frequencies of these anions. For a number of other hexahalogeno anions, see Refs. 625, 759, 767–770. For IO_6^{5-} and TeO_6^{6-}, see Refs. 771 and 772, respectively.

Vibrational spectra of octahedral XY_5Z-type molecules have been obtained for SF_5Cl(773), $SF_5(CF_3)$(774), IF_5O (775), and a number of anions of the type, $[MX_5O]^{2-}$, where M is Nb, Mo, or W and X is Cl or Br(776).

(3) XY_7 Molecules ($\mathbf{D}_{5h}$)

Molecules of this type are very rare. Lord and colleagues[777] have studied the infrared and Raman spectra of the IF_7 molecule. If it is pentagonal bipyramidal ($\mathbf{D}_{5h}$), five vibrations should be Raman active and five should be infrared active. In fact, this structure for IF_7 was confirmed through these vibrations: Raman—678 (A_1'), 635 (A_1'), 360 (E_1''), 511 (E_2'), and 313 (E_2') cm^{-1}; infrared—670 (A_2''), 368 (A_2''), 547 (E_1'), 426 (E_1') and 250 (E_1') cm^{-1}.

TABLE II-32b. VIBRATIONAL FREQUENCIES OF OCTAHEDRAL HEXACHLORIDES, HEXABROMIDES, AND HEXAIODIDES (cm^{-1})

Molecule	ν_1	ν_2	ν_3	ν_4	ν_5	ν_6	References
$[TlCl_6]^{3-}$	280	262	–	–	155	–	758
$[GeCl_6]^{2-}$	318	213	293	205	191	–	759, 760
$[SnCl_6]^{2-}$	314	235	313	–	157	–	761, 762
$[PbCl_6]^{2-}$	285	215	265	130	137	–	760
$[PCl_6]^-$	360	283	444	285	238	–	760
$[AsCl_6]^-$	337	289	333	220	202	–	760
$[SbCl_6]^-$	337	277	345	180	172	–	760
$[SeCl_6]^{2-}$	343	273	–	–	166	–	762
$[TeCl_6]^{2-}$	287	247	228	105	131	–	760
$[TiCl_6]^{2-}$	320	271	316	183	173	–	763
$[ReCl_6]^{2-}$	346	275	313	172	159	–	764
$[OsCl_6]^{2-}$	346	274	314	177	165	–	764
$[PdCl_6]^{2-}$	317	292	340	175	164	–	762, 765
$[PtCl_6]^{2-}$	344	320	343	182	162	–	762, 765
$[IrCl_6]^{2-}$	352	225	335	168	190	–	766
$[SnBr_6]^{2-}$	183	144	–	–	69	–	761, 762
$[ReBr_6]^{2-}$	213	174	217	118	104	–	764
$[PdBr_6]^{2-}$	198	176	253	130	100	–	766
$[PtBr_6]^{2-}$	207	190	244	90	97	–	765, 762
$[PtI_6]^{2-}$	–	–	186	46	–	–	625

The same conclusion has been obtained by nuclear magnetic resonance[778] and electron diffraction[779] studies of this molecule. Khanna[780] made a normal coordinate analysis of IF_7. Claassen and Selig have shown that the infrared and Raman spectra of ReF_7 can be accounted for on the basis of the pentagonal bipyramidal structure.[781] The infrared spectra of the $[NbF_7]^{2-}$ and $[TaF_7]^{2-}$ ions are reported.[743] Keller and Chetham-Strode[747,748] have studied the Raman spectra of these ions in aqueous HF solutions to identify the predominant species involved in equilibrium.

II-11. X_2Y_6, X_2Y_7, AND XY_8 MOLECULES

(1) Bridged X_2Y_6 Molecules (V_h)

Figure II-20 indicates the 18 normal modes of vibration[782] and band assignments for a bridged X_2Y_6 molecule. The vibrations of the A_g, B_{1g}, B_{2g}, and B_{3g} species are Raman active, whereas those of the B_{1u}, B_{2u}, and B_{3u} species are infrared active. Table II-33 lists the fundamental frequencies of five molecules of this type. The bridged XY′ stretching frequencies (ν_2, ν_6, ν_{13}, and ν_{17}) are always lower than the terminal XY stretching frequencies

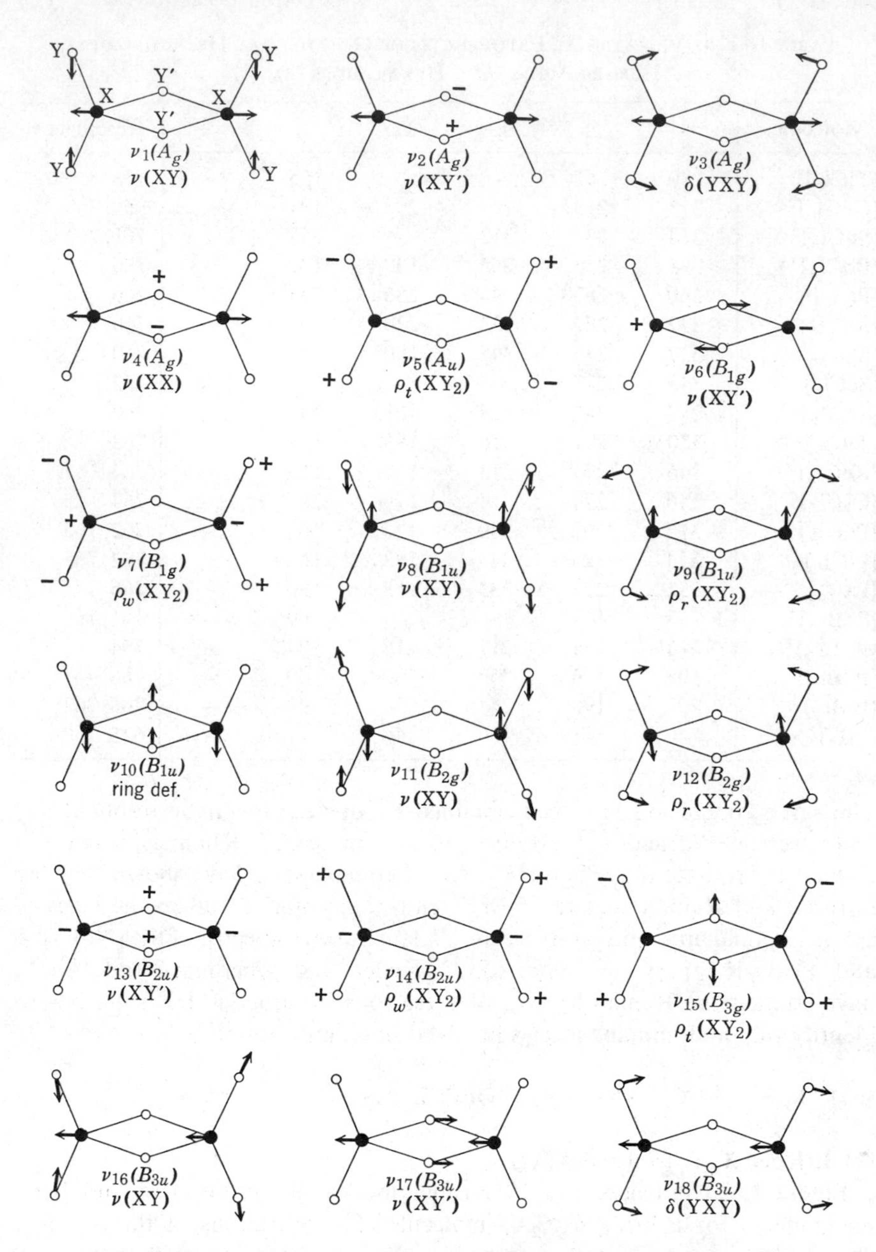

Fig. II-20. Normal modes of vibration of bridged X_2Y_6 molecules.[782]

TABLE II-33. VIBRATIONAL FREQUENCIES OF BRIDGED X_2Y_6 MOLECULES (cm^{-1})

	B_2H_6[a]	B_2D_6[b]	Al_2Cl_6[c]	Al_2Br_6[d]	Al_2I_6[d]
ν_1	2524	1850	506	491	406
ν_2	2104	1500	340	204	146
ν_3	1180	915	217	140	94
ν_4	794	700	112	73	53
ν_5	(829)	(592)	–	–	–
ν_6	1768	1270	284	(291)	(195)
ν_7	(1035)	(860)	–	176	–
ν_8	2612	1985	625	–	–
ν_9	(950)	730	–	–	–
ν_{10}	368	250	–	–	–
ν_{11}	2591	1975	606	407	344
ν_{12}	(920)	(725)	164	112	–
ν_{13}	1915	1460	420	–	–
ν_{14}	973	720	–	–	–
ν_{15}	1012	730	160	–	–
ν_{16}	2525	1840	484	–	–
ν_{17}	1606	1199	–	–	–
ν_{18}	1177	876	–	–	–

[a] Refs. 783–784
[b] Refs. 783, 785, 786
[c] Refs. 787, 788
[d] Ref. 789

(ν_1, ν_8, ν_{11}, and ν_{16}) in these compounds. Onishi and Shimanouchi[788] have carried out normal coordinate analyses of Al_2Cl_6, $Al_2(CH_3)_6$ and $Al_2(CH_3)_4Cl_2$ using a modified UBF field. Aluminum chloride exists as a dimer in the gaseous and liquid phases, whereas it consists of an $[AlCl_6]^{3-}$ unit in the crystalline state. Gallium chloride exists as a dimer even in the solid state.[790] For the dimer of $FeCl_3$, see Ref. 702. A normal coordinate analysis of B_2H_6 has been done by Brown and Longuet-Higgins[791] and by Venkateswarlu and Thirugnansambandam.[792]

A number of boron compounds have unusual structures. Only the references are cited for these compounds: $M(BH_4)$ (793); B_5H_9 (794, 795); $B_{10}H_{14}$ (796); $B_3H_6N_3$ (borazine) (797); BO_2, B_2O_2, and B_2O_3 (798, 799); HBO_2 (800); B_2F_4 (801, 802); B_2Cl_4 (803, 804): B_2S_3 (805); $(BOF)_3$ and $(BOCl)_3$ (806).

(2) Ethane-Type X_2Y_6 Molecules

The conformation of ethane-type molecules may be staggered ($\mathbf{D}_{3d}$) or eclipsed ($\mathbf{D}_{3h}$). The 12 normal modes of vibration of the staggered molecule are illustrated in Fig. II-21. The vibrations of the A_{1g} and E_g species are Raman active, and those of the A_{2u} and E_u species are infrared active. Table II-34 gives the fundamental frequencies and band assignments based on $\mathbf{D}_{3d}$ symmetry.

Vibrational spectra have been obtained for a large number of organosilicon compounds. Here references are given for relatively simple silicon compounds containing no Si-C bonds. Disiloxane, $(SiH_3)_2O$, has been studied by Lord et al.,[812] Curl and Pitzer,[813] and McKean.[814] Disilyl sulfide, $(SiH_3)_2S$, has been studied by Linton et al.[815] and Ebsworth et al.[816] According to electron diffraction studies,[817,818] the Si-O-Si angle in the former is 144°, whereas the Si-S-Si angle in the latter is 97.4°.

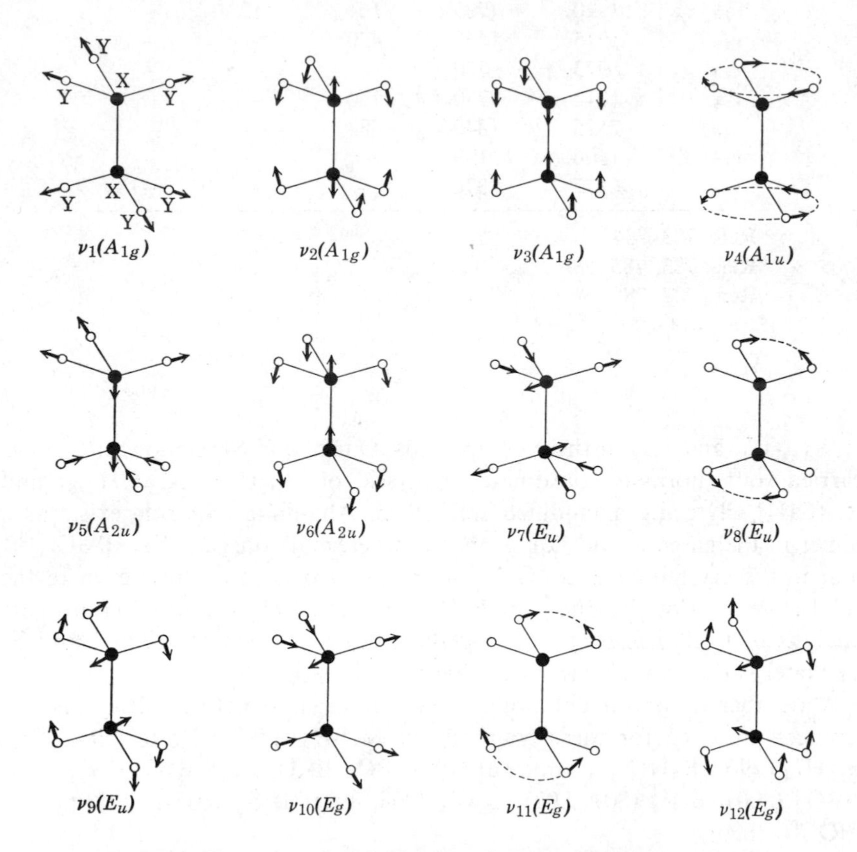

Fig. II-21. Normal modes of vibration of ethane-type X_2Y_6 molecules.

TABLE II-34. VIBRATIONAL FREQUENCIES OF ETHANE-TYPE X_2Y_6 MOLECULES (CM^{-1})

Molecule	C_2H_6[a]	C_2D_6[a]	Si_2H_6[b]	Si_2D_6[c]	Ge_2H_6[d]
ν_1 $\nu(XY)$	2899	2083	2152	1548	(2070)
ν_2 $\delta(XY_3)$	(1375)	1158	909	683	765
ν_3 $\nu(XX)$	993	852	434	408	229
ν_4 $\rho_t(XY_3)$	275	200	–	–	144
ν_5 $\nu(XY)$	2954	2111	2154	1549	2078
ν_6 $\delta(XY_3)$	1379	1072	844	625	755
ν_7 $\nu(XY)$	2994	2236	2179	1585	2114
ν_8 $\delta(XY_3)$	1486	1102	940	683	898
ν_9 $\rho_r(XY_3)$	821	601	379	277	407
ν_{10} $\nu(XY)$	2963	2225	2155	1569	2150
ν_{11} $\delta(XY_3)$	1460	1055	929	667	875
ν_{12} $\rho_r(XY_3)$	(1155)	970	625	475	417

[a] Ref. 807
[b] Refs. 808, 809
[c] Ref. 808
[d] Ref. 810, 811

Robinson[819] and Kriegsmann and Forster[820] have shown that the spectrum of trisilylamine, $(SiH_3)_3N$, can be explained by using a trigonal planar model. The vibrational spectra of trisilylphosphine. $(SiH_3)_3P$, can also be interpreted on the basis of the trigonal planar model.[821] The vibrational spectrum of $(SiH_3)_3As$ shows that the $AsSi_3$ skeleton is planar. However, the configuration of the $SbSi_3$ skeleton in $(SiH_3)_3Sb$ is not clear from its spectrum.[822] Spectral data are available for a number of other compounds containing the silyl (SiH_3) group; SiH_3X(X, halogen) (823); $(SiH_3)_2CN_2$ (824); $(SiH_3)PH_2$ (825). All the compounds containing the SiH_3, SiH_2 and SiH groups exhibit strong bands between 2200 and 2100 cm^{-1} due to the Si-H stretching modes. In addition, the former two groups exhibit several H-Si-H bending modes below 1000 cm^{-1}. For a correlation between the Si-H stretching frequency and the inductive effect or the structure, see Ref. 826 or 827.

(3) Y_3X-Y-XY_3 Molecules

Vibrational spectra have been obtained for the anions such as $[S_2O_7]^{2-}$ (828), $[Se_2O_7]^{2-}$ (829, 830), $[Cr_2O_7]^{2-}$ (831), $[P_2O_7]^{4-}$ (832) and for a neutral molecule such as Cl_2O_7 (833). For normal coordinate analysis, see Refs. 834 and 835. Gillespie and Robinson[836] have calculated the bond order of the X-Y-X bridge from the X-Y-X stretching force constant.

(4) XY_8 Molecules

The XY_8 molecule may take the form of (I) a cube ($\mathbf{O}_h$), (II) an Archimedean antiprism ($\mathbf{D}_{4h}$), (III) a dodecahedron ($\mathbf{D}_{2d}$) or (IV) a face-centered trigonal prism ($\mathbf{C}_{2v}$). Although XY_8 molecules are rare, x-ray analysis indicates that the $[TaF_8]^{3-}$ and $[CrO_8]^{3-}$ ions have structures II and III, respectively.[837,838] Weinstock and Malm[839] have shown that the compound previously suggested to be OsF_8 is OsF_6. Dove [840]has shown from infrared and x-ray studies that the compounds of the type $M_3HM^{IV}F_8$ (M = Na, K; M^{IV} = Nb, Pb, Ti, Sn) consist of M^+, $[M^{IV}F_6]^{2-}$, and $[HF_2]^-$ ions in the crystalline state. Little work has been done on the vibrational spectra of XY_8 molecules. Pistorius[841] has calculated the **G** and **F** matrix elements for a cubic XY_8 molecule, and Schlaefer and Wasgestian[842] have calculated those for an Archimedean antiprism XY_8 molecule.

II-12. MINERALS

A large amount of literature on the vibrational spectra of minerals is available. Some of these investigations are classified here, and only the references are given for each group.

(1) Theoretical Analysis

Vibrational spectra of minerals can be analyzed by the method of site group or factor group analysis (Part I, Sec. 16). Some references are: α-quartz (843); β-quartz (844); silicon carbide (845); rutile (846); silicates (847–849); sulfate minerals (850).

(2) Location of the Position of Hydrogen through Measurement of Infrared Dichroism

The position of hydrogen is rather difficult to determine from x-ray analysis. Studies of the infrared dichroism of, for example, the OH stretching band is useful in the elucidation of the orientation of the OH bond in minerals. Some references are: muscovite (851), afwillite (852), and silicates (853).

(3) Identification of Minerals

The differences in the spectra of calcite and aragonite have been discussed in Sec. I-16. Minerals having different crystal structures (modifications) can be distinguished by means of their spectra. Some references are: diamonds (854, 855); silicas (856–860); carbonates (861–863); hydrated aluminas (864); micas (865–866); serpentine minerals (867); and talcs (868). For a collection of infrared spectra of various minerals, see Ref. 869.

(4) Infrared Reflection Spectra

The reflection method is useful in obtaining the infrared spectra of minerals. Reflection spectra have been obtained for silicas (870, 871), carborundum (872), and titanates (873).

References

1. S. A. Rice and W. Klemperer, *J. Chem. Phys.*, **27**, 573 (1957); W. Klemperer and S. A. Rice, *ibid.*, **26**, 618 (1957); W. Klemperer, W. G. Norris, A Buchler and A. G. Emslie, *ibid.*, **33**, 1534 (1960); W. Klemperer and W. G. Norris, *ibid.*, **34**, 1071 (1961).
2. E. Whittle, D. A. Dows, and G. C. Pimentel, *J. Chem. Phys.*, **22**, 1943 (1954); E. D. Becker and G. C. Pimentel, *ibid.*, **25**, 224 (1956).
3. D. E. Milligan and M. E. Jacox, *J. Chem. Phys.*, **41**, 2838 (1964).
4. K. Rosengren and G. C. Pimentel, *J. Chem. Phys.*, **43**, 507 (1965).
5. D. E. Milligan and M. E. Jacox, *J. Chem. Phys.*, **40**, 2461 (1964).
6. J. F. Ogilvie, *Nature*, **204**, 572 (1964).
7. A. Arkell, R. R. Reinhard, and L. P. Larson, *J. Amer. Chem. Soc.*, **87**, 1016 (1965).
8. D. E. Mann, N. Acquista, and D. White, *J. Chem. Phys.*, **44**, 3453 (1966).
9. L. F. Keyser and G. W. Robinson, *J. Chem. Phys.*, **44**, 3225 and **45**, 1694 (1966).
10. D. F. Smith, *J. Chem. Phys.*, **28**, 1040 (1958).
11. J. L. Hollenberg, *J. Chem. Phys.*, **46**, 3271 (1967).
12. R. M. Adams and J. J. Katz, *J. Opt. Soc. Amer.*, **46**, 895 (1956).
13. P. A. Giguère and N. Zengin, *Can. J. Chem.*, **36**, 1013 (1958).
14. D. F. Hornig and W. E. Osberg, *J. Chem. Phys.*, **23**, 662 (1955); G. L. Hiebert and D. F. Hornig, *ibid.*, **28**, 316 (1958).
15. M. L. Josien, G. Sourisseau, and C. Castinel, *Bull. Soc. Chim. Fr.*, **1955**, 1539; M. L. Josien and G. Sourisseau, *ibid.*, **1955**, 178; P. Grange, J. Lascombe, and M. L. Josien, *Spectrochim. Acta*, **16**, 981 (1960).
16. J. Berkowitz, *J. Chem. Phys.*, **29**, 1386 (1958); **32**, 1519 (1960).
17. M. J. Linevsky, *J. Chem. Phys.*, **34**, 587 (1961).
18. R. L. Redington, *J. Chem. Phys.*, **44**, 1238 (1966).
19. A. Snelson, *J. Chem. Phys.*, **46**, 3652 (1967).
20. A. L. Smith, W. E. Keller, and H. L. Johnston, *J. Chem. Phys.*, **19**, 189 (1951).
21. W. G. Fateley, H. A. Bent, and B. L. Crawford, *J. Chem. Phys.*, **31**, 204 (1959).
22. D. J. Millen and D. Watson, *J. Chem. Soc.*, **1957**, 1369; J. D. S. Goulden and D. J. Millen, *ibid.*, **1950**, 2620.
23. V. Lorenzelli, *Compt. rend.*, **258**, 5386 (1964).
24. W. B. Person, R. E. Humphrey, W. A. Deskin, and A. I. Popov, *J. Amer. Chem. Soc.*, **80**, 2049 (1958); W. B. Person, R. E. Erickson, and R. E. Buckes, *ibid.*, **82**, 29 (1960); A. I. Popov, R. E. Humphrey, and W. B. Person, *ibid.*, 82, 1850 (1960).
25. W. D. Stalleup and D. Williams, *J. Chem. Phys.*, **10**, 199 (1942).
26. R. A. Penneman and L. H. Jones, *J. Chem. Phys.*, **24**, 293 (1956).
27. R. A. Penneman and L. H. Jones, *J. Chem. Phys.*, **28**, 169 (1958).
28. J. Goubeau, H. Haeberle, and H. Ulmer, *Z. Anorg. Allg. Chem.*, **311**, 110 (1961).
29. J. Guy and M. Cahigneau, *Bull. Soc. Chim. Fr.*, **1956**, 257.
30. F. A. Miller, S. G. Frankiss, and O. Sala, *Spectrochim. Acta*, **21**, 775 (1965); F. A. Miller and W. K. Baer, *Spectrochim. Acta*, **19**, 73 (1963).
31. D. A. Long and D. Steele, *Spectrochim. Acta*, **19**, 1731 (1963).
32. A. Langseth and C. K. Møller, *Acta. Chem. Scand.*, **4**, 725 (1950); *Nature*, **166**, 147 (1950).
33. L. H. Jones, *J. Chem. Phys.*, **22**, 217 (1954).
34. K. A. Wickersheim, *J. Chem. Phys.*, **31**, 863 (1959).
35. R. M. Hexter, *J. Chem. Phys.*, **34**, 941 (1961).
36. R. A. Buchanan, E. L. King, and H. H. Caspers, *J. Chem. Phys.*, **36**, 2665 (1962).
37. R. A. Buchanan, H. H. Caspers, and H. R. Marlin, *J. Chem. Phys.*, **40**, 1125 (1964).
38. E. Drouard, *Compt. rend.*, **249**, 665 (1959); **247**, 68 (1958).

39. W. R. Busing, *J. Chem. Phys.*, **23**, 933 (1955).
40. J. A. Ibers, J. Kumamoto, and R. G. Snyder, *J. Chem. Phys.*, **33**, 1164 (1960); R. G. Snyder, J. Kumamoto, and J. A. Ibers, *ibid.*, **33**, 1171 (1960).
41. R. A. Buchanan, *J. Chem. Phys.*, **31**, 870 (1959).
42. W. R. Busing and H. W. Morgan, *J. Chem. Phys.*, **28**, 998 (1958).
43. R. M. Hexter, *J. Opt. Soc. Amer.*, **48**, 770 (1958).
44. H. A. Benesi, *J. Chem. Phys.*, **30**, 852 (1959).
45. R. T. Mara and G. B. B. M. Sutherland, *J. Opt. Soc. Amer.*, **43**, 1100 (1953).
46. B. A. Phillips and W. R. Busing, *J. Phys. Chem.*, **61**, 502 (1957).
47. C. Cabannes-Ott, *Compt. rend.*, **242**, 355 (1956).
48. D. Krishnamurti, *Proc. Indian Acad. Sci.*, **50A**, 223 (1959).
49. C. Cabannes-Ott, *Compt. rend.*, **242**, 2825 (1956).
50. R. West and R. H. Baney, *J. Phys., Chem.*, **64**, 822 (1960).
51. A. V. Stuart and G. B. B. M. Sutherland, *J. Chem. Phys.*, **24**, 559 (1956).
52. M. Falk and E. Whalley, *J. Chem. Phys.*, **34**, 1554 (1961).
53. O. Glemser and E. Hartert, *Naturwissenschaften*, **40**, 552 (1953).
54. O. Glemser, *Nature*, **183**, 1476 (1959).
55. V. A. Kolesov and Y. I. Ryskin, *Opt. Spektrosk.*, **7**, 261 (1959).
56. T. Dupuis, *Rec. Trav. Chim., Pay -Bas.*, **79**, 518 (1960).
57. D. E. Bethell and N. Sheppard, *Trans. Faraday Soc.*, **51**, 9 (1955).
58. P. Natalis, *Ann. soc. sci. Bruxelles, Ser. I*, **73**, 261 (1959).
59. E. Hartert and O. Glemser, *Z. Elektrochem.*, **60**, 746 (1956).
60. R. L. Williams and R. J. Page, *J. Chem. Soc.*, **1957**, 4143.
61. J. H. Taylor, W. S. Benedict, and J. Strong, *J. Chem. Phys.*, **20**, 1884 (1952).
62. W. E. Osberg and D. F. Hornig, *J. Chem. Phys.*, **20**, 1345 (1952).
63. M. E. Jacox and D. E. Milligan, *Spectrochim. Acta*, **17**, 1196 (1961).
64. L. H. Jones and E. McLaren, *J. Chem. Phys.*, **28**, 995 (1958).
65. A. H. Nielsen and R. T. Lagemann, *J. Chem. Phys.*, **22**, 36 (1954).
66. T. Wentink, *J. Chem. Phys.* **29**, 188 (1958).
67. D. F. Smith, "Noble-Gas Compounds," edited by H. H. Hyman, Univ. of Chicago Press, 1963, p. 295.
68. H. H. Claassen, G. L. Goodman, J. C. Malm, and F. Schreiner, *J. Chem. Phys.*, **42**, 1229 (1965).
69. W. Klemperer and L. Lindeman, *J. Chem. Phys.*, **25**, 397 (1956); W. Klemperer, *ibid.*, **25**, 1066 (1956).
70. W. Klemperer, *J. Electrochem. Soc.*, **110**, 1023 (1963).
71. P. Gray and T. C. Waddington, *Trans. Faraday Soc.*, **53**, 901 (1957).
72. H. A. Papazian, *J. Chem. Phys.*, **34**, 1614 (1961).
72. D. A. Dows, E. Whittle, and G. C. Pimentel, *J. Chem. Phys.*, **23**, 1475 (1955); D. A. Dows, G. C. Pimentel, and E. Whittle, *ibid.*, **23**, 1606 (1955); D. E. Milligan, H. W. Brown, and G. C. Pimentel, *ibid.*, **25**, 1080 (1956).
74. J. I. Bryant, *J. Chem. Phys.*, **38**, 2845 (1963).
75. L. H. Jones and R. A. Penneman, *J. Chem. Phys.*, **22**, 781 (1954); R. Newman and R. M. Badger, *ibid.*, **19**, 1207 (1951).
76. A. Azman and A. Occirk, *Spectrochim. Acta*, **23A**, 1597 (1967).
77. J. C. Evans and G. Y-S. Lo., *J. Phys. Chem.*, **70**, 11 (1966).
78. T. C. Waddington, *J. Chem. Soc.*, **1958**, 1708.
79. J. C. Evans and G. Y-S. Lo, *J. Phys. Chem.*, **70**, 20 (1966).
80. J. C. Evans and G. Y-S. Lo, *J. Phys. Chem.*, **70**, 543 (1966).
81. J. R. Soulen and W. F. Schwartz, *J. Phys. Chem.*, **66**, 2066 (1962).

82. J. W. Nebgen, A. D. McElroy, and H. F. Klodowski, *Inorg. Chem.*, **4**, 1796 (1965).
83. R. A. Marcus and J. M. Fresco, *J. Chem. Phys.*, **27**, 564 (1957).
84. S. K. Deb and A. D. Yoffe, *Trans. Faraday Soc.*, **55**, 106 (1959).
85. G. K. T. Conn and C. K. Wu, *Trans. Faraday Soc.*, **34**, 1483 (1938).
86. D. E. Milligan, M. E. Jacox, and D. McKinley, *J. Chem. Phys.*, **42**, 902 (1965).
87. G. E. Leroi, T. C. James, J. T. Hougen, and W. Klemperer, *J. Chem. Phys.*, **36**, 2879 (1962).
88. A. Snelson, *J. Phys., Chem.*, **70**, 3208 (1966).
89. A. Büchler and W. Klemperer, *J. Chem. Phys.*, **29**, 121 (1958).
90. R. E. Rundle and P. H. Lewis, *J. Chem. Phys.*, **20**, 132 (1952).
91. D. White, K. S. Seshadri, and D. F. Dever, *J. Chem. Phys.*, **39**, 2463 (1963).
92. J. A. Salthouse and T. C. Waddington, *J. Chem. Soc.*, **A,1966**, 28.
93. J. C. Evans and G. Y-S. Lo, *J. Chem. Phys.*, **44**, 3638 (1966).
94. J. C. Evans and G. Y-S. Lo, *Inorg. Chem.*, **6**, 1483 (1967).
95. A. G. Maki and R. Forneris, *Spectrochim. Acta*, **23A**, 867 (1967).
96. G. C. Haywood and P. J. Hendra, *Spectrochim. Acta*, **23A**, 2309 (1967).
97. K. O. Christe, W. Sawodny, and J. P. Guertin, *Inorg. Chem.*, **6**, 1159 (1967).
98. D. A. Long, F. S. Murfin, and R. L. Williams, *Proc. Roy. Soc.*, **A223**, 251 (1954).
99. H. D. Rix, *J. Chem. Phys.*, **22**, 429 (1954).
100. F. A. Miller and W. G. Fateley, *Spectrochim. Acta*, **20**, 253 (1964).
101. W. H. Smith and G. E. Leroi, *J. Chem. Phys.*, **45**, 1778 (1966).
102. C. K. Ingold, D. J. Millen, and H. G. Poole, *Nature*, **158**, 480 (1946); *J. Chem. Soc.*, **1950**, 2576; C. K. Ingold and D. J. Millen, *ibid.*, **1950**, 2612.
103. J. Prigent, *Compt. rend.*, **247**, 1739 (1958).
104. J. Sutton, *Nature*, **169**, 235 (1952).
105. L. H. Jones and R. A. Penneman, *J. Chem. Phys.* **21**, 542 (1953); L. H. Jones, *ibid.*, **23**, 2105 (1955).
106. L. H. Jones, *Spectrochim. Acta*, **15**, 409 (1959).
107. B. M. Gatehouse and A. E. Comyns, *J. Chem. Soc.*, **1958**, 3965.
108. G. L. Caldow, A. B. V. Cleave, and R. L. Eager, *Can. J. Chem.*, **38**, 772 (1960).
109. S. P. McGlynn, J. K. Smith, and W. C. Neely, *J. Chem. Phys.*, **35**, 105 (1961).
110. R. Tsuchida, *Bull. Chem. Soc. Jap.*, **13**, 388, 438 (1938).
111. H. C. Allen, E. D. Tidwell, and E. K. Plyler, *J. Chem. Phys.*, **25**, 302 (1956).
112. W. S. Richardson, *J. Chem. Phys.*, **19**, 1213 (1951).
113. P. A. Staats, H. W. Morgan, and J. H. Goldstein, *J. Chem. Phys.*, **25**, 582 (1956).
114. R. E. Dodd and R. Little, *Spectrochim. Acta*, **16**, 1083 (1960).
115. W. J. Orville-Thomas, *J. Chem. Phys.*, **20**, 920 (1952).
116. W. O. Freitag and E. R. Nixon, *J. Chem. Phys.*, **24**, 109 (1956).
117. A. Fadini, *Z. Angew. Math. u. Mech.*, **45**, 29 (1965); **46**, 52 (1966).
118. W. J. Lafferty, D. R. Lide, and R. A. Toth, *J. Chem. Phys.*, **43**, 2063 (1965).
119. J. Wagner, *Z. Phys. Chem.*, **B48**, 309 (1941).
120. A. Fadini, *Z. Naturforsch.* ,**A21**, 426 (1966).
121. J. Wagner, *Z. Phys. Chem.*, **A193**, 55 (1943).
122. W. West and M. Farnsworth, *J. Chem. Phys.*, **1**, 402 (1933).
123. G. M. Begun and W. H. Fletcher, *J. Chem. Phys.*, **28**, 414 (1958).
124. D. A. Dows, *J. Chem. Phys.*, **26**, 745 (1957).
125. H. J. Callomon, D. C. McKean, and H. W. Thompson, *Proc. Roy. Soc.*, **A208**, 341 (1951).
126. P. O. Kinell and B. Strandberg, *Acta Chem. Scand.*, **13**, 1607 (1959).
127. L. H. Jones, *J. Chem. Phys.*, **25**, 1069 (1956); **28**, 1234 (1958).

128. H. W. Morgan, *J. Inorg. Nucl. Chem.*, **16**, 368 (1960).
129. A. Maki and J. C. Decius, *J. Chem. Phys.*, **31**, 772 (1959).
130. T. C. Waddington, *J. Chem. Soc.*, **1959**, 2499.
131. D. E. Milligan and M. E. Jacox, *J. Chem. Phys.*, **39**, 712 (1963).
132. D. E. Milligan and M. E. Jacox, *J. Chem. Phys.*, **41**, 3032 (1964).
133. M. E. Jacox, D. E. Milligan, N. C. Moll, and W. E. Thompson, *J. Chem. Phys.*, **43** 3734 (1965).
134. D. E. Milligan and M. E. Jacox, *J. Chem. Phys.*, **44**, 2850 (1966).
135. R. L. Webb, S. Frank, and W. C. Schneider, *J. Amer. Chem. Soc.*, **77**, 3491 (1955).
136. T. Wadsten and S. Andersson, *Acta Chem. Scand.*, **13**, 1069 (1959).
137. B. R. Penfold and W. N. Lipscomb, *Acta Cryst.*, **14**, 589 (1961).
138. R. Savote and M. Pezolet, *Can. J. Chem.*, **45**, 1677 (1967).
139. C. B. Baddiel and G. J. Janz, *Trans. Faraday Soc.*, **60**, 2009 (1964).
140. F. Fehér and H. Weber, *Z. Naturforsch.*, **B11**, 426 (1956).
141. R. G. R. Bacon, R. S. Irwin, J. M. Pollock, and A. D. E. Pullin, *J. Chem. Soc.*, **1958** 764.
142. M. J. Nelson and A. D. E. Pullin, *J. Chem. Soc.*, **1960**, 604.
143. F. A. Miller and W. K. Baer, *Spectrochim. Acta*, **18**, 1311 (1963).
144. K. Oba, F. Watari and K. Aida, *Spectrochim. Acta*, **23A**, 1515 (1967).
145. K. E. Hjartaas, *Acta Chem. Scand.*, **21**, 1381 (1967).
146. F. A. Miller and G. L. Carlson, *Spectrochim. Acta*, **17**, 977 (1961).
147. T. Wentink, *J. Chem. Phys.*, **30**, 105 (1959).
148. W. J. Jones, W. J. Orville-Thomas, and U. Opik, *J. Chem. Soc.*, **1959**, 1625.
149. W. B. Person, G. R. Anderson, J. N. Fordemwalt, H. Stammreich, and R. Forneris, *J. Chem. Phys.*, **35**, 908 (1961).
150. W. S. Benedict, N. Gailar, and E. K. Plyler, *J. Chem. Phys.*, **24**, 1139 (1956).
151. J. H. Hibben, *J. Chem. Phys.*, **5**, 166 (1937).
152. C. Haas and D. F. Hornig, *J. Chem. Phys.*, **32**, 1763 (1960); D. F. Hornig, H. F. White, and F. P. Reding, *Spectrochim. Acta*, **12**, 338 (1958).
153. G. Gompertz and W. J. Orville-Thomas, *J. Phys. Chem.*, **63**, 1331 (1959).
154. R. D. Waldron, *J. Chem. Phys.*, **26**, 809 (1957).
155. S. Pinchas and M. Halmann, *J. Chem. Phys.*, **31**, 1692 (1959).
156. P. A. Staats, H. W. Morgan, and J. H. Goldstein, *J. Chem. Phys.*, **24**, 916 (1956).
157. H. C. Allen and E. K. Plyler, *J. Chem. Phys.*, **25**, 1132 (1956).
158. F. P. Reding and D. F. Hornig, *J. Chem. Phys.*, **27** 1024 (1957).
159. A. H. Nielsen and H. H. Nielsen, *J. Chem. Phys.*, **5**, 277 (1937).
160. D. M. Cameron, W. C. Sears, and H. H. Nielsen, *J. Chem. Phys.*, **7**, 994 (1939).
161. M. K. Wilson and R. M. Badger, *J. Chem. Phys.*, **16**, 741 (1948).
162. K. Herman and P. A. Giguère, *Can. J. Chem.*, **43**, 1746 (1965).
163. H. J. Bernstein and J. Powling, *J. Chem. Phys.*, **18**, 685 (1950).
164. M. M. Rochkind and G. C. Pimentel, *J. Chem. Phys.*, **42**, 1361 (1965).
165. A. H. Nielsen and P. J. H. Woltz, *J. Chem. Phys.*, **20**, 1878 (1952).
166. K. O. Criste and W. Sawodny, *Inorg. Chem.*, **6**, 313 (1967).
167. J. P. Mathieu, *Compt. rend.*, **234**, 2272 (1952).
168. E. T. Arakawa and A. H. Nielsen, *J. Mol. Spectry.*, **2**, 413 (1958).
169. R. E. Weston and T. F. Brodasky, *J. Chem. Phys.*, **27**, 683 (1957).
170. R. D. Shelton, A. H. Nielsen, and W. H. Fletcher, *J. Chem. Phys.*, **21**, 2178 (1953).
171. E. C. M. Grigg and G. R. Johnston, *Aust. J. Chem.*, **19**, 1147 (1966).
172. R. N. Wiener and E. R. Nixon, *J. Chem. Phys.*, **25**, 175 (1956).
173. S. R. Polo and M. K. Wilson, *J. Chem. Phys.*, **22**, 900 (1954).

174. P. A. Giguère and M. Falk, *Spectrochim. Acta*, **16**, 1 (1960).
175. H. Stammreich, R. Forneris, and K. Sone, *J. Chem. Phys.*, **23**, 972 (1955).
176. D. E. Milligan and M. E. Jacox, *J. Chem. Phys.*, **43**, 4487 (1965).
177. D. E. Milligan, D. E. Mann, and R. A. Mitsch, *J. Chem. Phys.*, **41**, 1199 (1964).
178. V. M. Khanna, R. Hauge, R. F. Curl, and J. L. Margrave, *J. Chem. Phys.*, **47**, 5031 (1967).
179. J. W. Hastie, R. Hauge, and J. L. Margrave, *J. Phys. Chem.*, **72**, 4492 (1966).
180. D. E. Milligan and M. E. Jacox, *J. Chem. Phys.*, **47**, 703 (1967).
181. R. V. St. Louis and B. L. Crawford, *J. Chem. Phys.*, **42**, 857 (1965).
182. M. J. Linevsky, D. White, and D. E. Mann, *J. Chem. Phys.*, **41**, 542 (1964).
183. J. E. Bertie and E. Whalley, *J. Chem. Phys.*, **40**, 1637, 1646, 1660 (1964); **41**, 1450 (1964)
184. E. Greinacher, W. Lüttke, and R. Mecke, *Z. Elektrochem.*, **59**, 23 (1955).
185. M. L. Joisen and P. Saumagne, *Bull. Soc. Chim. Fr.*, **1956**, 937.
186. M. Anbar, M. Halmann, and S. Pinchas, *J. Chem. Soc.*, **1960**, 1242.
187. C. Duval, J. Lecomte, and M. J. Morandat, *Bull. Soc. Chim. Fr.*, **1951**, 745.
188. K. Hedberg and R. M. Badger, *J. Chem. Phys.*, **19**, 508 (1951).
189. P. J. H. Woltz, E. A. Jones, and A. H. Nielsen, *J. Chem. Phys.*, **20**, 378 (1952).
190. L. Landau, *J. Mol. Spectry.* **4**, 276 (1960).
191. W. H. Eberhardt, *J. Chem. Phys.*, **20**, 529 (1952).
192. W. G. Burns and H. J. Bernstein, *J. Chem. Phys.*, **18**, 1669 (1950).
193. A. Müller, G. Nagarajan, O. Glemser, and J. Wegener, *Spectrochim. Acta*, **23A** 2683 (1967).
194. H. Richert and O. Glemser, *Z. Anorg. Allg. Chem.*, **307**, 328 (1961).
195. D. E. Milligan and M. E. Jacox, *J. Chem. Phys.*, **38**, 2627 (1963).
196. R. D. Spratley, J. J. Turner, and G. C. Pimentel, *J. Chem. Phys.*, **44**, 2063 (1966).
197. A. Arkell, *J. Amer. Chem. Soc.*, **87**, 4057 (1965).
198. D. E. Milligan, M. E. Jacox, A. M. Bass, J. J. Comeford, and D. E. Mann, *J. Chem. Phys.*, **42**, 3187 (1965).
199. M. E. Jacox and D. E. Milligan, *J. Chem. Phys.*, **43**, 866 (1965).
200. H. Y. Sheng, E. F. Barker, and D. M. Dennison, *Phys. Rev.*, **60**, 786 (1941).
201. C. M. Lewis and W. V. Houston, *Phys. Rev.*, **44**, 903 (1933).
202. F. P. Reding and D. F. Hornig, *J. Chem. Phys.*, **19**, 594 (1951); **22**, 1926 (1954).
203. M. V. Migeotte and E. F. Barker, *Phys. Rev.*, **50**, 418 (1936).
204. H. W. Morgan, P. A. Staats, and J. H. Goldstein, *Phys. Rev.*, **27**, 1212 (1957).
205. S. Sundaram and F. F. Cleveland, *J. Mol. Spectry.*, **5**, 61 (1960).
206. E. Lee and C. K. Wu, *Trans. Faraday Soc.*, **35**, 1366 (1939).
207. W. H. Haynie and H. H. Nielsen, *J. Chem. Phys.*, **21**, 1839 (1953).
208. R. C. Taylor and G. L. Vidale, *J. Amer. Chem. Soc.*, **78**, 5999 (1956).
209. R. Savoie and P. A. Giguère, *J. Chem. Phys.*, **41**, 2698 (1964).
210. R. E. Weston, *J. Amer. Chem. Soc.*, **76**, 2645 (1954).
211. C. C. Feriso and D. F. Hornig, *J. Chem. Phys.*, **23**, 1464 (1955); J. T. Mulhaupt and D. F. Hornig, *ibid.*, **24**, 169 (1956).
212. D. E. Bethell and N. Sheppard, *J. Chem. Phys.*, **21**, 1421 (1953).
213. D. J. Millen and E. G. Vaal, *J. Chem. Soc.*, **1956**, 2913.
214. W. J. Biermann and J. B. Gilmour, *Can. J. Chem.*, **37**, 1249 (1959).
215. R. D. Gillard and G. Wilkinson, *J. Chem. Soc.*, **1964**, 1640.
216. J. Kinumaki and K. Aida, *Sci. Repts. Research Insts., Tohoku Univ., Ser. A*, **6**, 186 (1954); K. Aida, *Bull. Chem. Research Inst. Non-aqueous Solutions, Tohoku Univ.*, **4**, 126 (1954).
217. C. A. Plint, R. M. B. Small, and H. L. Welsh, *Can. J. Phys.*, **32**, 653 (1954).

218. R. D. Waldron and D. F. Hornig, *J. Amer. Chem. Soc.*, **75**, 6079 (1953).
219. F. P. Reding and D. F. Hornig, *J. Chem. Phys.*, **23**, 1053 (1955).
220. J. S. Burgess, *Phys. Rev.*, **76**, 1267 (1949).
221. R. E. Weston, *J. Chem. Phys.*, **20**, 1820 (1952).
222. S. Sundaram, F. Suszek, and F. F. Cleveland, *J. Chem. Phys.*, **32**, 251 (1960).
223. G. DeAlti, G. Costa, and V. Galasso, *Spectrochim. Acta*, **20**, 965 (1964).
224. M. Pariseau, E. Wu, and J. Overend, *J. Chem. Phys.*, **39**, 217 (1963).
225. E. L. Pace and L. Pierce, *J. Chem. Phys.*, **23**, 1248 (1955).
226. J. Shamir and H. H. Hyman, *Spectrochim. Acta*, **23A**, 1899 (1967).
227. M. K. Wilson and S. R. Polo, *J. Chem. Phys.*, **20**, 1716 (1952); **21**, 1426 (1953).
228. H. S. Gutowsky and A. D. Liehr, *J. Chem. Phys.*, **20**, 1652 (1952).
229. D. M. Yost and J. E. Sherborne, *J. Chem. Phys.*, **2**, 125 (1934).
230. P. W. Davis and R. A. Oetjen, *J. Mol. Spectry.*, **2**, 253 (1958).
231. J. K. Wilmshurst, *J. Mol. Spectrosc.*, **5**, 343 (1960).
232. K. Venkateswarlu and S. Sundaram, *Proc. Phys. Soc.* (*London*), **A69**, 180 (1956).
233. M. L. Delwaulle, *Compt. rend.*, **228**, 1585 (1949).
234. L. A. Woodward and M. J. Taylor, *J. Chem. Soc.*, **1962**, 407.
235. F. A. Miller and W. K. Baer, *Spectrochim. Acta*, **17**, 114 (1961).
236. J. C. Evans, *J. Mol. Spectry.*, **4**, 435 (1960).
237. T. R. Manley and D. A. Williams, *Spectrochim. Acta*, **21**, 1773 (1965).
238. H. Stammreich, R. Forneris, and Y. Tavares, *J. Chem. Phys.*, **25**, 580 (1956).
239. H. Gerding and H. Houtgraaf, *Rec. Trav. Chim. Pays-Bas*, **73**, 737 and 759 (1954).
240. G. E. Moore, *J. Amer. Chem. Soc.*, **74**, 6076 (1952).
241. P. N. Schatz, *J. Chem. Phys.*, **29**, 481 (1958).
242. A. K. Ramdas, *Proc. Indian Acad. Sci.*, **37A**, 451 (1953); **36A**, 55 (1952).
243. C. Rocchiccioli, *Compt. rend.*, **242**, 2922 (1956).
244. J. L. Hollenberg and D. A. Dows, *Spectrochim. Acta*, **16**, 1155 (1960).
245. M. Rolla, *Gazz. Cheim. Ital.*, **69**, 779 (1939).
246. C. Rocchiccioli, *Compt. rend.*, **249**, 236 (1959).
247. W. E. Dasent and T. C. Waddington, *J. Chem. Soc.*, **1960**, 2429, 3350.
248. J. R. Durig, O. D. Bonner, and W. H. Breazeale, *J. Phys. Chem.*, **69**, 3886 (1965).
249. J. C. Evans and H. J. Bernstein, *Can. J. Chem.*, **33**, 1270 (1955).
250. A. Simon and K. Waldmann, *Z. Phys. Chem.*, **204**, 235 (1955).
251. C. Rocchiccioli, *Compt. rend.*, **244**, 2704 (1957).
252. H. Siebert, *Z. Anorg. Allg. Chem.*, **275**, 225 (1955).
253. C. Rocchiccioli, *Compt. rend.*, **247**, 1108 (1958).
254. H. H. Claassen and G. Knapp, *J. Amer. Chem. Soc.*, **86**, 2341 (1964).
255. A. Depaigne-Delay, C. Duval, and J. Lecomte, *Bull. Soc. Chim. Fr.*, **1946**, 54.
256. J. T. Last, *Phys. Rev.*, **105**, 1740 (1957).
257. L. J. Schoen and D. R. Lide, *J. Chem. Phys.*, **38**, 461 (1963).
258. G. E. Moore and R. M. Badger, *J. Amer. Chem. Soc.*, **74**, 6076 (1952).
259. R. Ettinger, *J. Chem. Phys.*, **38**, 2427 (1963).
260. J. J. Comeford, *J. Chem. Phys.*, **45**, 3463 (1966).
261. A. Müller, O. Glemser, and E. Nicke, *Z. Naturforsch.*, **B21**, 732 (1966).
262. R. J. Gillespie and E. A. Robinson, *Can. J. Chem.*, **39**, 2171 (1961).
263. J. K. O'Loane and M. K. Wilson, *J. Chem. Phys.*, **23**, 1313 (1955).
264. E. L. Pace and H. V. Samuelson, *J. Chem. Phys.*, **44**, 3682 (1965).
265. C. A. McDowell, *Trans. Faraday Soc.*, **49**, 371 (1953).
266. G. Allen and C. A. McDowell, *J. Chem. Phys.*, **23**, 209 (1955).
267. H. Stammreich, R. Forneris, and Y. Tavares, *J. Chem. Phys.*, **25**, 1277 (1956).

268. R. Paetzold and K. Aurich, *Z. Anorg. Allg. Chem.*, **335**, 281 (1965).
269. J. A. Rolfe and L. A. Woodward, *Trans. Faraday Soc.*, **51**, 779 (1955).
270. A. Simon and R. Paetzold, *Z. Anorg. Allg. Chem.*, **301**, 246 (1959); *Naturwissenschaften*, **44** 108 (1957).
271. M. Falk and P. A. Giguère, *Can. J. Chem.*, **34**, 1680 (1958).
272. D. F. Smith, G. M. Begun, and W. H. Fletcher, *Spectrochin. Acta*, **20**, 1763 (1964).
273. A. J. Arvia and P. J. Aymonino, *Spectrochim. Acta*, **19**, 1449 (1963).
274. A. Simon and R. Paetzold, *Z. Anorg. Allg. Chem.*, **303**, 39, 46, 53, 72, 79 (1960); *Z. Elektrochem.*, **64**, 209 (1960).
275. F. A. Cotton and W. D. Horrocks, *Spectrochim. Acta*, **16**, 358 (1960).
276. J. Vanderryn, *J. Chem. Phys.*, **30**, 331 (1959).
277. D. A. Dows, *J. Chem. Phys.*, **31**, 1637 (1959).
278. A. H. Nielsen, *J. Chem. Phys.*, **22**, 659 (1954).
279. R. E. Scruby, J. R. Lacher, and J. D. Park, *J. Chem. Phys.*, **19**, 386 (1951).
280. D. A. Dows and G. Bottger, *J. Chem. Phys.*, **34**, 689 (1961).
281. T. Wentink and V. H. Tiensuu, *J. Chem. Phys.*, **28**, 826 (1958).
282. A. Snelson, *J. Phys. Chem.*, **71**, 3202 (1967).
283. R. W. Lovejoy, J. H. Colwell, D. F. Eggers, and G. D. Halsey, J. *Chem. Phys.*, **36**, 612 (1962).
284. R. Bent and W. R. Ladner, *Spectrochim. Acta*, **19**, 931 (1963).
285. W. C. Steele and J. C. Decius, *J. Chem. Phys.*, **25**, 1184 (1956).
286. B. Krebs, G. Gattow, and A. Müller, *Z. Anorg. Allg. Chem.*, **337**, 279 (1965); *Z. Naturforsch.*, **B20**, 1017 (1965); **A20**, 1124, 1242 and 1664 (1965).
287. B. Krebs and A. Müller, *Spectrochim. Acta*, **22**, 1532 (1966).
288. A Müller, G. Gattow, and H. Seidel, *Z. Anorg. Allg. Chem.*, **347**, 24 (1966); G. Nagarajan and A. Müller, *Z. Naturforsch.*, **B21**, 393 (1966).
289. L. P. Lindeman and M. K. Wilson, *J. Chem. Phys.*, **24**, 242 (1956).
290. J. Goubeau, D. E. Richter, and H. J. Becher, *Z. Anorg. Allg. Chem.*, **278**, 12 (1955)
291. M. Perec and L. N. Becka, *J. Chem. Phys.*, **43**, 721 (1965).
292. L. Lynds, *J. Chem. Phys.*, **42**, 1124 (1965).
293. C. D. Bass, L. Lynds, T. Wolfram, and R. E. DeWames, *Inorg. Chem.*, **3**, 1063 (1964).
294. L. Lynds and C. D. Bass, *J. Chem. Phys.*, **41**, 3165 (1964).
295. W. R. Heslop and J. W. Linnett, *Trans. Faraday Soc.*, **49**, 1262 (1953).
296. K. Shimizu and H. Shingu, *Spectrochim. Acta*, **22**, 1999 (1966).
297. S. Konaka, Y. Murata, K. Kuchitsu, and Y. Morino, *Bull. Chem. Soc. Jap.*, **39**, 1134 (1966).
298. L. Beckmann, L. Gutjahr, and R. Mecke, *Spectrochim. Acta*, **21**, 141 (1965).
299. I. W. Levin and S. Abramowitz, *J. Chem. Phys.*, **43**, 4213 (1965).
300. J. L. Duncan, *J. Mol. Spectry.*, **13**, 338 (1964).
301. C. D. Bass, L. Lynds, T. Wolfram, and R. E. DeWames, *J. Chem. Phys.*, **40**, 3611 (1964).
302. H. Gerding and W. J. Nijveld, *Rec. Trav. Chim. Pays-Bas*, **59**, 1206 (1940); *Z. Phys. Chem.*, **B35**, 193 (1937).
303. R. J. Gillespie and E. A. Robinson, *Can. J. Chem.*, **39**, 2189 (1961).
304. C. Duval and J. Lecomte, *Bull. Soc. Chim. Fr.*, **1952**, 101.
305. J. P. Laperches and P. Tarte, *Spectrochim. Acta*, **22**, 1201 (1966).
306. I. C. Hisatsune and N. H. Suarez, *Inorg. Chem.*, **3**, 168 (1964).
307. R. R. Servoss and H. M. Clark, *J. Chem. Phys.*, **26**, 1175 (1957).
308. D. F. Hornig and R. C. Plumb, *J. Chem. Phys.*, **26**, 637 (1957).
309. C. W. F. T. Pistorius, *J. Chem. Phys.*, **31**, 1454 (1959).

310. J. L. Parsons, *J. Chem. Phys.*, **33**, 1860 (1960).
311. S. Bhagavantum and T. Venkatarayudu, *Proc. Indian Acad. Sci.*, **9A**, 224 (1939).
312. D. Krishnamurti, *Proc. Indian Acad. Sci.*, **43A**, 210 (1956).
313. T. S. Krishnan, *Proc. Indian Acad. Sci.*, **44A**, 96 (1956).
314. M. Anbar, M. Halmann, and S. Pinchas, *J. Chem. Soc.*, **1960**, 1242, 1246.
315. B. S. R. Rao, *Proc. Indian Acad. Sci.*, **10A**, 167 (1939); **19A**, 93 (1944).
316. R. Newman and R. S. Halford, *J. Chem. Phys.*, **18**, 1276, 1291 (1950).
317. W. E. Keller and R. S. Halford, *J. Chem. Phys.*, **17**, 26 (1949).
318. J. Louisfert, *Compt. rend.*, **233**, 381 (1951); **235**, 287 (1952).
319. K. Buijs and C. J. H. Schutte, *Spectrochim. Acta*, **17**, 917, 921, 927 (1961); **18**, 307 (1962).
320. J. C. Decius, *J. Chem. Phys.*, **22**, 1941 (1954); **23**, 1290 (1955).
321. J. R. Ferraro, *J. Mol. Spectry.*, **4**, 99 (1960).
322. C. C. Addison and B. M. Gatehouse, *J. Chem. Soc.*, **1960**, 613.
323. G. E. Walrafen and D. E. Irish, *J. Chem. Phys.*, **40**, 911 (1964).
324. G. J. Janz and T. R. Kozlowski, *J. Chem. Phys.*, **40**, 1699 (1964).
325. H. Cohn, C. K. Ingold, and H. G. Poole, *J. Chem. Soc.*, **1952**, 4272.
326. R. E. Dodd, J. A. Rolfe, and L. A. Woodward, *Trans. Faraday Soc.*, **52**, 145 (1956)
327. R. Ryason and M. K. Wilson, *J. Chem. Phys.*, **22**, 2000 (1954).
328. T. A. Hariharan, *Proc. Indian Acad. Sci.*, **48A**, 49 (1958).
329. J. Overend and J. C. Evans, *Trans. Faraday Soc.*, **55**, 1817 (1959).
330. A. H. Nielsen, T. G. Burke, P. J. H. Woltz, and E. A. Jones, *J. Chem. Phys.*, **20**, 596 (1952).
331. E. Catalano and K. S. Pitzer, *J. Amer. Chem. Soc.*, **80**, 1054 (1958).
322. E. A. Jones and T. G. Burke, *J. Chem. Phys.*, **18**, 1308 (1950).
333. R. R. Patty and R. T. Lagemann, *Spectrochim. Acta*, **15**, 60 (1959).
334. R. F. Stratton and A. H. Nielsen, *J. Mol. Spectry.*, **4**, 373 (1960).
335. A. J. Downs, *Spectrochim. Acta*, **19**, 1165 (1963).
336. H. H. Claassen, B. Weinstock, and J. G. Malm, *J. Chem. Phys.*, **28**, 285 (1958).
337. K. Venkateswarlu and S. Sundaram, *J. Chim. Phys.*, **54**, 202 (1957).
338. J. W. Linnett and D. F. Heath, *Trans. Faraday Soc.*, **48**, 592 (1952).
339. G. J. Janz and Y. Mikawa, *J. Mol. Spectry.*, **5**, 92 (1960).
340. C. W. F. T. Pistorius, *J. Chem. Phys.*, **30**, 332 (1959).
341. J. P. Devlin and I. C. Hisatsune, *Spectrochim. Acta*, **17**, 206 (1961).
342. L. Beckmann, L. Gutjahr, and R. Mecke, *Spectrochim. Acta*, **20**, 1295 (1964).
343. W. R. Busing and H. A. Levy, *J. Chem. Phys.*, **42**, 3054 (1965).
344. O. Bain and P. A. Giguère, *Can. J. Chem.*, **33**, 527 (1955).
345. P. A. Giguère and O. Bain, *J. Phys. Chem.*, **56**, 340 (1952).
346. P. A. Giguère, *J. Chem. Phys.*, **18**, 88 (1950).
347. R. C. Taylor, *J. Chem. Phys.*, **18**, 898 (1950).
348. R. C. Taylor and P. C. Cross, *J. Chem. Phys.*, **24**, 41 (1956).
349. A. Simon and H. Kriegsmann, *Naturwissenschaften*, **42**, 12 (1955).
350. R. L. Miller and D. F. Hornig, *J. Chem. Phys.*, **35**, 265 (1961).
351. P. A. Giguère and C. Chapados, *Spectrochim. Acta*, **22**, 1131 (1966).
352. N. Zengin and P. A. Giguère, *Can. J. Chem.*, **37**, 632 (1959).
353. M. K. Wilson and R. M. Badger, *J. Chem. Phys.*, **17**, 1232 (1949).
354. F. Fehér, W. Lane, and G. Winkhaus, *Z. Anorg. Allg. Chem.*, **288**, 113 (1956).
355. J. R. B. Mututano and C. Otero, *An. Real Soc. Espan. Fis. Quim., Ser. B*, **51**, 223 (1955).

356. J. A. A. Ketelaar, F. N. Hooge, and G. Blasse, *Rec. Trav. Chim. Pays-Bas*, **75**, 220 (1956).
357. H. J. Bernstein and J. Powling, *J. Chem. Phys.*, **18**, 1018 (1950).
358. F. N. Hooge and J. A. A. Ketelaar, *Rec. Trav. Chim. Pays-Bas*, **77**, 902 (1958).
359. H. Gerding and R. Westrik, *Rec. Trav. Chim. Pays-Bas*, **60**, 701 (1941).
360. H. Stammreich and R. Forneris, *Spectrochim. Acta*, **8**, 46 (1956).
361. E. B. Bradley, C. R. Bennett, and E. A. Jones, *Spectrochim. Acta*, **21**, 1505 (1965).
362. S. T. King and J. Overend, *Spectrochim. Acta.*, **23A**, 61 (1967).
363. R. H. Sanborn, *J. Chem. Phys.*, **33**, 1855 (1960).
364. S. T. King and J. Overend, *Spectrochim Acta*, **22**, 689 (1966).
365. A. Cornelis-Benoit, *Spectrochim. Acta*, **21**, 623 (1965).
366. A. Simon and H. Kriegsmann, *Naturwissenschaften*, **42**, 14 (1955).
367. O. Knop and P. A. Giguère, *Can. J. Chem.*, **37**, 1794 (1959).
368. F. Fehér and M. Baudler, *Z. Anorg. Allg. Chem.*, **254**, 289 (1947); **258**, 132 (1949); F. Fehér and G. Winkhaus, *ibid.*, **288**, 123 (1956).
369. E. G. Brame, S. Cohen, J. L. Margrave, and V. W. Meloche, *J. Inorg. Nucl. Chem.*, **4**, 90 (1957).
370. G. E. McGraw, D. L. Bernitt, and I. C. Hisatsune, *Spectrochim. Acta*, **23A**, 25 (1967)
371. D. A. Dows and G. C. Pimentel, *J. Chem. Phys.*, **23**, 1258 (1955).
372. G. Herzberg and C. Reid, *Disc. Faraday Soc.*, **9**, 92 (1952).
373. L. H. Jones, J. N. Shoolery, and R. G. Shulman, *J. Chem. Phys.*, **18**, 990 (1950).
374. C. Reid, *J. Chem. Phys.*, **18**, 1512 (1950).
375. L. H. Jones and R. M. Badger, *J. Chem. Phys.*, **18**, 1511 (1950).
376. J. R. Durig and D. W. Wertz, *J. Chem. Phys.*, **46**, 3069 (1967).
377. M. E. Jacox and D. E. Milligan, *J. Chem. Phys.*, **40**, 2457 (1964).
378. W. Beck and K. Feldl, *Angew. Chem.*, **78**, 746 (1966).
379. C. Reid, *J. Chem. Phys.*, **18**, 1544 (1950).
380. R. A. Ashby and R. L. Werner, *J. Mol. Spectry.*, **18**, 184 (1965).
381. R. A. Ashby and R. L. Werner, *Spectrochim. Acta*, **22**, 1345 (1966).
382. L. H. Jones, R. M. Badger, and G. E. Moore, *J. Chem. Phys.*, **19**, 1599 (1951).
383. L. d'Or and P. Tarte, *Bull. Soc. Roy. Sci. Liege*, **20**, 478 (1951).
384. P. Tarte, *Bull. Soc. Roy. Sci. Liege*, **20**, 16 (1951).
385. W. T. Thompson and W. H. Fletcher, *Spectrochim. Acta*, **22**, 1907 (1966).
386. K. Venkateswarlu and P. Thirugnanasambandam, *Z. Phys. Chem.* (*Leipzig*), **218**, 1 (1961).
387. A. Palm, *J. Chem. Phys.*, **26**, 855 (1957).
388. I. C. Hisatsune and J. P. Devlin, *Spectrochim. Acta*, **16**, 401 (1960); **17**, 218 (1961).
389. G. M. Begun and W. H. Fletcher, *J. Mol. Spectry.*, **4**, 388 (1960).
390. R. N. Wiener and E. R. Nixon, *J. Chem. Phys.*, **26**, 906 (1957).
391. R. G. Snyder and I. C. Hisatsune, *J. Mol. Spectry.*, **1**, 139 (1957); *J. Chem. Phys.*, **26**, 960 (1957).
392. I. C. Hisatsune and R. V. Fitzsimmons, *Spectrochim. Acta*, **15**, 206 (1959).
393. I. C. Hisatsune, J. P. Devlin, and S. Califano, *Spectrochim. Acta*, **16**, 450 (1960).
394. I. C. Hisatsune, J. P. Devlin, and Y. Wada, *J. Chem. Phys.*, **33**, 714 (1960).
395. R. Teranishi and J. C. Decius, *J. Chem. Phys.*, **22**, 896 (1954); **21**, 116 (1953).
396. D. J. Millen, C. N. Polydoropoulous, and D. Watson, *J. Chem. Soc.*, **1960**, 687.
397. R. J. W. LeFèvre, W. T. Oh, I. H. Reece, and R. L. Werner, *Aust. J. Chem.*, **10**, 361 (1957).
398. I. C. Hisatsune, J. P. Devlin, and Y. Wada, *Spectrochim. Acta*, **18**, 1641 (1962).
399. G. E. McGraw, D. L. Bernitt, and I. C. Hisatsune, *J. Chem. Phys.*, **42**, 237 (1965).

400. P. A. Giguère and I. D. Liu, *J. Chem. Phys.*, **20**, 136 (1952).
401. E. L. Wagner and E. L. Bulgozdy, *J. Chem. Phys.*, **19**, 1210 (1951).
402. A. Yamaguchi, *J. Chem. Soc. Japan*, **80**, 1109 (1959).
403. J. S. Ziomek and M. D. Zeidler, *J. Mol. Spectry.*, **11**, 163 (1963).
404. R. E. Nightingale and E. L. Wagner, *J. Chem. Phys.*, **22**, 203 (1954).
405. P. A. Giguère and I. D. Liu, *Can. J. Chem.*, **30**, 948 (1952).
406. J. R. Durig and R. C. Lord, *Spectrochim. Acta*, **19**, 1877 (1963).
407. R. G. Snyder and J. C. Decius, *Spectrochim. Acta*, **13**, 280 (1959).
408. D. L. Frasco and E. L. Wagner, *J. Chem. Phys.*, **30**, 1124 (1959).
409. Landolt-Börnstein, *Physikalisch-chemische Tabellen*, 2 Teil, 1951.
410. I. F. Kovalev, *Opt. Spektrosk.*, **2**, 310 (1957).
411. L. P. Lindeman and M. K. Wilson, *Z. Phys. Chem.*, **9**, 29 (1956).
412. I. W. Levin and H. Ziffer, *J. Chem. Phys.*, **43**, 4023 (1965).
413. G. E. MacWood and H. C. Urey, *J. Chem. Phys.*, **4**, 402 (1936).
414. H. M. Kaylor and A. H. Nielsen, *J. Chem. Phys.*, **23**, 2139 (1955).
415. J. H. Meal and M. K. Wilson, *J. Chem. Phys.*, **24**, 385 (1956).
416. H. W. Morgan, P. A. Staats, and J. H. Goldstein, *J. Chem. Phys.*, **27**, 1212 (1957).
417. J. V. Martinez and E. L. Wagner, *J. Chem. Phys.*, **27**, 1110 (1957).
418. A. Heineman, *Ber. Bunsenges. Phys. Chem.*, **68**, 280, 287 (1964).
419. A. R. Emery and R. C. Taylor, *J. Chem. Phys.*, **28**, 1029 (1958).
420. C. J. H. Schutte, *Spectrochim. Acta*, **16**, 1054 (1960); J. A. A. Ketelaar and C. J. H. Schutte, *ibid.*, **17**, 1240 (1961).
421. L. A. Woodward and H. L. Roberts, *Trans. Faraday Soc.*, **52**, 1458 (1956)
422. H. C. Longuet-Higgins and D. A. Brown, *J. Inorg. Nucl. Chem.*, **1**, 60 (1955); D. A. Brown, *J. Chem. Phys.*, **29**, 451 (1958).
423. C. W. F. T. Pistorius, *J. Chem. Phys.*, **27**, 965 (1957).
424. M. Radhakrishnan, *Z. Phys. Chem. (Frankfurt)*, **36**, 227 (1963); **41**, 197 and 201 (1964).
425. J. L. Duncan and I. M. Mills, *Spectrochim. Acta*, **20**, 523 and 1089 (1964).
426. A. Müller and B. Krebs, *J. Mol. Spectry.*, **24**, 180, 198 (1967).
427. E. L. Wagner and D. F. Hornig, *J. Chem. Phys.*, **18**, 296, 305 (1950).
428. R. C. Plumb and D. F. Hornig, *J. Chem. Phys.*, **21**, 366 (1953).
429. R. C. Plumb and D. F. Hornig, *J. Chem. Phys.*, **23**, 947 (1955).
430. W. Vedder and D. F. Hornig, *J. Chem. Phys.*, **35**, 1560 (1961).
431. T. C. Waddington, *J. Chem. Soc.*, **1958**, 4340.
432. J. P. Mathieu, R. M. Aguirre, and L. C. Mathieu, *Compt. rend.*, **232**, 318 (1951); J. P. Mathieu, *ibid* , **233**, 1595 (1951).
433. J. P. Mathieu and H. Poulet, *Spectrochim. Acta*, **16**, 696 (1960).
434. L. C. Mathieu and J. P. Mathieu, *J. Chim. Phys.*, **49**, 226 (1952).
435. J. P. Mathieu, *Compt. rend.*, **240**, 2508 (1955).
436. D. Penot, H. Poulet, and J. P. Mathieu, *Compt. rend.*, **243**, 1303 (1956).
437. J. Goubeau, W. Bues, and F. W. Kampmann, *Z. Anorg. Allg. Chem.*, **283**, 123 (1956).
438. P. J. H. Woltz and A. H. Nielsen, *J. Chem. Phys.*, **20**, 307 (1952).
439. E. A. Jones, J. S. Kirby-Smith, P. J. H. Woltz, and A. H. Nielsen, *J. Chem. Phys.*, **19**, 242 (1951).
440. J. Heicklen and V. Knight, *Spectrochim. Acta*, **20**, 295 (1964).
441. A. D. Caunt, L. N. Short, and L. A. Woodward, *Trans. Faraday, Soc.* **48**, 873 (1952).
442. A. D. Caunt, L. N. Short, and L. A. Woodward, *Nature*, **168**, 557 (1951).
443. A. Büchler, J. B. Berkowitz-Mattuck, and D. H. Dugre, *J. Chem. Phys.*, **34**, 2202 (1961).
444. H. L. Welsh, M. F. Crawford, and G. D. Scott, *J. Chem. Phys.*, **16**, 97 (1948).

445. D. A. Long, T. V. Spencer, D. N. Waters, and L. A. Woodward, *Proc. Roy. Soc.*, **A240**, 499 (1957).
446. N. J. Hawkins and D. R. Carpenter, *J. Chem. Phys.*, **23**, 1700 (1955).
447. M. F. A. Dove, J. A. Creighton, and L. A. Woodward, *Spectrochim. Acta*, **18**, 267 (1962).
448. L. P. Lindeman and M. K. Wilson, *Spectrochim. Acta*, **9**, 47 (1957).
449. M. L. Delwaulle, F. François, M. B. Delhaye-Buisset, and M. Delhaye, *J. Phys. Radium*, **15**, 206 (1954).
450. J. T. Neu and W. D. Gwinn, *J. Amer. Chem. Soc.*, **70**, 3463 (1948).
451. M. L. Delwaulle and F. François, *Compt. rend.*, **219**, 64 (1944).
452. R. R. Haun and W. D. Harkins, *J. Amer. Chem. Soc.*, **54**, 3917 (1932).
453. F. A. Miller and G. L. Carlson, *Spectrochim. Acta*, **16**, 6 (1960).
454. H. Stammreich, Y. Tavares, and D. Bassi, *Spectrochim. Acta*, **17**, 661 (1961).
455. H. Stammreich, R. Forneris, and Y. Tavares, *J. Chem. Phys.*, **25**, 1278 (1956).
456. D. P. Stevenson and V. Schomaker, *J. Amer. Chem. Soc.*, **62**, 1267 (1940).
457. R. E. Dodd, L. A. Woodward, and H. L. Roberts, *Trans. Faraday Soc.*, **52**, 1052 (1956).
458. I. W. Levin and C. V. Berney, *J. Chem. Phys.*, **44**, 2557 (1966).
459. J. A. Rolfe, L. A. Woodward, and D. A. Long, *Trans. Faraday Soc.*, **49**, 1388 (1953).
460. P. L. Goggin, H. L. Roberts, and L. A. Woodward, *Trans. Faraday Soc.*, **57**, 1877 (1961).
461. J. W. George N. Katsaros, and K. J. Wynne, *Inorg. Chem.*, **6**, 903 (1967).
462. D. M. Adams and P. J. Lock, *J. Chem. Soc.*, **A**, **1967**, 145.
463. G. C. Haywood and P. J. Hendra, *J. Chem. Soc.*, **A**, **1967** 643.
464. J. Goubeau and W. Bues, *Z. Anorg. Allg. Chem.*, **268**, 221 (1952).
465. N. N. Greenwood, *J. Chem. Soc.*, **1959**, 3811.
466. G. L. Carlson, *Spectrochim. Acta*, **19**, 1291 (1963).
467. L. A. Woodward and A. A. Nord, *J. Chem. Soc.*, **1956**, 3721.
468. L. A. Woodward and M. J. Taylor, *J. Chem. Soc.*, **1960**, 4473.
469. T. G. Spiro, *Inorg. Chem.*, **6**, 569 (1967).
470. A. Müller and A. Fadini, *Z. Anorg. Allg. Chem.*, **349**, 164 (1967).
471. J. Weidlein and K. Dehnicke, *Z. Anorg. Allg. Chem.*, **337**, 113 (1965).
472. K. Dehnicke and J. Weidlein, *Ber. Bunsenges. Phys. Chem.*, **69**, 1087 (1965).
473. A. Sabatini and L. Sacconi, *J. Amer. Chem. Soc.*, **86**, 17 (1964).
474. R. J H. Clark, and T. M. Dunn, *J. Chem. Soc.*, **1963**, 1198.
475. G. J. Janz and D. W. James, *J. Chem. Phys.*, **38**, 905 (1963).
476. L. A. Woodward and A. A. Nord, *J. Chem. Soc.*, **1955**, 2655.
477. L. A. Woodward and P. T. Bill, *J. Chem. Soc.*, **1955**, 1699.
478. M. L. Delwaulle, *Compt. rend.*, **238**, 2522 (1954).
479. H. Gerding and P. C. Nobel, *Rec. Trav. Chem. Pays-Bas*, **77**, 472 (1958).
480. M. L. Delwaulle, *Compt. rend.*, **240**, 2132 (1955).
481. J. A. Rolfe, D. E. Sheppard, and L. A. Woodward, *Trans. Faraday Soc.*, **50**, 1275 (1954).
482. L. A. Woodward and G. H. Singer, *J. Chem. Soc.*, **1958**, 716.
483. D. M. Adams, J. Chatt, J. M. Davidson, and J. Gerratt, *J. Chem. Soc.*, **1963**, 2189.
484. T. C. Waddington and F. Klanberg, *J. Chem. Soc.*, **1960**, 2339.
485. W. Kynaston, B. E. Larcombe, and H. S. Turner, *J. Chem. Soc.*, **1960**, 1772.
486. D. Fortnum and J. O. Edwards, *J. Inorg. Nucl. Chem.*, **2**, 264 (1956).
487. E. Steger and K. Herzog, *Z. Anorg. Allg. Chem.*, **331**, 169 (1964).
488. E. Steger and W. Schmidt, *Ber. Bunsenges. Phys. Chem.*, **68**, 102 (1964).
489. H. Siebert, *Z. Anorg. Allg. Chem.*, **275**, 225 (1954).

490. H. Colm, *J. Chem. Soc.*, **1952**, 4282.
491. S. D. Ross, *Spectrochim Acta*, **18**, 225 (1962).
492. A. Müller, B. Krebs, W. Rittner, and M. Stockburger, *Ber. Bunsenges. Phys. Chem.* **71**, 182 (1967).
493. A. Müller, R. Kebabcioglu, M. J. F. Leroy, and G. Kaufmann, *Z. Naturforsch*, **B23**, 740 (1968).
494. F. A. Miller and C. H. Wilkins, *Anal. Chem.*, **24**, 1253 (1952).
495. A. Müller and B. Krebs, *Z. Naturforsch.*, **B21**, 3, (1966); **A21**, 433 (1966); **B20**, 1127 (1965); **A20**, 745 (1965).
496. R. H. Busey and O. L. Keller, *J. Chem. Phys.*, **41**, 215 (1964).
497. Refs. 33 and 34 of Part I.
498. G. Nagarajan, *Bull. Soc. Chim. Belges*, **71**, 119 (1962).
499. R. E. Dodd, *Trans. Faraday Soc.*, **55**, 1480 (1959).
500. H. Siebert, *Z. Anorg. Allg. Chem.*, **273**, 21 (1953).
501. G. J. Janz and D. W. James, *J. Chem. Phys.*, **38**, 902 (1963).
502. H. H. Claassen and A. J. Zielen, *J. Chem. Phys.*, **22**, 707 (1954).
503. L. A. Woodward and H. L. Roberts, *Trans. Faraday Soc.*, **52**, 615 (1956).
504. N. J. Hawkins and W. W. Sabol, *J. Chem. Phys.*, **25**, 775 (1956).
505. A. Müller and G. Nagarajan, *Z. Naturforsch.*, **B21**, 508 (1966); A. Müller and A. Fidini, *ibid.*, **B21**, 585 (1966); A. Müller and B. Krebs, *Z. Anorg. Allg. Chem.*, **344**, 56 (1966); G. Gattow, A. Franke and A. Müller, *Naturwissenschaften*, **62**, 428 (1965),
506. A. Müller, B. Krebs, R. Kebabcioglu, M. Stockburger, and O. Glemser, *Spectrochim. Acta*, **24A**, 1831 (1968); A. Müller, B. Krebs and G. Gattow, *Spectrochim. Acta*, **23A**, 2809 (1967).
507. J. O. Edwards, G. C. Morrison, V. F. Ross, and J. W. Schultz, J. *Amer. Chem. Soc.*, **77**, 266 (1955).
508. E. R. Lippincott, J. A. Psellos, and M. C. Tobin, *J. Chem. Phys.*, **20**, 536 (1952).
509. R. S. Krishnan, *Proc. Indian. Acad. Sci.*, **23A**, 288 (1946).
510. M. Hass and G. B. B. M. Sutherland, *Proc. Roy. Soc.*, **A236**, 427 (1956).
511. D. Krishnamurti, *Proc. Indian Acad. Sci.*, **48A**, 355 (1958).
512. E. Steger and W. Schmidt, *Ber. Bunsenges. Phys. Chem.*, **68**, 102 (1964).
513. C. Shantakumari, *Proc. Indian Acad. Sci.*, **37A**, 393 (1953).
514. C. Duval and J. Lecomte, *Compt. rend.*, **248**, 1977 (1959).
515. T. S. Krishnan and P. S. Narayanan, *Proc. Indian Acad. Sci.*, **41A**, 121 (1955).
516. J. P. Mathieu, L. C. Mathieu, and H. Poulet, *J. Phys. Radium*, **16**, 781 (1955).
517. D. Krishnamurti, *Proc. Indian Acad. Sci.*, **42A**, 77 (1955).
518. T. S. Krishnan, *J. Indian Inst. Sci.*, **38A**, 207 (1956).
519. C. Rocchiccioli, *Compt. rend.*, **257**, 3851 (1963).
520. R. Lafont, *Compt. rend.*, **242**, 1154 (1956).
521. P. A. Giguère and R. Savoie, *Can. J. Chem.*, **38**, 2467 (1961).
522. R. J. Gillespie and E. A. Robinson, *Can. J. Chem.*, **40**, 644 (1962).
523. A. Weil and J. P. Mathieu, *Compt. rend.*, **238**, 2510 (1954).
524. A. J. Dahl, J. C. Trowbridge, and R. C. Taylor, *Inorg. Chem.*, **2**, 654 (1963).
525. R. M. Ansidei, *Boll. Sci. Fac. Chim. Ind. Bologna*, **18**, 116 (1940).
526. P. A. Giguère and R. Savoie, *Can. J. Chem.*, **40**, 495 (1962).
527. O. Redlich, E. K. Holt, and J. Bigeleisen, *J. Amer. Chem. Soc.*, **55**, 13 (1944).
528. S. D. Ross, *Sprectrochim. Acta*, **18**, 225 (1962).
529. A. C. Chapman and L. E. Thrilwell, *Spectrochim. Acta*, **20**, 937 (1964).
530. J. V. Pustinger, W. T. Cave, and M. L. Nielsen, *Spectrochim. Acta*, **11**, 909 (1959).
531. J. A. A. Ketelaar, *Acta Cryst.*, **7**, 691 (1954).

532. M. Tsuboi, *J. Amer. Chem. Soc.*, **79**, 1351 (1957).
533. J. S. Ziomek, J. R. Ferraro, and D. F. Peppard, *J. Mol. Spectry.*, **8**, 212 (1962).
534. A. Mutschin and K. Maenuchen, *Z. Anal. Chem.*, **156**, 241 (1957).
535. C. Duval and J. Lecomte, *Mikrochim. Acta*, **1956**, 454; *Compt. rend.*, **240**, 66 (1955).
536. J. Lecomte, A. Boulle and M. Lang-Dupont, *Compt., rend.*, **241**, 1927 (1955).
537. C. Duval and J. Lecomte, *Z. Elektrochem.*, **64**, 582 (1960).
538. C. Duval and J. Lecomte, *Compt. rend.*, **239**, 249 (1954).
539. W. S. Benedict, K. Morikawa, R. B. Barnes, and H. S. Taylor, *J. Chem. Phys.*, **5**, 1 (1937).
540. G. E. MacWood and H. C. Urey, *J. Chem. Phys.*, **4**, 402 (1936).
541. C. Newman, J. K. O'Loane, S. R. Polo, and M. K. Wilson, *J. Chem. Phys.*, **25**, 855 (1956).
542. R. N. Dixon and N. Sheppard, *Trans. Faraday Soc.*, **53**, 282 (1957).
543. M. L. Delwaulle and F. François, *Compt. rend.*, **219**, 335 (1944); **220**, 173 (1945).
544. F. François and M. B. Delhaye-Buisset, *Compt. rend.*, **230**, 1946 (1950).
545. D. W. Mayo, H. E. Opitz, and J. S. Peake, *J. Chem. Phys.*, **23**, 1344 (1955).
546. A. Monfils, *J. Chem. Phys.*, **19**, 138 (1951).
547. A. Monfils, *Compt. rend.*, **236**, 795 (1953).
548. C. Newman, *Spectrochim. Acta*, **15**, 793 (1959).
549. J. A. Hawkins, S. R. Polo, and M. K. Wilson, *J. Chem. Phys.*, **21**, 1122 (1953).
550. F. A. Andersen and B. Bak, *Acta Chem. Scand.*, **8**, 738 (1954).
551. J. A. Hawkins, *J. Chem. Phys.*, **21**, 360 (1953).
552. T. G. Gibian and D. S. McKinney, *J. Amer. Chem. Soc.*, **73**, 1431 (1951).
553. Y. Kakiuchi, *Bull. Chem. Soc. Jap.*, **26**, 260 (1953).
554. R. C. Lord and C. M. Steese, *J. Chem. Phys.*, **22**, 542 (1954).
555. S. L. N. G. Krishnamachari, *Indian J. Phys.*, **29**, 384 (1955).
556. M. L. Delwaulle, *Compt. rend.*, **230**, 1945 (1950).
557. M. L. Delwaulle and F. François, *Compt. rend.*, **230**, 743 (1950).
558. J. E. Griffiiths, T. N. Srivastava, and M. Onyszchuk, *Can. J. Chem.*, **40**, 579 (1962)
559. M. L. Delwaulle, *Compt. rend.*, **238**, 84 (1954).
560. I. N. Godnev, A. S. Sverklin, and N. I. Ushanova, *Opt. Spektrosk.* **2**, 704 (1957).
561. R. Dupeyrat, *Compt. rend.*, **241**, 932 (1955).
562. O. Háová, *Chem. Listy*, **49**, 640 (1955).
563. M. L. Delwaulle and F. François, *Compt. rend.*, **224**, 1422 (1947).
564. D. A. Long and R. T. Bailey, *Spectrochim Acta*, **19**, 1607 (1963)
565. R. P. Madden and W. S. Benedict, *J. Chem. Phys.*, **25**, 594 (1956).
566. F. X. Powell and E. R. Lippincott, *J. Chem. Phys.*, **32**, 1883 (1960).
567. D. R. Linde and D. E. Mann, *J. Chem. Phys.*, **25**, 1128 (1956).
568. R. Duval and J. Lecomte, *Compt. rend.*, **213**, 998 (1941).
569. E. Steger, I. C. Ciurea, and A. Fadini, *Z. Anorg. Allg. Chem.*, **350**, 225 (1967).
570. H. Siebert, *Z. Anorg. Allg. Chem.*, **289**, 15 (1957).
571. D. W. A. Sharp, *J. Chem. Soc.*, **1957**, 3761.
572. H. Stammreich, O. Sala, and D. Bassi, *Spectrochim. Acta*, **19**, 593 (1963).
573. H. Stammreich, O. Sala, and K. Kawai, *Spectrochim. Acta*, **17**, 226 (1961).
574. L. A. Woodward, J. A. Creighton, and K. A. Taylor, *Trans. Faraday Soc.*, **56**, 1267 (1960).
575. J. Lewis and G. Wilkinson, *J. Inorg. Nucl. Chem.*, **6**, 12 (1958).
576. F. A. Miller and G. L. Carlson, *Spectrochim. Acta*, **16**, 1148 (1960).
577. A. Müller, B. Krebs, and W. Höltje, *Spectrochim. Acta*, **23A**, 2753 (1967).

578. A. Müller and B. Krebs, *Z. Anorg. Allg. Chem.*, **342**, 182 (1966); A. Müller, H. W. Roesky, and B. Krebs, *Z. Chem.*, **7**, 159 (1967).
579. H. Selig and H. H. Claassen, *J. Chem. Phys.*, **44**, 1404 (1966).
580. F. A. Miller and L. R. Cousins, *J. Chem. Phys.*, **26**, 329 (1957).
581. H. S. Gutowsky and A. D. Liehr, *J. Chem. Phys.*, **20**, 1652 (1952).
582. M. L. Delwaulle and F. François, *Compt. rend.*, **220**, 817 (1945).
583. H. Gerding and M. van Driel, *Rec. Trav. Chim. Pays-Bas*, **61**, 419 (1942).
584. H. Horn, A. Müller, and O. Glemser, *Z. Naturforsch.*, **A20**, 746 (1965).
585. M. L. Delwaulle and F. François, *Compt. rend.*, **226**, 896 (1948).
586. H. Gerding and R. Westrik, *Rec. Trav. Chim. Pays-Bas*, **61**, 842 (1942).
587. A. Müller, E. Diemann, B. Krebs, and M. J. F. Leroy, *Angew. Chem.*, **20**, 846 (1968).
588. A. Müller and E. Diemann, *Z. Naturforsch.*, **B23**, 1607 (1968).
589. C. G. Barraclough, J. Lewis, and R. S. Nyholm, *J. Chem. Soc.*, **1959**, 3552.
590. F. A. Cotton and R. M. Wing, *Inorg. Chem.*, **4**, 867 (1965).
591. F. S. Martin, J. M. Fletcher, P. G. M. Brown, and B. M. Gatehouse, *J. Chem. Soc.*, **1959**, 76.
592. W. P. Griffith, *J. Chem. Soc.*, **1964**, 245.
593. P. Bender and J. M. Wood, *J. Chem. Phys.*, **23**, 1316 (1955).
594. W. D. Perkins and M. K. Wilson, *J. Chem. Phys.*, **20**, 1791 (1952); G. R. Hunt and M. K. Wilson, *Spectrochim. Acta*, **16**, 570 (1960).
595. R. Vogel-Hogler, *Acta Phys. Austriaca*, **1**, 311 (1947).
596. D. E. Martz and R. T. Langeman, *J. Chem. Phys.*, **22**, 1193 (1954).
597. W. E. Hobbs, *J. Chem. Phys.*, **28**, 1220 (1958).
598. H. Stammreich, K. Kawai, and Y. Tavares, *Spectrochim. Acta*, **15**, 438 (1959).
599. F. A. Miller, G. L. Carlson, and W. B. White, *Spectrochim. Acta*, **15**, 709 (1959).
600. G. Mitra, *J. Amer. Chem. Soc.*, **80**, 5639 (1958).
601. T. Dupuis, *Compt. rend.*, **246**, 3332 (1958).
602. E. Steger and K. Martin, *Z. Anorg. Allg. Chem.*, **308**, 330 (1960).
603. T. T. Crow and R. T. Lagemann, *Spectrochim. Acta*, **12**, 143 (1958).
604. A. Müller, H. Horn, and O. Glemser, *Z. Naturforsch.*, **B20**, 1150 (1965).
605. J. R. Durig and J. W. Clark, *J. Chem. Phys.*, **46**, 3057 (1967).
606. A. Müller and H. W. Roesky, *Z. Phys. Chem.*, **55**, 218 (1966); A. Müller, B. Krebs, E. Niecke, and A. Ruoff, *Ber. Bunsenges. Phys. Chem.*, **71**, 571 (1967).
607. M. L. Delwaulle and F. François, *Compt. rend.*, **222**, 1193 (1946).
608. A. Müller, E. Niecke, and O. Glemser, *Z. Anorg. Allg. Chem.*, **350**, 246 (1967).
609. G. D. Flesch and H. J. Svec, *J. Amer. Chem. Soc.*, **80**, 3189 (1958).
610. J. W. Linnett, *Quart. Rev.* (*London*), **1**, 73 (1947).
611. L. A. Woodward, *Trans. Faraday Soc.*, **54**, 1271 (1958).
612. C. W. F. T. Pistorius, *J. Chem. Phys.*, **28**, 514 (1958).
613. K. Venkateswarlu and S. Sundaram, *J. Chem. Phys.*, **23**, 2365 (1955).
614. S. Sundaram, *J. Chem. Phys.*, **33**, 708 (1960).
615. H. Murata and K. Kawai, *J. Chem. Phys.*, **23**, 2451 (1955).
616. K. Kawai and H. Murata, *J. Chem. Soc. Japan*, **77**, 504 (1956).
617. V. Galasso, G. de Alti, and G. Costa, *Spectrochim. Acta*, **21**, 669 (1965).
618. H. Siebert, *Z. Anorg. Allg. Chem.*, **275**, 210 (1954).
619. D. A. Long, R. B. Gravenor, and D. T. L. Jones, *Trans. Faraday Soc.*, **60**, 1509 (1964).
620. L. S. Mayants, E. M. Popov, and M. I. Kabachnik, *Opt. Spektrosk.*, **6**, 589 (1959).
621. J. S. Ziomek and E. A. Piotrowski, *J. Chem. Phys.*, **34**, 1087 (1961).
622. H. H. Claassen, C. L. Chernick, and J. G. Malm, *J. Amer. Chem. Soc.*, **85**, 1927 (1963).
623. H. Stammreich and R. Forneris, *Spectrochim. Acta*, **16**, 363 (1960).

624. A. Sabatini, L. Sacconi, and V. Schettino, *Inorg. Chem.*, **3**, 1775 (1964).
625. D. M. Adams and H. A. Gebbie, *Spectrochim. Acta*, **19**, 925 (1963).
626. J. D. S. Goulden, A. Maccoll, and D. J. Millen, *J. Chem. Soc.*, **1950**, 1635.
627. C. H. Perry, D. P. Athans, E. F. Young, J. R. Durig, and B. R. Mitchell, *Spectrochim. Acta*, **23A**, 1137 (1967).
628. P. J. Hendra, *J. Chem. Soc.*, **A**, **1967**, 1298.
629. A. Fadini and A. Müller, *Mol. Phys.*, **12**, 145 (1967).
630. J. Hiraishi and T. Shimanouchi, *Spectrochim. Acta*, **22**, 1483 (1966).
631. D. E. C. Corbridge and E. J. Lowe, *J. Chem. Soc.*, **1954**, 493, 4555.
632. H. J. Bernstein and J. Powling, *J. Chem. Phys.*, **18**, 1018 (1950).
633. N. B. Slater, *Trans. Faraday Soc.*, **50**, 207 (1954).
634. C. W. F. T. Pistorius, *J. Chem. Phys.*, **29**, 1421 (1958).
635. E. R. Nixon, *J. Phys. Chem.*, **60**, 1054 (1956).
636. M. Baudler and L. Schmidt, *Z. Anorg. Allg. Chem.*, **289**, 219 (1957).
637. S. G. Frankiss and F. A. Miller, *Spectrochim. Acta*, **21**, 1235 (1965).
638. S. G. Fankiss and F. A. Miller, *Spectrochim. Acta*, **23A**, 543 (1967).
639. M. Baudler and G. Fricke, *Z. Anorg. Allg. Chem.*, **345**, 129 (1966).
640. H. Gerding, H. van Brederode, and H. C. J. de Decker, *Rec. Trav. Chim. Pays-Bas*, **61**, 549 (1942).
641. H. Gerding and H. C. J. de Decker, *Rec. Trav. Chim. Pays-Bas*, **64**, 191 (1945).
642. M. Baudler, *Z. Anorg. Allg. Chem.*, **279**, 115 (1955).
643. W. G. Palmer, *J. Chem. Soc.*, **1961**, 1552.
644. M. Baudler, *Z. Anorg. Allg. Chem.*, **292**, 325 (1957).
645. A. Simon and H. Richter, *Z. Anorg. Allg. Chem.*, **301**, 154 (1959).
646. E. Steger and G. Leukroth, *Z. Anorg. Allg. Chem.*, **303**, 169 (1960).
647. A. Simon and H. Richter, *Z. Anorg. Allg. Chem.*, **304**, 1 (1960).
648. E. Steger, *Z. Anorg. Allg. Chem.*, **296**, 305 (1958).
649. A. Simon and E. Steger, *Z. Anorg. Allg. Chem.*, **277**, 209 (1954).
650. E. Steger and A. Simon, *Z. Anorg. Allg. Chem.*, **291**, 76 (1957).
651. E. Steger and A. Simon, *Z. Anorg. Allg. Chem.*, **294**, 1, 147 (1958).
652. H. Gerding and H. van Brederode, *Rec. Trav. Chim. Pays-Bas*, **64**, 183 (1945).
653. H. Gerding, J. W. Maarsen, and P. C. Nobel, *Rec. Trav. Chim. Pays-Bas*, **76**, 757 (1957).
654. E. Steger, *Z. Elektrochem.*, **61**, 1004 (1957); *Z. Anorg. Allg. Chem.*, **310**, 114 (1961).
655. E. Steger, *Z. Anorg. Allg. Chem.*, **309**, 304 (1961).
656. H. J. Becher and F. Seel, *Z. Anorg. Allg. Chem.*, **305**, 148 (1960).
657. A. C. Chapman and N. L. Paddock, *J. Chem. Soc.*, **1962**, 635.
658. I. C. Hisatsune, *Spectrochim. Acta*, **21**, 1899 (1965).
659. T. R. Manley and D. A. Williams, *Spectrochim. Acta*, **23A**, 149 (1967)
660. A. Simon and H. Kriegsmann, *Z. phys. Chem.*, **204**, 369 (1955).
661. H. Siebert, *Z. Anorg. Allg. Chem.*, **275**, 225 (1954).
662. L. A. Nimon, V. D. Neff, R. E. Cantley, and R. O. Buttlar, *J. Mol. Spectry.*, **22**, 105 (1967).
663. K. Venkateswarlu and P. Thirungnanasambandam, *Trans. Faraday Soc.*, **55**, 1993 (1959).
664. G. M. Barrow, *J. Chem. Phys.*, **21**, 219 (1953).
665. D. W. Scott and J. P. McCullough, *J. Mol. Spectry.*, **6**, 372 (1961).
666. G. W. Chantry, A. Anderson, and H. A. Gebbie, *Spectrochim. Acta*, **20**, 1223 (1964).
667. D. W. Scott, J. P. McCullough, and F. H. Kruse, *J. Mol. Spectry.*, **13**, 313 (1964).
668. A. Simon, *Z. Anorg. Allg. Chem.*, **260**, 161 (1949).
669. A. Simon and K. Waldmann, *Z. Anorg. Allg. Chem.*, **281**, 135 (1955).

670. A. Simon, K. Waldmann, and E. Steger, *Z. Anorg. Allg. Chem.*, **288**, 131 (1956).
671. A. Simon and K. Waldmann, *Z. Anorg. Allg. Chem.*, **283**, 359 (1956); **284**, 36 (1956).
672. A. Simon and G. Kratzsch, *Z. Anorg. Allg. Chem.*, **242**, 369 (1939).
673. A. Simon and H. Richter, *Naturwissenschaften*, **44**, 178 (1957).
674. A. Meuwsen and G. Heinze, *Z. Anorg. Allg. Chem.*, **269**, 86 (1952).
675. R. J. Gillespie and E. A. Robinson, *Can. J. Chem.*, **39**, 2179 (1961).
676. A. Simon and R. Lehmann, *Z. Anorg. Allg. Chem.*, **311**, 212, 224 (1961).
677. F. B. Dudley and G. H. Cady, *J. Amer. Chem. Soc.*, **79**, 513 (1957).
678. A. V. Jones, *J. Chem. Phys.*, **18**, 1263 (1950).
679. I. Nakagawa, S. Mizushima, A. J. Saraceno, T. J. Lane, and J. V. Quagliano, *Spectrochim. Acta*, **12**, 239 (1958).
680. A. M. Vuagnat and E. L. Wagner, *J. Chem. Phys.*, **26**, 77 (1957).
681. T. Dupuis, *Compt. rend.*, **243**, 1621 (1956).
682. V. Wannagat and R. Pfeiffenschneider, *Z. Anorg. Allg. Chem.*, **297**, 151 (1958).
683. E. R. Lippincott and M. C. Tobin, *J. Chem. Phys.*, **21**, 1559 (1953); *J. Amer. Chem. Soc.*, **73**, 4990 (1951).
684. H. Garcia-Fernandy, *Bull. Soc. Chim. Fr.*, **1959**, 760.
685. G. A. Gallup and J. L. Koenig, *J. Phys. Chem.*, **64**, 395 (1960).
686. O. Glemser and H. Richert, *Z. Anorg. Allg. Chem.*, **307**, 313, 328 (1960).
687. K. Ramaswamy, K. Sathianandan, and F. F. Cleveland, *J. Mol. Spectry.*, **9**, 107 (1962).
688. R. E. Dodd, L. A. Woodward, and H. L. Roberts, *Trans. Faraday Soc.*, **53**, 1545, 1557 (1957).
689. J. K. Wilmshurst and H. J. Bernstein, *Can. J. Chem.*, **35**, 191 (1957).
690. A. Müller and B. Krebs, *Z. Anorg. Allg. Chem.*, **345**, 164 (1966).
691. W. K. Busfield, M. J. Taylor, and E. Whalley, *Can. J. Chem.*, **42**, 2107 (1964).
692. R. J. Gillespie and E. A. Robinson, *Can. J. Chem.*, **41**, 2074 (1963).
693. R. P. Carter and R. R. Holmes, *J. Chem. Phys.*, **41**, 863 (1964); **42**, 2632 (1965).
694. J. P. Pemsler and W. G. Planet, *J. Chem. Phys.*, **24**, 920 (1956).
695. L. C. Hoskins and R. C. Lord, *J. Chem. Phys.*, **46**, 2402 (1967).
696. J. Gaunt and J. B. Ainscough, *Spectrochim. Acta*, **10**, 57 (1957).
697. H. H. Claassen and H. Seilig, *J. Chem. Phys.*, **44**, 4039 (1965).
698. R. G. Cavell and H. C. Clark, *Inorg. Chem.*, **3**, 1789 (1964).
699. J. K. Wilmshurst and H. J. Bernstein, *J. Chem. Phys.*, **27**, 661 (1957).
700. G. L. Carlson, *Spectrochim. Acta*, **19**, 1291 (1963).
701. P. C. Van der Voorn, K. F. Purcell, and R. S. Drago, *J. Chem. Phys.*, **43**, 3457 (1965).
702. J. K. Wilmshurst, *J. Mol. Spectry.*, **5**, 343 (1960).
703. J. Gaunt and J. B. Ainscough, *Spectrochim. Acta*, **10**, 52 (1957).
704. H. Gerding and H. Houtgraaf, *Rec. Trav. Chim. Pays-Bas*, **74**, 5 (1955).
705. J. S. Ziomek and C. B. Mast, *J. Chem. Phys.*, **21**, 862 (1953); J. S. Ziomek, *ibid.*, **22**, 1001 (1954).
706. P. C. Haarhoff and C. W. F. T. Pistorius, *Z. Naturforsch.*, **A14**, 972 (1959).
707. R. A. Condrate and K. Nakamoto, *Bull. Chem. Soc. Jap.*, **39**, 1108 (1966).
708. A. J. Edwards, *J. Chem. Soc.*, **1964**, 3714.
709. D. E. Sands and A. Zalkin, *Acta Cryst.*, **12**, 723 (1959)
710. A. Zalkin and D. E. Sands, *Acta Cryst.*, **11**, 615 (1958).
711. R. R. Holmes, *J. Chem. Phys.*, **46**, 3724, 3730, 3718 (1967).
712. J. A. Salthouse and T. C. Waddington, *Spectrochim. Acta*, **23A**, 1069 (1967).
713. J. Goubeau, R. Baumgärtner, and H. Weiss, *Z. Anorg. Allg. Chem.*, **348**, 286 (1966).
714. K. Dehnicke and J. Weidlein, *Z. Anorg. Allg. Chem.*, **323**, 267 (1963).

715. G. M. Begun, W. H. Fletcher, and D. F. Smith, *J. Chem. Phys.*, **42**, 2236 (1965).
716. R. J. Gillespie and H. J. Clase, *J. Chem. Phys.*, **47**, 1071 (1967).
717. N. N. Greenwood, A. C. Sarma, and B. P. Straughan, *J. Chem. Soc.*, **A,1966**, 1446.
718. G. Nagarajan, *Bull. Soc. Chim. Belges*, **72**, 5 (1962).
719. R. T. Lagemann and E. A. Jones, *J. Chem. Phys.*, **19**, 534 (1951).
720. A. de Lattre, *J. Chem. Phys.*, **20**, 520 (1952).
721. D. Edelson and K. B. McAfee, *J. Chem. Phys.*, **19**, 1311 (1951).
722. C. W. Gullikson, J. R. Nielsen, and A. T. Stair, *J. Mol. Spectry.*, **1**, 151 (1957).
723. B. Weinstock and G. L. Goodman, *Advan. Chem. Phys.*, **11**, 169 (1965).
724. J. Gaunt, *Trans. Faraday Soc.*, **49**, 1122 (1953).
725. T. G. Burke, *J. Chem. Phys.*, **25**, 791 (1956).
726. J. Gaunt, *Trans. Faraday Soc.*, **51**, 893 (1955).
727. R. B. Badachhape, G. Hunter, L. D. McCory, and J. L. Margrave, *Inorg. Chem.*, **5**. 929 (1966).
728. H. H. Claassen, H. Selig, and J. G. Malm, *J. Chem. Phys.*, **36**, 2888 (1962).
729. K. N. Tanner and A. B. F. Duncan, *J. Amer. Chem. Soc.*, **73**, 1164 (1951).
730. T. G. Burke, D. F. Smith, and A. H. Nielsen, *J. Chem. Phys.*, **20**, 447 (1952).
731. H. H. Claassen, J. G. Malm, and H. Seig, *J. Chem. Phys.*, **36**, 2890 (1962).
732. J. Gaunt, *Trans. Faraday Soc.*, **50**, 209 (1954).
733. B. Weinstock, H. H. Claassen, and C. L. Chernick, *J. Chem. Phys.*, **38**, 1470 (1963).
734. B. Weinstock, H. H. Claassen, and J. G. Malm, *J. Chem. Physl*, **32**, 181 (1960).
735. H. C. Mattraw, N. J. Hawkins, D. R. Carpenter, and W. W. Sabol, *J. Chem. Phys.*, **23**, 985 (1955).
736. H. H. Claassen, B. Weinstock, and J. G. Malm, *J. Chem. Phys.*, **25**, 426 (1956).
737. J. G. Malm, B. Weinstock, and H. H. Claassen, *J. Chem. Phys.*, **23**, 2192 (1955).
738. N. J. Hawkins, H. C. Mattraw, and W. W. Sabol, *J. Chem. Phys.*, **23**, 2191 (1955).
739. B. Weinstock and J. G. Malm, *J. Inorg. Nucl. Chem.*, **2**, 380 (1956).
740. J. E. Griffiths and D. E. Irish, *Inorg. Chem.*, **3**, 1134 (1964).
741. J. Weidlein and K. Dehnicke, *Z. Anorg. Allg. Chem.*, **337**, 113 (1965).
742. L. A. Woodward and M. J. Ware, *Spectrochim. Acta*, **19**, 775 (1963).
743. R. D. Peacock and D. W. A. Sharp, *J. Chem. Soc.*, **1959**, 2762.
744. A. D. Meyers and F. A. Cotton, *J. Amer. Chem. Soc.*, **82**, 5027 (1960).
745. D. H. Brown, K. R. Dixon, C. M. Livingston, R. H. Nuttall, and D. W. A. Sharp, *J. Chem. Soc.*, **A,1967**, 100.
746. D. F. Evans and P. A. W. Dean, *J. Chem. Soc.*, **A,1967**, 698.
747. O. L. Keller, *Inorg. Chem.*, **2**, 783 (1963).
748. O. L. Keller and A. Chetham-Strode, *Inorg. Chem.*, **5**, 367 (1966).
749. H. H. Hyman, "Noble-Gas Compounds," Univ. of Chicago Press, 1963.
750. H. Kim, H. H. Claassen, and E. Pearson, *Inorg. Chem.*, **7**, 616 (1968).
751. J. Gaunt, *Trans. Faraday Soc.*, **50**, 546 (1954).
752. K. Venkateswarlu and S. Sundaram, *Z. Phys. Chem.*, **9**, 174 (1956).
753. C. W. F. T. Pistorius, *J. Chem. Phys.*, **29**, 1328 (1958).
754. H. H. Claassen, *J. Chem. Phys.*, **30**, 968 (1959).
755. J. W. Linnett and C. J. S. M. Simpson, *Trans. Faraday Soc.*, **55**, 857 (1959).
756. O. N. Singh and D. K. Rai, *Can. J. Phys.*, **43**, 167, 378 (1965).
757. S. Sundaram, *Proc. Phys. Soc.*, **91**, 764 (1967).
758. T. G. Spiro, *Inorg. Chem.*, **4**, 1290 (1965).
759. D. M. Adams, J. Chatt, J. M. Davidson, and J. Gerratt, *J. Chem. Soc.*, **1963**, 2189.
760. I. R. Beattie, T. Gilson, K. Livingston, V. Fawcett, and G. A. Ozin *J. Chem. Soc.*, **A,1967**, 712.

761. L. A. Woodward and L. E. Anderson, *J. Chem. Soc.*, **1957**, 1284.
762. J. Hiraishi, I. Nakagawa, and T. Shimanouchi, *Spectrochim. Acta*, **20**, 819 (1964).
763. R. J. H. Clark, private communication.
764. L. A. Woodward and M. J. Ware, *Spectrochim. Acta*, **20**, 711 (1964).
765. L. A. Woodward and J. A. Creighton, *Spectrochim. Acta*, **17**, 594 (1961).
766. P. J. Hendra and P. J. D. Park, *Spectrochim. Acta*, **23A**, 1635 (1967).
767. D. M. Adams, H. A. Gebbie, and R. D. Peacock, *Nature*, **199**, 278 (1963).
768. D. M. Adams, *Proc. Chem. Soc.*, 335 (1961).
769. R. A. Walton and B. J. Brisdon, *Spectrochim. Acta*, **23A**, 2222 (1967).
770. N. N. Greenwood and B. P. Stranghan, *J. Chem. Soc.*, **A,1966**, 962.
771. H. Siebert, *Z. Anorg. Allg. Chem.*, **303**, 162 (1960).
772. H. Siebert, *Z. Anorg. Allg. Chem.*, **301**, 161 (1959).
773. L. H. Cross, H. L. Roberts, P. Goggin, and L. A. Woodward, *Trans. Faraday Soc.*, **56**, 945 (1960); *Spectrochim. Acta*, **17**, 344 (1961).
774. D. F. Eggers, H. E. Wright, and D. W. Robinson, *J. Chem. Phys.*, **35**, 1045 (1961).
775. D. F. Smith and G. M. Begun, *J. Chem. Phys.*, **43**, 2001 (1965).
776. A. Sabatini and I. Bertini, *Inorg. Chem.*, **5**, 204 (1966).
777. R. C. Lord, M. A. Lynch, W. C. Schumb, and E. Slowinski, *J. Amer. Chem. Soc.*, **72**, 522 (1950).
778. H. S. Gutowsky and C. J. Hoffmann, *J. Chem. Phys.*, **19**, 1259 (1951).
779. R. E. LaVilla and S. H. Bauer, *J. Chem. Phys.*, **33**, 182 (1960).
780. R. K. Khanna, *J. Mol. Spectry.*, **8**, 134 (1962).
781. H. H. Claassen and H. Selig, *J. Chem. Phys.*, **43**, 103 (1965).
782. R. P. Bell and H. C. Longuet-Higgins, *Proc. Roy. Soc.*, **A183**, 357 (1945).
783. W. J. Lehmann, J. F. Ditter, and I. Shapiro, *J. Chem. Phys.*, **29**, 1248 (1958).
784. W. E. Anderson and E. F. Barker, *J. Chem. Phys.*, **18**, 698 (1950).
785. R. C. Taylor and A. R. Emery, *Spectrochim. Acta*, **10**, 419 (1958).
786. R. C. Lord and E. Nielsen, *J. Chem. Phys.*, **19**, 1 (1951).
787. W. Klemperer, *J. Chem. Phys.*, **24**, 353 (1956).
788. T. Onishi and T. Shimanouchi, *Spectrochim. Acta*, **20**, 325 (1964).
789. B. Rice and K. C. Bald, *Spectrochim. Acta*, **20**, 721 (1964).
790. H. Gerding, H. G. Haring, and P. A. Renes, *Rec. Trav. Chim. Pays-Bas*, **72**, 78 (1953).
791. D. A. Brown and H. C. Longuet-Higgins, *J. Inorg. Nucl. Chem.*, **1**, 352 (1955).
792. K. Venkateswarlu and P. Thirugnanasambandam, *Proc. Indian Acad. Sci.*, **48A,** 344 (1958).
793. W. C. Price, *J. Chem. Phys.*, **17**, 1044 (1949); W. C. Price, H. C. Longuet-Higgins B. Rice, and T. F. Young, *ibid.*, **17**, 217 (1949).
794. H. J. Hrostowski and G. C. Pimentel, *J. Amer. Chem. Soc.*, **76**, 998 (1954).
795. W. J. Taylor, C. W. Beckett, J. Y. Tung, R. B. Holden, and H. L. Johnston, *Phys. Rev.*. **79**, 234 (1950).
796. W. E. Keller and H. L. Johnston, *J. Chem. Phys.*, **20**, 1749 (1952).
797. W. C. Price, R. D. B. Fraser, T. S. Robinson, and H. C. Longuet-Higgins, *Disc. Faraday Soc.*, **9**, 131 (1950).
798. D. White, D. E. Mann, P. N. Walsh, and A. Sommer, *J. Chem. Phys.*, **32**, 481 (1960); D. White, P. N. Walsh, and D. E. Mann, *ibid.*, **28**, 508 (1958).
799. A. Sommer, D. White, M. J. Linevsky, and D. E. Mann, *J. Chem. Phys.*, **38**, 87 (1963)
800. D. White, D. E. Mann, P. N. Walsh, and A. Sommer, *J. Chem. Phys.*, **32**, 488 (1960).
801. J. N. Gayles and J. Self, *J. Chem. Phys.*, **40**, 3530 (1964).
802. A. Finch, I. Hyams, and D. Steele, *Spectrochim. Acta*, **21**, 1423 (1965).

803. M. J. Linevsky, E. R. Shull, D. E. Mann, and T. Wartik, *J. Amer. Chem. Soc.*, **75**, 3287 (1953).
804. D. E. Mann and L. Fano, *J. Chem. Phys.*, **26**, 1665 (1957).
805. F. T. Green and J. L. Margrave, *J. Amer. Chem. Soc.*, **81**, 5555 (1959).
806. B. Latimer and J. P. Devlin, *Spectrochim. Acta*, **21**, 1437 (1965).
807. B. L. Crawford, W. H. Avery, and J. W. Linnett, *J. Chem. Phys.*, **6**, 682 (1938).
808. G. W. Bethke and M. K. Wilson, *J. Chem. Phys.*, **26**, 1107 (1957).
809. H. S. Gutowsky and E. O. Stejskal, *J. Chem. Phys.*, **22**, 939 (1954).
810. D. A. Dows and R. M. Hexter, *J. Chem. Phys.*, **24**, 1029, 117 (1956).
811. V. A. Crawford, K. H. Rhee, and M. K. Wilson, *J. Chem. Phys.*, **37**, 2377 (1962).
812. R. C. Lord, D. W. Robinson, and W. C. Schumb, *J. Amer. Chem. Soc.*, **78**, 1327 (1956); J. R. Aronson, R. C. Lord, and D. W. Robinson, *J. Chem. Phys.*, **33**, 1004 (1960).
813. R. F. Curl and K. S. Pitzer, *J. Amer. Chem. Soc.*, **80**, 2371 (1958).
814. D. C. McKean, *Spectrochim. Acta*, **13**, 38 (1958).
815. H. R. Linton and E. R. Nixon, *J. Chem. Phys.*, **29**, 921 (1958).
816. E. A. V. Ebsworth, R. Taylor, and L. A. Woodward, *Trans. Faraday Soc.*, **55**, 211 (1959).
817. A. Almenningen, O. Bastiansen, V. Ewing, K. Hedberg, and M. Traetteberg, *Acta Chem. Scand.*, **17**, 2455 (1963).
818. A. Almenningen, K. Hedberg, and R. Seip, *Acta Chem. Scand.*, **17**, 2264 (1963).
819. D. W. Robinson, *J. Amer. Chem. Soc.*, **80**, 5924 (1958).
820. H. Kriegsmann and W. Forster, *Z. Anorg. Allg. Chem.*, **298**, 212 (1959); **299**, 78, 223, 232 (1959).
821. G. Davidson, E. A. V. Ebsworth, G. M. Sheldrick, and L. A. Woodward, *Spectrochim. Acta*, **22**, 67 (1966).
822. G. Davidson, L. A. Woodward, E. A. V. Ebsworth, and G. M. Sheldrick, *Spectrochim. Acta*, **23A**, 2609 (1967).
823. C. Newman, J. K. O'Loane, S. R. Polo, and M. K. Wilson, *J. Chem. Phys.*, **25**, 855 (1956).
824. E. A. V. Ebsworth and M. J. Mays, *Spectrochim. Acta*, **19**, 1127 (1963).
825. H. R. Linton and E. R. Nixon, *Spectrochim. Acta*, **15**, 146 (1959).
826. H. W. Thompson, *Spectrochim. Acta*, **16**, 239 (1960).
827. A. L. Smith and N. C. Angelotti, *Spectrochim. Acta*, **15**, 412 (1959).
828. A. Simon and H. Wagner, *Z. Anorg. Allg. Chem.*, **311**, 102 (1961).
829. R. Paetzold, Z. Anorg. Allg. Chem., **337**, 225 (1965)
830. R. Paetzoid, *Z. Anorg. Allg. Chem.*, **336**, 278 (1965).
831. H. Stammreich, D. Bassi, O. Sala, and H. Siebert, *Spectrochim. Acta*, **13**, 192 (1958).
832. E. Steger and G. Leukroth, *Z. Anorg. Allg. Chem.*, **303**, 169 (1960).
833. R. Savoie and P. A. Giguère, *Can. J. Chem.*, **40**, 991 (1962).
834. R. W. Mooney and S. Z. Toma, *Spectrochim. Acta*, **23A**, 1541 (1967).
835. B. Beagley, *Trans. Faraday Soc.*, **61**, 1821 (1965).
836. R. J. Gillespie and E. A. Robinson, *Can. J. Chem.*, **42**, 2496 (1964).
837. J. L. Hoard, W. G. Martin, M. E. Smith, and J. E. Whitney, *J. Amer. Chem. Soc.*, **76**, 3820 (1954).
838. R. Stomberg and C. Brosset, *Acta Chem. Scand.*, **14**, 441 (1960).
839. B. Weinstock and J. G. Malm, *J. Amer. Chem. Soc.*, **80**, 4466 (1958).
840. M. F. A. Dove, *J. Chem. Soc.*, **1959**, 3722.
841. C. W. F. T. Pistorius, *Bull. Soc. Chim. Belges*, **68**, 630 (1959).
842. H. L. Schlaefer and H. F. Wasgestian, *Theor. Chim. Acta*, **1**, 369 (1963).
843. B. D. Saksena, *Proc. Indian Acad. Sci.*, **12A**, 93 (1940); **30A**, 308 (1949).

844. B. D. Saksena, *Proc. Indian Acad. Sci.*, **30A**, 128 (1949).
845. A. K. Ramdas, *Proc. Indian Acad. Sci.*, **37A**, 571 (1953).
846. F. Matossi, *J. Chem. Phys.*, **19**, 1543 (1951).
847. F. Matossi, *J. Chem. Phys.*, **17**, 679 (1949).
848. B. D. Saksena, *Trans. Faraday Soc.*, **57**, 242 (1961).
849. V. Stubičan and B. Roy, *Z. Krist.*, **115**, 200 (1961).
850. H. H. Adler and P. F. Kerr, *Amer. Mineralogist*, **50**, 132 (1965).
851. M. Tsuboi, *Bull. Chem. Soc. Jap.*, **23**, 83 (1950).
852. H. E. Petch, *Acta Cryst.*, **9**, 29 (1956).
853. J. M. Serratosa and W. F. Bradley, *J. Phys. Chem.*, **62**, 1164 (1958).
854. G. B. B. M. Sutherland and H. A. Willis, *Trans. Faraday Soc.*, **41**, 289 (1945); D. E. Blackwell and G. B. B. M. Sutherland, *J. Chim. Phys.*, **46**, 9 (1949).
855. J. J. Charett, *J. Chem. Phys.*, **35**, 1906 (1961).
856. M. Haccuria, *Bull. Soc. Chim. Belges*, **62**, 428 (1953).
857. P. J. Launer, *Amer. Mineralogist*, **37**, 764 (1952).
858. W. D. Keller and E. E. Pickett, *Amer. Mineralogist*, **34**, 855 (1949); **37**, 764 (1952); **38**, 725 (1953); **39**, 256 (1954).
859. J. J. Kirkland, *Anal. Chem.*, **27**, 1537 (1955).
860. R. Soda, *Bull. Chem. Soc. Jap.*, **34**, 1491 (1961).
861. C. K. Huang and P. F. Kerr, *Amer. Mineralogist*, **45**, 311 (1960).
862. J. Louisfert and T. Pobegiun, *Compt. rend.*, **235**, 287 (1953).
863. H. H. Adler and P. F. Kerr, *Amer. Mineralogist*, **47**, 700 (1962); **48**, 124 (1963).
864. L. D. Fredrickson, *Anal. Chem.*, **26**, 1883 (1954).
865. J. M. Serratosa and W. F. Bradley, *Nature*, **181**, 111 (1958).
866. A. M. Vergnoux, S. Théron, and M. Pouzol, *Compt. rend.*, **238**, 467 (1954).
867. G. W. Brindley and J. Zussman, *Amer. Mineralogist*, **44**, 185 (1959).
868. V. C. Farmer, *Mineral. Mag.*, **31**, 829 (1958).
869. J. M. Hunt, M. P. Wisherd, and L. C. Bonham, *Anal. Chem.*, **22**, 1478 (1950).
870. I. Simon and H. O. McMahon, *J. Chem. Phys.*, **21**, 23 (1953); I. Simon, *J. Opt. Soc. Amer.*, **41**, 336 (1951).
871. J. Reitzel, *J. Chem. Phys.*, **23**, 2407 (1955).
872. J. P. Mathieu and H. Poulet, *Compt. rend.*, **244**, 2794 (1957).
873. P. Turlier, L. Eyraud, and C. Eyraud, *Compt. rend.*, **243**, 659 (1956).

Coordination Compounds

Part III

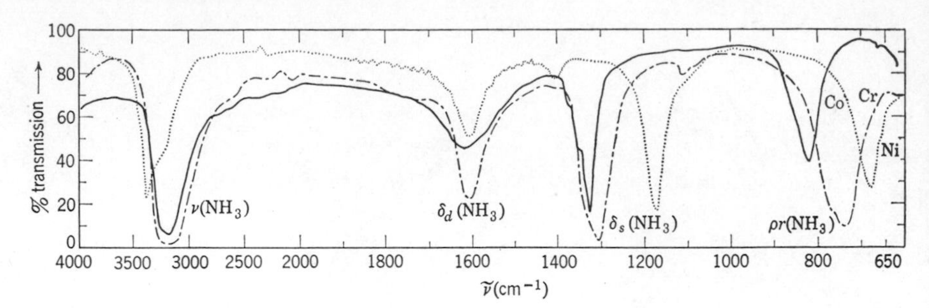

Fig. III-1. Infrared spectra of hexammine complexes: $[Co(NH_3)_6]Cl_3$, solid line; $[Cr(NH_3)_6]Cl_3$, dot-dash line; $[Ni(NH_3)_6]Cl_2$, dotted line.

III-1. AMMINE, AMIDO AND RELATED COMPLEXES

(1) Ammine (NH_3) Complexes

Figure III-1 shows the infrared spectra of some typical ammine complexes. Although the structure of the ammine complex as a whole is highly complicated, the vibrational spectrum may be understood if a simple 1 : 1 model (i.e., $M—NH_3$) is used. The normal modes of vibration of such a simple model may be represented by those of a tetrahedral ZXY_3 molecule, shown in Fig. III-2. Thus the following six vibrations are expected for the

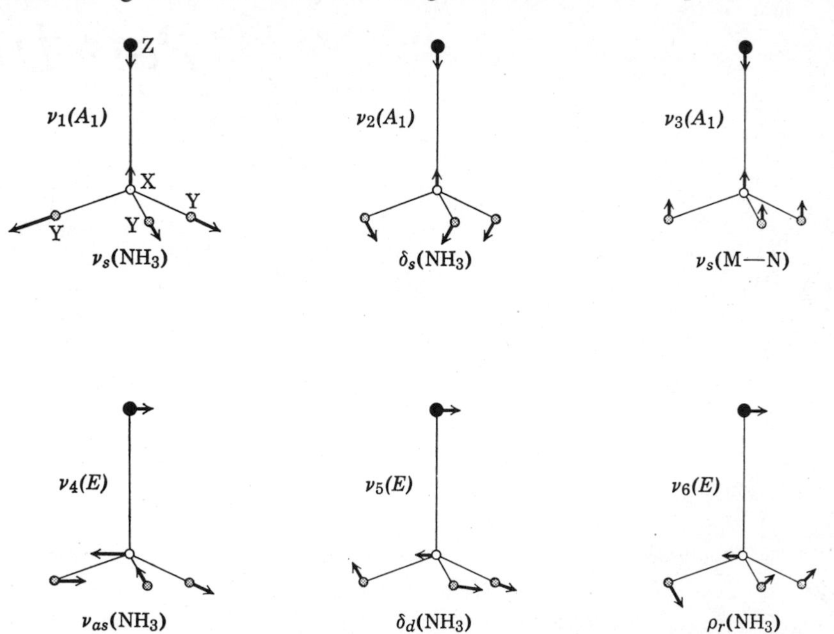

Fig. III-2. Normal modes of vibration of tetrahedral ZXY_3 molecules. (The band assignment is given for an $M—NH_3$ group.)

1 : 1 complex model: antisymmetric and symmetric NH_3 stretching, NH_3 degenerate deformation, NH_3 symmetric deformation, NH_3 rocking and M—N stretching. According to the selection rule for the point group $\mathbf{C}_{3v}$, these vibrations are both infrared and Raman active. Of these six vibrations, the NH_3 stretching bands can be assigned empirically, since they always appear between 3400 and 3000 cm^{-1}. The three bending modes were first assigned by Nakagawa and Mizushima[1,2] from a normal coordinate analysis based on a 1 : 1 complex model like that in Fig. III-2.

(a) NH_3 Stretching Bands

Table III-1 lists the vibrational frequencies of typical ammine complexes. The NH_3 stretching bands in the complexes are usually broader, and their frequencies lower, than those of the free NH_3 molecule. Several factors

TABLE III-1. VIBRATIONAL FREQUENCIES OF TYPICAL AMMINE COMPLEXES (CM^{-1})[3]

Compound	$\nu(NH_3)$	$\delta_d(NH_3)$	$\delta_s(NH_3)$	$\rho_r(NH_3)$
NH_3	3414, 3336	1628	950	–
$[Ni(NH_3)_6](ClO_4)_2$	3397, 3312	1618	1236	(620)
$[Cr(NH_3)_6](ClO_4)_3$	3330, 3280	1622	1334	718
$[Co(NH_3)_6]ClO_4)_3$	3320, 3240	1630	1352	803
$[Co(NH_3)_6](NO_3)_3$	3290, 3200	1618	–	–
$[Co(NH_3)_6]I_3$	3150	1590	1323	792
$[Co(NH_3)_6]Br_3$	3120	1578	1318	797
$[Co(NH_3)_6]Cl_3$	3070	1603	1325	818
$[NH_4]Cl$	3138, 3041	1710	1403	–

may be responsible for the frequency shift. One is the *effect of coordination.*[3] The NH_3 stretching frequency decreases in the series NH_3 > $[Ni(NH_3)_6]^{2+}$ > $[Cr(NH_3)_6]^{3+}$ > $[Co(NH_3)_6]^{3+}$ > $[NH_4]^+$, if the perchlorates are compared. On the other hand, the stability order of these compounds is known to be Co(III) > Cr(III) > Ni(II). Therefore it is reasonable to conclude that the N—H bond order decreases (and the NH_3 stretching frequency decreases) as the M—N bond order increases in the stability order mentioned. Since the NH_3 stretching frequencies of the ammine complexes are intermediate between those of free NH_3 and the $[NH_4]^+$ ion, the M—N bond of the ammine complexes is partially ionic.

The NH_3 stretching frequency is also sensitive to changes in the anion. As seen in Table III-1, the frequency of the NH_3 stretching band of the $[Co(NH_3)_6]^{3+}$ ion decreases by about 230 cm^{-1} when $[ClO_4]^-$ is replaced by Cl^-. There is ample evidence of the presence of hydrogen bonding between the N—H of ammine complex ions and anions such as halogens.[4]

Therefore it is evident that *hydrogen bonding*[3] weakens the N—H bond and shifts the band to a lower frequency. It should be noted that the magnitude of the shift due to hydrogen bonding is even greater than the difference in the NH_3 stretching frequencies between $[Ni(NH_3)_6](ClO_4)_2$ and $[Co(NH_3)_6](ClO_4)_3$. In addition to these two factors, Svatos et al.[5] cite the effects of *hydration* and *configuration* (*cis, trans*, etc.).

The broadening of the NH_3 stretching band in Fig. III-1 may also be the result of hydrogen bonding as well as overlapping of the individual N—H stretching bands of the whole complex ion.

(b) NH_3 Deformation and Rocking Bands

As Table III-1 shows, the effects of coordination and hydrogen bonding shift the three bending bands to higher frequencies. The direction of the band shifts of these bending modes is opposite to that of the stretching modes. Svatos and co-workers[6] have examined the frequencies of these three bending modes in a series of ammine complexes and have found that the NH_3 rocking frequency is most sensitive and the degenerate deformation frequency least sensitive to the metal. These results can be reasonably explained by the conclusion drawn from normal coordinate analysis.[1,2] Wilmshurst[7] has found a linear relation between the electronegativity of the metal and the square of either the symmetric deformation or the rocking frequency.

(c) Metal-Nitrogen (M—N) Stretching Bands

The M—N stretching frequency is of particular interest, since it provides direct information about the coordinate bond. Because of the relatively heavy mass of the metal and the low bond order of the coordinate bond, the M—N stretching vibration may appear in the lower frequency region.

The band assignments for the M—N stretching vibrations of the Co(III) ammine complexes have been controversial. Originally, Powell and Sheppard[8] have assigned an extremely weak band at 502 cm^{-1} to the M—N stretching vibration in $[Co(NH_3)_6]Cl_3$. Although the reason for the weakness of this band is not obvious, the corresponding bands are clearly seen in $[Co(NH_3)_5Cl]Cl_2$ (493 cm^{-1}), $[Pd(NH_3)_4]Cl_2 \cdot H_2O$ (498 cm^{-1}), and $[Pt(NH_3)_4]Cl_2$ (511 cm^{-1}). Figure III-3 shows the infrared spectra of $[Co(NH_3)_6]Cl_3$ and $[Co(NH_3)_5X]X_2$ (X: a halogen) in the CsBr region.[9] Additional evidence to support this assignment was the existence of a linear relation between the Raman active M—N stretching frequency or the NH_3 rocking frequency and the stability constant in a series of ammine complexes of various metals.[8] This assignment was followed by Nakamoto et al[10] who attributed the bands near 500 cm^{-1} of nitroammine complexes to Co(III)—N stretching modes.

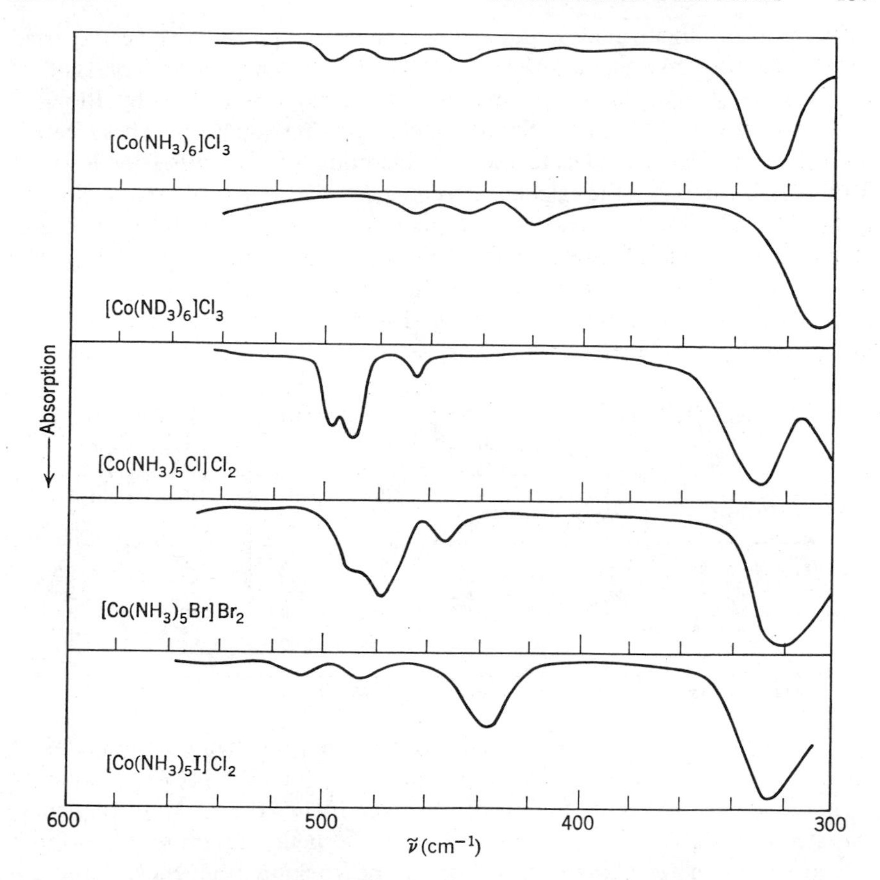

Fig. III-3. Infrared spectra of Co(III) ammine complexes.[9]

Several investigators[11,12] preferred to assign a strong band at 330 cm^{-1} of $[Co(NH_3)_6]Cl_3$ to the Co(III)—N stretching mode. However, this assignment seems to be contradictory with the following findings: (1) According to this interpretation, both Co(III)—N and Co(II)—N stretching bands must be assigned in the same region (near 330 cm^{-1}). If so, the strength of these two bonds should be similar. On the contrary, the heat of formation of the Co(III) ammine complex is twice that of the Co(II) ammine complex.[13] (2) The relationship between stability and M—N stretching frequency[8] will break down if the 328 cm^{-1} band of the Co(III) complex is assigned to the Co(III)—N stretching mode. (3) Normal coordinate analyses on the $[Co(NH_3)_6]^{3+}$ ion yield a fairly reasonable value for the Co(III)-N stretching force constant if the bands near 500 cm^{-1} are interpreted as the infrared active (F_{1u}) Co—N stretching mode.

Normal coordinate analyses on metal ammine complexes have been carried out by several investigators. The Co(III)—N stretching force constant of the $[Co(NH_3)_6]^{3+}$ ion was estimated to be 2 mdyn/A (GVF) by Block,[14] and 1.07 mdyn/A (UBF) by Shimanouchi and Nakagawa.[15] These calculations have been extended to the ammine complexes of other metals.[16,17] Table III-2 compares the calculated frequencies with those observed. For the $[Rh(NH_3)_6]^{3+}$ ion, the agreement is quite satisfactory for both infrared and Raman active modes. However, the results of calculation on the $[Co(NH_3)_6]^{3+}$ ion are not in good agreement with the observed spectra. The table also includes the observed frequencies and band assignments for the $[Ru(NH_3)_6]^{3+}$ and $[Ir(NH_3)_6]^{3+}$ ions.

TABLE III-2. SKELETAL FREQUENCIES OF OCTAHEDRAL HEXAMINE COMPLEXES (CM^{-1})

Skeletal Vibration[a]	$[Co(NH_3)_6]^{3+}$			$[Rh(NH_3)_6]^{3+}$		$[Ru(NH_3)_6]^{3+}$	$[Ir(NH_3)_6]^{3+}$
	Calc.[14]	Calc.[15]	Obs.[18]	Calc.[17]	Obs.[17]	Obs.[19]	Obs.[19]
ν_1, A_{1g}(R), stretching	498	383	495	512	515	500	527
ν_2, E_g(R), stretching	413	337	440	478	480	475	500
ν_3, F_{1u}(IR), stretching	542	464	464	470	470	463	475
ν_4, F_{1u}(IR), bending	-	329	325	310	310	283, 263	279, 264
ν_5, F_{2g}(R), bending	-	370	-	234	240	248	262
ν_6, F_{2u}(ia), bending	-	261	-	-	-	-	-

[a] Based on O_h symmetry of the octahedral XY_6-type molecule (see Sec. II-10).

Table III-3 summarizes the infrared frequencies and band assignments of metal ammine complexes. As stated previously[1,2], the NH_3 rocking frequency is most sensitive to the nature of a metal; the higher the rocking frequency, the stronger the M–N bond. In Table III-3, metal hexammine chlorides are arranged in the decreasing order of the rocking frequency. Thus the order of the strength of the M–N bond may be

$$\text{Pt(IV)} > \text{Ir(III)} > \text{Rh(III)} > \text{Co(III)} > \text{Ru(III)} > \text{Cr(III)} >$$
$$\text{Ni(II)} > \text{Co(II)} \approx \text{Zn(II)} \approx \text{Fe(II)} > \text{Mn(II)} \approx \text{Cd(II)}$$

The M–N stretching force constant also follows the same order of metals:[16,20,24]

$$\begin{array}{ccccccccc} \text{Pt(IV)} & > & \text{Co(III)} & > & \text{Cr(III)} & \gg & \text{Ni(II)} & \approx & \text{Co(II)} \\ 2.13 & & 1.05 & & 0.94 & & 0.34 & & 0.33 \quad \text{mdyn/A(UBF)} \end{array}$$

This order of metals is in good agreement with that of the ligand field stabilization energy. The Co(III)—N stretching band (F_{1u}) splits into three peaks in the chloride and bromide, but not in the iodide. The former two crystals are monoclinic and the latter is cubic.[25] The splitting observed for the former may indicate that the symmetry of the $[Co(NH_3)_6]^{3+}$ ion in these crystals is much lower than that in the iodide crystal.

Table III-3 also lists the infrared frequencies of square-planar tetrammine complexes (Cu(II), Pt(II), and Pd(II)). Normal coordinate analyses have been made on the Cu(II)[16] and Pt(II)[26,27] complexes. Table III-4 gives the skeletal frequencies of the $[Pt(NH_3)_4]^{2+}$ and $[Pd(NH_3)_4]^{2+}$ ions. The Pt—N stretching force constant was estimated to be 2.804(UBF)[26] or 2.1(UBF)[27] mdyn/A. Only one tetrahedral complex (Zn(II)) is included in this table. The existence of the $[Na(NH_3)_4]^+$ ion was demonstrated by Leonard et al[28] from a study of Raman spectra. The vibrational spectrum of the linear $[Hg(NH_3)_2]^{2+}$ ion has been assigned by Bertin et al[12] based on $\mathbf{D}'_{3h}$ symmetry. The Hg—N stretching force constant obtained was 1.695 mdyn/A(UBF) and the corresponding stretching frequency 493 cm^{-1}.

TABLE III-3. INFRARED FREQUENCIES AND BAND ASSIGNMENTS OF METAL AMMINE COMPLEXES (CM^{-1})

Metal Ammine Complex	$\delta_d(NH_3)$	$\delta_s(NH_3)$	$\rho_r(NH_3)$	$\nu(MN)$	$\delta(NMN)$	References
$[Pt(NH_3)_6]Cl_4$	1578	1370	945	536	317	20
$[Ir(NH_3)_6]Cl_3$	1587	1350 1323	857	475	279 264	19
$[Rh(NH_3)_6]Cl_3$	1618	1352 1318	845	472	302 287	19
$[Co(NH_3)_6]Cl_3$	1610	1330	830	499 476 449	332	21
$[Ru(NH_3)_6]Cl_3$	1612	1362 1338 1316	788	474 464 452	283 268	22,19
$[Cr(NH_3)_6]Cl_3$	1630	1307	748	495 473 456	270	23,16
$[Ni(NH_3)_6]Cl_2$	1607	1175	680	334	215	21
$[Co(NH_3)_6]Cl_2$	1602	1163	654	327	192	21
$[Zn(NH_3)_6]Cl_2$	1596	1145	645	300	–	21
$[Fe(NH_3)_6]Cl_2$	1597	1151	641	321	–	21
$[Mn(NH_3)_6]Cl_2$	1592	1134	617	307	–	21
$[Cd(NH_3)_6]Cl_2$	1585	1091	613	298	–	21
$[Zn(NH_3)_4]Cl_2$	1610	1255 1215	702	437	–	21
$[Cu(NH_3)_4]SO_4$	1610	1270 1240	713	420	250	16
$[Pt(NH_3)_4]Cl_2$	1563	1325	842	510	297	24
$[Pd(NH_3)_4]Cl_2$	1601	1285	797	491	295	24
$[Hg(NH_3)_2]Cl_2$	1605	1270	719	513	–	16

TABLE III-4. SKELETAL FREQUENCIES OF SQUARE-PLANAR TETRAMMINE COMPLEXES (cm^{-1})

Skeletal Vibration[a]	$[Pt(NH_3)_4]Cl_2$ Calc.[26]	$[Pt(NH_3)_4]Cl_2$ Obs.[26]	$[Pt(NH_3)_4]Cl_2$ Obs.[27]	$[Pd(NH_3)_4]Cl_2$ Obs.[27]
ν_1, A_{1g}(R), stretching	520	524	538	510
ν_2, B_{1g}(R), in-plane bending	260	265	270	305
ν_3, A_{2u}(IR), out-of-plane bending	–	–	–	–
ν_4, B_{2g}(R), stretching	500	508	526	468
ν_5, B_{2u}(ia), out-of-plane bending	–	–	–	–
ν_6, E_u(IR), stretching	508	510	577	496
ν_7, E_u(IR), in-plane bending	245	236	204	245

[a] Based on $\mathbf{D}_{4h}$ symmetry of XY_4-type ions (see Sec. II-8).

(d) Lattice Vibrations

Infrared spectra of metal hexammine complexes obtained in the solid state exhibit lattice modes below 150 cm^{-1}. Nakagawa and Shimanouchi[29] have carried out a detailed analysis of these lattice modes based on crystal structure.

(e) Halogenoammine Complexes

The infrared spectra of halogenoammine complexes of Co(III) have been studied extensively.[15,16,24] Table III-5 lists the skeletal frequencies and band assignments for pentammine and *trans*-tetrammine complexes. Nakagawa and Shimanouchi[16] obtained the following UBF stretching force constants: K(Co–N),1.05; K(Co—F),0.99; K(Co–Cl),0.91; K(Co—Br), 1.03 and K(Co—I), 0.62 mdyn/A. Although the Co—Cl stretching of the $[Co(NH_3)_5Cl]^{2+}$ ion was originally assigned at 484 cm^{-1},[15] it was later assigned at 272 cm^{-1} (Table III-5).[16,25] The infrared spectra of halogenoammine complexes of Cr(III) have been obtained by Tanaka et al.[23] and Schreiner and McLean.[30]

The infrared spectra of $M(NH_3)_2X_2$-type complexes (M, Pt(II), or Pd(II); X, halogen) have been studied by many investigators. Table III-6 summarizes the band assignments and observed frequencies of skelatal vibrations. Figure III-4 shows the infrared spectra of *cis*- and *trans*-$Pd(NH_3)_2Cl_2$ obtained by Layton et al.[32] It is seen that both the Pd—N and Pd—Cl stretching bands split into two peaks in the *cis*-isomer. This is expected, since the symmetry of the *cis*-planar PdN_2Cl_2 skeleton is $\mathbf{C}_{2v}$, which makes all stretching vibrations infrared active. Durig et al.[33] have found that the Pd—N stretching frequency ranges from 528 to 436 cm^{-1} depending on the nature of other ligands in a complex. It has been known that the Pt—N stretching band is shifted to a lower frequency as a ligand of stronger *trans*-

TABLE III-5. SKELETAL VIBRATIONS OF PENTAMMINE AND *Trans*-TETRAMINE Co(III) COMPLEXES (CM^{-1})[16, 24]

Pentammine Complex (C_{4v} symmetry)		ν(Co—N)	ν(Co—X)	Skeletal Bending
$[Co(NH_3)_5F]^{2+}$	A_1	480, 438	343	308
	E	498	–	345, 290, 219
$[Co(NH_3)_5Cl]^{2+}$	A_1	476, 416	272	310
	E	498	–	292, 287, 188
$[Co(NH_3)_5Br]^{2+}$	A_1	475, 410	215	287
	E	497	–	290, 263, 146
$[Co(NH_3)_5I]^{2+}$	A_1	473, 406	168	271
	E	498	–	290, 259, 132
Trans-Tetrammine Complex (D_{4h} symmetry)				
$[Co(NH_3)_4Cl_2]^+$	A_{2u}	–	353	186
	E_u	501	–	290, 167
$[Co(NH_3)_4Br_2]^+$	A_{2u}	–	317	227
	E_u	497	–	280, 120

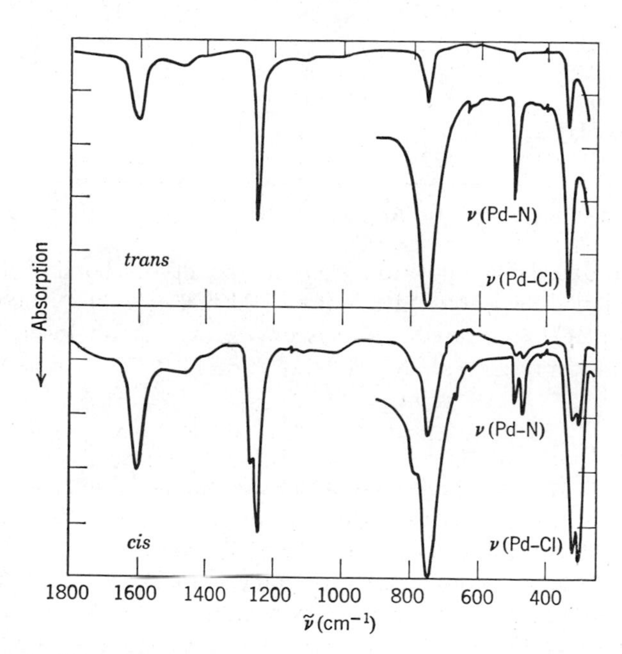

Fig. III-4. Infrared spectra of *trans*- and *cis*-$[Pd(NH_3)_2Cl_2]$.[32]

TABLE III-6. SKELETAL FREQUENCIES OF SQUARE-PLANAR $M(NH_3)_2X_2$-TYPE COMPLEXES (CM^{-1})*

Complex		ν(M—N)	ν(M—X)	Bending	References
trans-$[Pd(NH_3)_2Cl_2]$					
	IR	496	333	245, 222, 162, 137	31
	R	494	295		
cis-$[Pd(NH_3)_2Cl_2]$					
	IR	495, 476	327, 306	245, 218, 160, 135	31
trans-$[Pd(NH_3)_2Br_2]$					
	IR	490	–	220, 220, 122, 101	31
cis-$[Pd(NH_3)_2Br_2]$					
	IR	480, 460	258	225, 225, 120, 100	31
trans-$[Pd(NH_3)_2I_2]$					
	IR	480	191	263, 218, 109	31
trans-$[Pt(NH_3)_2Cl_2]$					
	IR	572	365	220, 195	27
	R	529	318	–	27
cis-$[Pt(NH_3)_2Cl_2]$					
	IR	507	328	240, 132	26
trans-$[Pt(NH_3)_2Br_2]$					
	IR	504	206	230	27
	R	535	206	–	27
trans-$[Pt(NH_3)_2I_2]$					
	R	532	153	–	27

* For band assignments, see also Ref. 36.

effect is introduced in a position *trans* to the Pt—N bond.[34] Durig and Mitchell[35] studied the isomerization of *cis*-$[Pd(NH_3)_2X_2]$ to the corresponding *trans*-isomer by infrared spectroscopy. For metal-halogen stretching bands of these complexes, see Sec. III-12.

(f) Rotation of NH_3 Group

Leech et al.[37] studied the band contours of the NH_3 vibrations of *trans*-$[Pd(NH_3)_2X_2]$(X, Cl, or I) as a function of temperature (90–470°K). They concluded that the NH_3 groups of the iodo complex are rotating with gradually decreasing freedom over this temperature range, and that those of the chloro complex are not rotating appreciably at 298°K and below and begin to rotate with limited freedom at temperatures above 298°K. Nakamoto et al.[38] have postulated that, in *cis*- and *trans*-$[Pt(NH_3)_2X_2]$-type complexes, the NH_3 groups are probably not rotating because the NH_3 hydrogens interact with filled *d*-orbitals of the platinum.

(2) Amido (NH_2) Complexes

The vibrational spectra of amido complexes may be interpreted in terms of the normal modes of vibration of a pyramidal ZXY_2-type molecule. Mizushima and colleagues[39] have carried out a normal coordinate analysis of the $[Hg(NH_2)_2]^+_\infty$ ion (an infinite chain polymer). The results are given in Table III-7. Brodersen and Becher[40] have also studied the infrared spectra of a number of compounds containing Hg—N bonds and have assigned the Hg—N stretchings in the range 700 to 400 cm^{-1}. The infrared spectrum of the NH_2^- ion in alkali metal salts has been measured.[41]

TABLE III 7. VIBRATIONAL FREQUENCIES AND BAND ASSIGNMENTS OF AMIDO COMPLEXES (CM^{-1})[39]

Compound	$\delta(NH_2)$	$\rho_w(NH_2)$	$\rho_t(NH_2)$	$\rho_r(NH_2)$	ν(Hg—N)
$[Hg(NH_2)]^+_\infty(Cl^-)_\infty$	1534	1022	(978)	668	573
$[Hg(NH_2)]^+_\infty(Br^-)_\infty$	1528	1005	(978)	647	560
	1505	950	(978)	620	510

(3) Alkylamine Complexes

Infrared spectra of methylamine complexes of the type, $[Pt(CH_3NH_2)_2X_2]$ (X, a halogen) have been studied by Watt et al.[42] and Kharitonov et al.[43] Far-infrared spectra of complexes of the type, $[M(R_2NH)_2X_2]$(R, ethyl or *n*-propyl; M, Zn(II), or Cd(II); X, Cl, or Br) are also reported.[44] Chatt and co-workers[45] have studied the effect of hydrogen bonding on the N—H stretching frequencies of $[Pt(RNH_2)Cl_2L]$-type complexes (R, methyl, ethyl, etc.; L, ethylene, triphenylphosphine, etc.) in organic solvents such as chloroform and dioxane. Their study revealed that the complexes of primary amines have a strong tendency to associate through intermolecular hydrogen bonds of the NH···Cl type, whereas those of secondary amines have little tendency to associate. Later, this difference was explained on the basis of steric repulsion and intramolecular interaction between the NH hydrogen and the nonbonding *d*-electrons of the metal.[46]

(4) Complexes of Hydroxylamine and Hydrazine

Vibrational spectra of hydroxylamine (NH_2OH) complexes have been studied by Kharitonov et al.[47] Sacconi and Sabatini[48] have reported the infrared spectra of hydrazine (NH_2NH_2) complexes of the type, $M(N_2H_4)_2Cl_2$, where M is a divalent metal. They assigned the M—N stretching bands between 440 and 330 cm^{-1}. Brodersen[49] reported the infrared spectra of hydrazine complexes of mercury.

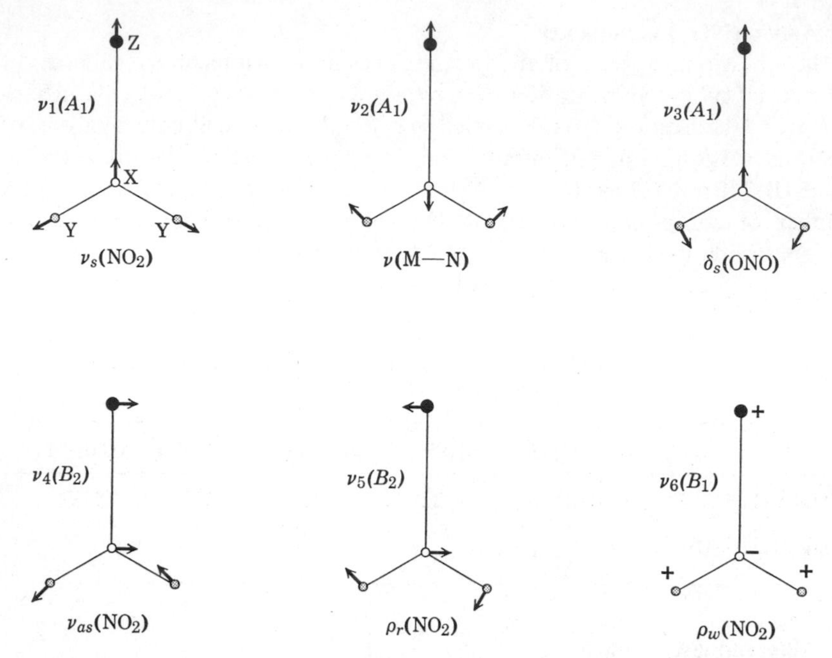

Fig. III-5. Normal modes of vibration of planar ZXY_2 molecules. (The band assignment is given for an M—NO_2 group.)

III-2. NITRO AND NITRITO COMPLEXES

(1) Nitro (NO_2) Complexes

The normal vibrations of the nitro group coordinated to a metal may be approximated by those of a planar ZXY_2 molecule, as illustrated in Fig. III-5. In addition, the NO_2 twisting mode may become Raman active. Recently, Nakagawa and Shimanouchi[29,50] have carried out a normal coordinate analysis to assign the vibrational spectra of hexanitrocobaltic salts; both internal and lattice modes were assigned completely from factor group analysis. The results indicate that the complex anion takes the $\mathbf{T_h}$ symmetry in the crystals of the potassium, rubidium, and cesium salts, but the $\mathbf{S_6}$ symmetry in the crystal of the sodium salt. Figure III-6 illustrates the $\mathbf{T_h}$ and $\mathbf{S_6}$ symmetry of the $[Co(NO_2)_6]^{3-}$ ion. Figure III-7 shows the infrared spectra of the potassium and sodium salts obtained by these authors. Infrared spectra of $M_3[Co(NO_2)_6]$ type compounds have been studied by other investigators.[51–53] Elliott et al.[54] and LePostollec et al.[55] have obtained the infrared spectra of nitro complexes of other metals. Table III-8 summaries the

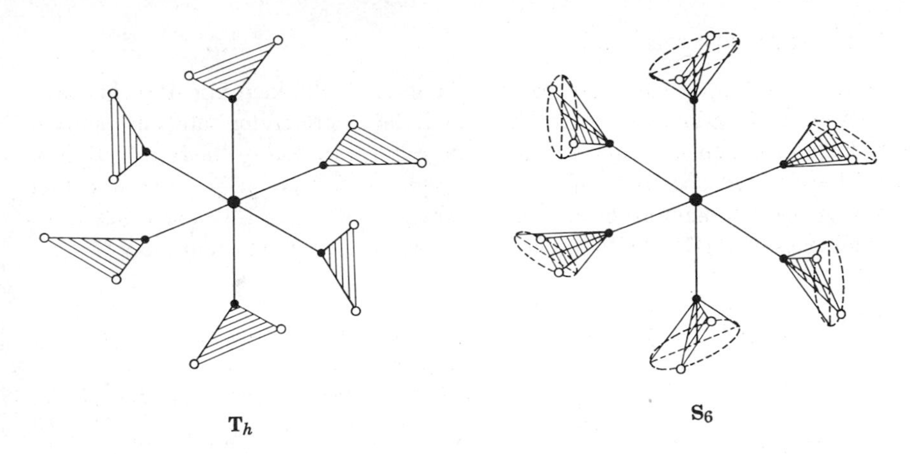

Fig. III-6. Possible structures of the $[Co(NO_2)_6]^{3-}$ ion.

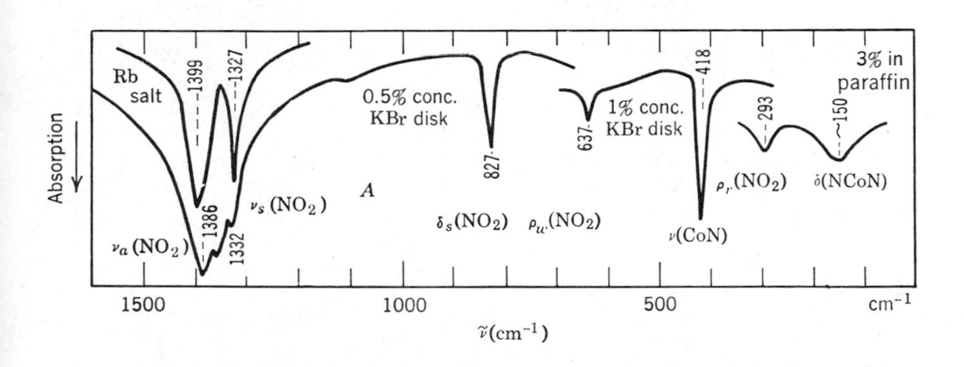

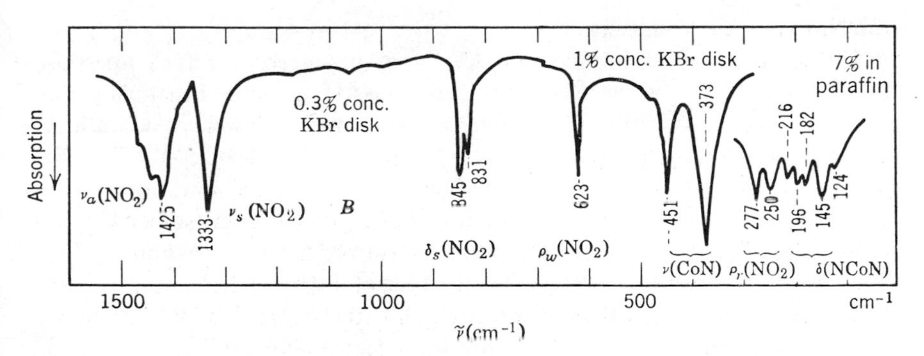

Fig. III-7. Infrared spectra of $K_3[Co(NO_2)_6]$ (A) and $Na_3[Co(NO_2)_6]$ (B).[50]

vibrational frequencies and band assignments obtained for typical nitro complexes. In all cases, the antisymmetric NO_2 stretching bands are shifted to higher frequencies upon coordination, relative to that of the free NO_2^- ion.

The infrared spectrum of the square-planar $[Pt(NO_2)_4]^{2-}$ ion was first analyzed by Nakamoto et al.[10] The infrared spectra of this and other nitro complexes of Pt(II), Pt(IV), and Pd(II) have been studied extensively.[56–58]

TABLE III-8. VIBRATIONAL FREQUENCIES OF NITRO COMPLEXES (cm^{-1})

Complex	$\nu_{as}(NO_2)$	$\nu_s(NO_2)$	$\delta(ONO)$	$\rho_w(NO_2)$	$\nu(M—N)$	$\rho_r(NO_2)$	References
NO_2^-	1250	1335	830	–	–	–	10
$K_3[Co(NO_2)_6]$	1386	1332	827	637	418	293	50
$Na_3[Co(NO_2)_6]$	1425	1333	845 831	623	451 373	277 250	50
$K_2Ba[Co(NO_2)_6]$	1650 1400	1330 1292 1250	826	–	418 384	326 293	54
$K_2Ba[Fe(NO_2)_6]$	1630	1332 1184	820	644 607	408	292	54
$K_2Ba[Ni(NO_2)_6]$	1640 1405	1348 1341	836 810	–	435	296 285	54
$K_2Ba[Cu(NO_2)_6]$	1630 1420	1332 1260	816 800	556	447	315 288	54
$K_2[Pt(NO_2)_4]$	1436 1410 1386	1350	838 832 828	636 613	450		10
$K_2[Pd(NO_2)_4]$	1434 1400 1374	1340	827 832				56

(2) Nitroammine Complexes

The spectra of nitroammine complexes may be interpreted as a superposition of the spectra of the $M—NO_2$ and $M—NH_3$ groups. Practically, this approach cannot be utilized, since band overlapping as well as vibrational coupling between these two groups is extensive. In fact, Beattie and Tyrrell[59] found it difficult to correlate the stereochemistry and the number of nitro groups with the spectra. It was noted, however, that the *cis* isomer exhibits more bands than the *trans* isomer, since the former has lower symmetry than the latter.[60,61] In dinitroammine complexes of Pt(II), Pd(II), and Co(III) Chatt et al.[62] noted that the *cis* isomer exhibits the symmetric NO_2 stretching at a lower frequency than the *trans* isomer, with an additional band at ca. 1350 cm^{-1} and shows the ONO bending band split into two distinct peaks. *Cis* and *trans* isomers of nitroethylenediamine complexes can also be distinguished on a similar basis.[63]

O'Connor,[64] using infrared and ultraviolet spectra, has shown that $[Co(NH_3)_3(NO_2)_3]$ is in the *cis-cis* configuration. Nakamoto et al.[10] suggested that the $M—NH_3$ stretching bands near 500 cm^{-1} are useful in distinguishing these stereoisomers. Nakagawa and Shimanouchi[65] have carried out normal coordinate analyses on various Co(III) nitroammine complexes; the $Co—NO_2$ and $Co—NH_3$ stretching force constants (UBF) were estimated to be 1.16 and 1.05 mdyn/A, respectively. Far-infrared spectra of these complexes are reported by Blyholder and Kittila.[66] Cleare and Griffith[67] confirmed the band assignments of nitro, nitrito, and nitrosyl complexes by comparing the spectra of normal and N^{15}-substituted complexes.

(3) Bridging Nitro Group

The nitro group is known to form a bridge between two metal atoms. Nakamoto et al.[10] suggested that, of the three possible structures;

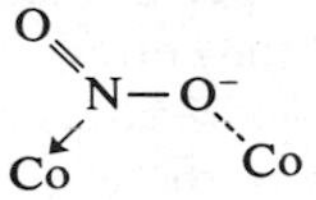

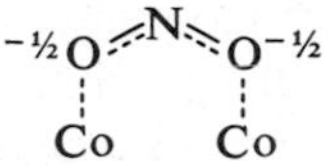

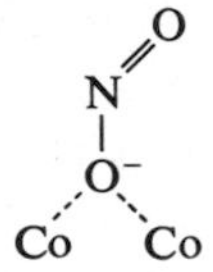

O=N(→Co)—O⁻···Co (I)

$^{-1/2}$O≐N≐O$^{-1/2}$ with each O···Co (II)

O=N—O⁻ with O···Co and O···Co (III)

I II III

I is most probable, since the NO_2 stretching frequencies of the

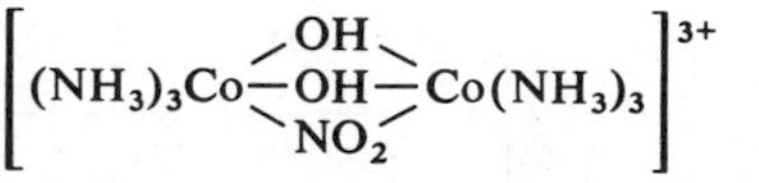

$$\left[(NH_3)_3Co\begin{matrix}OH \\ OH \\ NO_2\end{matrix}Co(NH_3)_3\right]^{3+}$$

ion (1516 and 1200 cm^{-1}) are markedly different from those of nitro and nitrito complexes. This result rules out structure III because III is expected to show some similarity to that of the nitrito complexes. The high NO_2 stretching frequency of the bridged complex (1516 cm^{-1}) is also difficult to explain on the basis of structure II. Bridged complexes such as

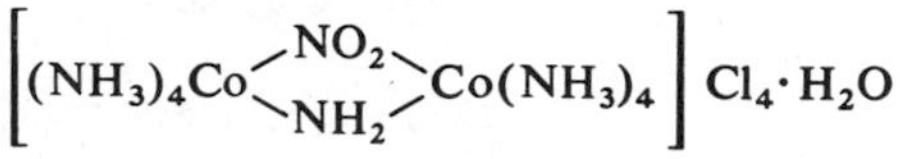

$$\left[(NH_3)_4Co\begin{matrix}NO_2 \\ NH_2\end{matrix}Co(NH_3)_4\right]Cl_4\cdot H_2O$$

and $[(P(n\text{-}Bu)_3)_2Pd_2(NO_2)_4]$ exhibit the NO_2 stretching bands at ca. 1480 and 1210 cm^{-1}.[62] Thus the nitro bridges in these complexes may take structure I. These frequencies are markedly different from those of the $M—NO_2$ group, since the bond orders of two NO bonds in structure I are quite different.

(4) Nitrito (ONO) Complexes

The nitro group coordinates to a metal through one of its oxygen atoms (nitrito complex):

$$M{-}O{-}N{=}O$$

Table III-9 lists the NO stretching frequencies of nitrito complexes. Similar to the case of the bridging NO_2 group, the two NO_2 stretching frequencies are well separated in nitrito complexes, since one NO bond is almost a single and the other a double bond. In fact, the N—O stretching frequency is lower than that found in bridging complexes. Thus the distinction of nitro- and nitrito complexes can be made readily by comparing their infrared spectra. It was also noted that nitrito complexes lack rocking vibrations near 620 cm^{-1} which appear in nitro complexes.[10]

The red nitritopentammine complex, $[Co(NH_3)_5(ONO)]Cl_2$, is unstable and is gradually converted to the yellow nitro complex. Penland et al.[70] followed this conversion by observng the disappearance of the nitrito bands in KBr pellets. Nakamoto et al.[10] have shown that the corresponding Cr(III) complex exists as a nitrito and not as a nitro complex. Beattie and Satchell[71] studied the kinetics of the nitrito-nitro conversion of the Co(III) complexes using infrared spectra. Basolo and Hammaker[72] made a similar study for the

TABLE III 9. VIBRATIONAL FREQUENCIES OF NITRITO COMPLEXES (CM^{-1})

Compound	ν(N=O)	ν(N—O)	δ(ONO)	References
$[Co(NH_3)_5(ONO)]Cl_2$	1468	1065	825	10
$[Cr(NH_3)_5(ONO)]Cl_2$	1460	1048	839	10
$[Rh(NH_3)_5(ONO)]Cl_2$	1461, 1445	1063	830	67
$[Rh(NH_3)_5(ON^{15}O)]Cl_2$	1412, 1382	1049	825	67
$Ni(N,N\text{-diethyl-en})_2(ONO)_2$	1361	1163	824	68
$Ni(py)_4(ONO)_2$	1393	1114	825	68
Ni(2-(methylaminomethyl)-pyridine)$_2$(ONO)$_2$	1375	1180	817, 822	69
Ni(2-(aminomethyl)-6-methyl-piperidine)$_2$(ONO)$_2$	1341	1206	827, 813	69

pentammine complexes of Ir(III), Rh(III), and Pt(IV). Goodgame and Hitchman[68] obtained several nitrito complexes of Ni(II). For example, $Ni(N,N\text{-diethyl-en})(ONO)_2$ exhibits the N=O and N—O stretchings at 1361 and 1163 cm^{-1}, respectively. If this compound is heated at 116°C in vacuo, these bands are replaced by strong bands at 1429 and 1220 cm^{-1} without any change in chemical composition. The observed high frequency shifts may suggest that the nitrito complex is converted into a bridging nitro complex in this case.

(5) Chelating Nitro Group

Goodgame and Hitchman[73] proposed the bidentate, chelating structure for the nitro groups in complexes of the type, $M(NO_2)_2L_2$, where M is Co(II) or Ni(II), and L is an uncharged ligand such as $(C_6H_5)_3PO$. Table III-10 lists the frequencies of the bidentate, chelating nitro groups. It is to be noted that both the antisymmetric and symmetric NO_2 stretching frequencies are lower and the ONO bending frequency is higher in the bidentate, chelating nitro complexes than in the unidentate (N-bonded) nitro complexes.

TABLE III-10. VIBRATIONAL FREQUENCIES OF CHELATING NITRO GROUPS (CM^{-1})[73]

Complex	$\nu_{as}(NO_2)$	$\nu_s(NO_2)$	$\delta(ONO)$
$Co(Ph_3PO)_2(NO_2)_2$	1266	1199, 1176	856
$Co(Ph_3AsO)_2(NO_2)_2$	1314	1193, 1183	847
$Ni(\alpha\text{-picoline})_2(NO_2)_2$	1272	1199	866, 862
$Ni(quinoline)_2(NO_2)_2$	1299, 1289	1203, 1193	867

(6) Complexes Containing Both Nitro and Nitrito Groups

All the six nitro groups of $K_4[Ni(NO_2)_6] \cdot H_2O$ are coordinated to the metal through their nitrogen atoms. However, its anhydrous salt, $K_4[Ni(NO_2)_6]$, exhibits the bands characteristic of the nitro as well as nitrito coordination. From UV spectral evidence, Goodgame and Hitchman[74] suggested the structure, $K_4[Ni(NO_2)_4(ONO)_2]$, for the anhydrous salt. The infrared spectrum of $Cs_3[Ni(NO_2)_5]$ is similar to that of

$$K_4[Ni(NO_2)_4(ONO)_2].$$

The $[Ni(NO_2)_5]^{3-}$ ion may consist of four unidentate (N-bonded) nitro groups and one bidentate, chelating nitro group. A chain polymer structure in which the Ni atom is coordinated by four unidentate (N-bonded) nitro groups and two bridging nitro groups (Ni—O$\overset{N}{\diagup\;\diagdown}$O—Ni type) is also probable. A novel Ni complex, $((CH_3)_4N)[Ni(NO_2)_3]$, exhibits the NO_2 stretchings at 1435 and 1202 cm^{-1}. They suggested that this compound is a polymer containing a three-dimensional net work of the —Ni—N(=O)—O—Ni—

linkages. El-Sayed and Ragsdale[69] suggested that Ni(2-(aminomethyl) pyridine)$_2$(NO_2)$_2$ may contain both nitro and nitrito groups. Table III-11 summarizes the vibrational frequencies of two complexes that contain both groups.

TABLE III-11. VIBRATIONAL FREQUENCIES OF Ni(II) COMPLEXES CONTAINING NITRO AND NITRITO GROUPS (cm^{-1})

Complex	Nitro Group			Nitrito Group		References
	$\nu_{as}(NO_2)$	$\nu_s(NO_2)$	δ(ONO)	ν(N=O)	ν(N—O)	
$K_4[Ni(NO_2)_6]\cdot H_2O$	1346	1319	427	–	–	74
$K_4[Ni(NO_2)_4(ONO)_2]$	1347	1325	423 414	1387	1206	74
$Cs_3[Ni(NO_2)_5]$	1348	1325	433 406	1372*	1208*	74
Ni(2-(aminomethyl)-pyridine)$_2$(NO_2)(ONO)	1338	1318		1368	1251	69

* Chelating or bridging nitro group.

III-3. LATTICE WATER, AQUO AND HYDROXO COMPLEXES

Water in inorganic salts may be classified as lattice water or coordinated water. There is, however, no definite borderline between the two. The former term denotes water molecules trapped in the crystalline lattice, either by weak hydrogen bonds to the anion or by weak ionic bonds to the metal, or by both,

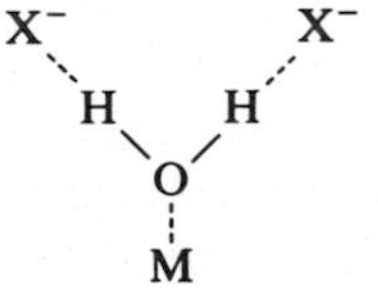

$$
\begin{array}{ccccc}
X^- & & & & X^- \\
 & \ddots & & \cdot^{\cdot^{\cdot}} & \\
 & H & & H & \\
 & & \diagdown\;\diagup & & \\
 & & O & & \\
 & & \vdots & & \\
 & & M & &
\end{array}
$$

whereas the latter denotes water molecules bonded to the metal through partially covalent bonds. Although bond distances and angles obtained from x-ray and neutron diffraction data provide direct information concerning the geometry of the water molecule in the crystal lattice, studies of vibrational spectra are also useful for this purpose.

(1) Lattice water

In general, lattice water absorbs at 3550–3200 cm^{-1} (antisymmetric and symmetric OH stretchings)[75] and at 1630–1600 cm^{-1} (HOH bending). If the spectrum is examined under high resolution, the fine structure of these bands is observed. For example, Hass and Sutherland (Ref. II-510) found eight peaks between 3500 and 3400 cm^{-1} in $CaSO_4 \cdot 2H_2O$. Detailed vibrational studies have been made for a number of inorganic salts having lattice water: $Li_2SO_4 \cdot H_2O$(76,77), $K_2HgCl_4 \cdot H_2O$(77), $LiClO_4 \cdot 3H_2O$(78), $MnCl_2 \cdot 4H_2O$(79), $CaCl_2 \cdot 6H_2O$(80), $MgSO_4 \cdot 7H_2O$(81), and $WO_3 \cdot nH_2O$(82).

In the low frequency region, these crystals exhibit *librational modes* that are due to rotational oscillations of the water molecule restricted by interactions with neighboring atoms. According to van der Elsken and Robinson[83], they appear in the region from 600 to 300 cm^{-1} in the hydrates of alkali and alkaline earth halides.

(2) Aquo (H_2O) Complexes

In addition to the three fundamental modes of the free water molecule, coordinated water is expected to show other modes such as those in Fig. III-5. The rocking, and metal-oxygen stretching modes will become infrared active if the metal-oxygen bond is sufficiently covalent. The presence of these bands in aquo complexes was first suggested by Fujita et al.[84]. Gamo[85] assigned the bands at 880–650 cm^{-1} of inorganic salts to the rocking mode of coordinated water. Sartori and co-workers[86] attempted to calculate the frequencies of these modes using the model shown in Fig. III-5. Nakagawa and Shimanouchi[87] have carried out normal coordinate analyses on the $[M(H_2O)_6]$ ($\mathbf{T}_h$ symmetry) and $[M(H_2O)_4]$ ($\mathbf{D}_{4h}$ symmetry)-type ions to assign these low frequency modes. Table III-12 lists the frequencies and band assignments obtained by these authors. Figure III-8 illustrates the far-infrared spectra of four aquo complexes.

TABLE III-12. OBSERVED FREQUENCIES, BAND ASSIGNMENTS AND M—O STRETCHING FORCE CONSTANTS OF AQUO COMPLEXES[87]

Compound	$\rho_r(H_2O)$	$\rho_w(H_2O)$	ν(M—O)	K(M—O)[a]
$[Cr(H_2O)_6]Cl_3$	800	541	490	1.31
$[Ni(H_2O)_6]SiF_6$	(755)[b]	645	405	0.84
$[Ni(D_2O)_6]SiF_6$	–	450	389	0.84
$[Mn(H_2O)_6]SiF_6$	(655)[c]	560	395	0.80
$[Fe(H_2O)_6]SiF_6$	–	575	389	0.76
$[Cu(H_2O)_4]SO_4 \cdot H_2O$	887, 855	535	440	0.67
$[Zn(H_2O)_6]SO_4 \cdot H_2O$	–	541	364	0.64
$[Zn(D_2O)_6]SO_4 \cdot D_2O$	467	392	358	0.64
$[Mg(H_2O)_6]SO_4 \cdot H_2O$	–	460	310	0.32
$[Mg(D_2O)_6]SO_4 \cdot D_2O$	474	391	–	0.32

[a] UBF field(mdyn/*A*)
[b] $Ni(H_2O)_4Cl_2$
[c] $Mn(H_2O)_4Cl_2$

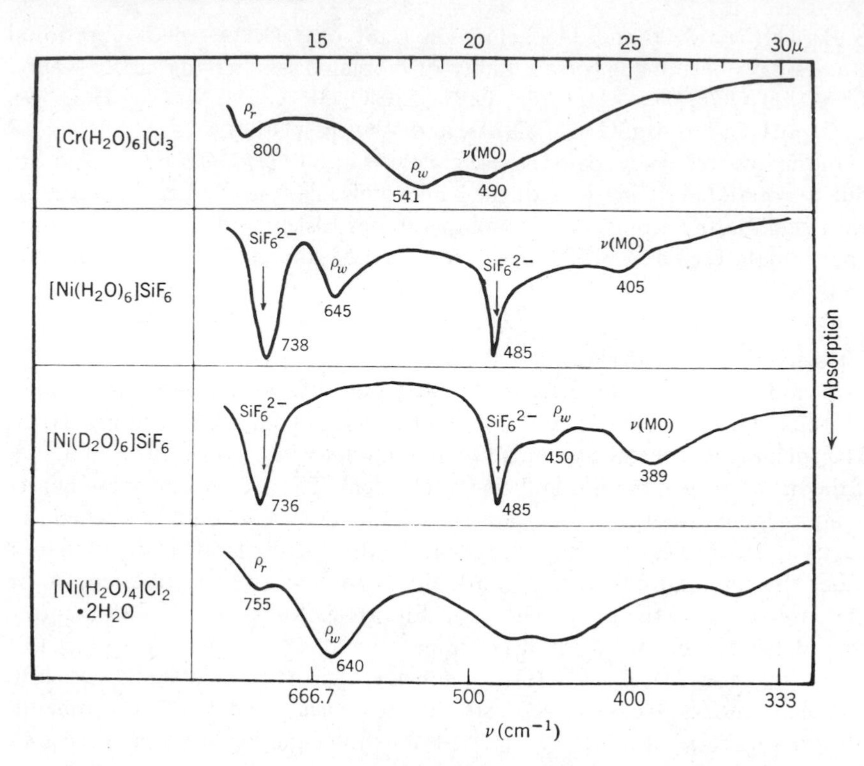

Fig. III-8. Infrared spectra of aquo complexes in the low frequency region.[87]

Concentrated aqueous solutions of $Cu(NO_3)_2$, $CuSO_4$ and $Cu(ClO_4)_2$ exhibit Raman lines at 440 cm^{-1}. Similar Raman lines appear at 400–390 cm^{-1} in Zn(II) salts and at 370–360 cm^{-1} in Mg(II) salts. They are probably due to totally symmetric metal-oxygen stretching modes of the hydrated ions since they are highly polarized[88]. It is possible to observe these low frequency modes by inelastic scattering of low energy neutrons. Boutin and co-workers[89] have carried out an extensive study on lattice water and aquo complexes by using this method. Durig et al.[90] have assigned the low frequency modes of coordinated water in *trans*- and *cis*-$[Ru(H_2O)_3Cl_3]$; the rocking, wagging, and Ru–OH_2 stretching are at 780, 545, and 470–439 cm^{-1} for the *trans*-, and at 870, 578, and 440–432 cm^{-1} for the *cis*-isomer. Far-infrared spectra of Cr(III) aquoammine complexes have been studied by Blyholder and Vergez.[91]

(3) Hydroxo (OH) Complexes

The spectra of hydroxo complexes are expected to be similar to those of the metal hydroxides discussed previously (Sec. II-1). The hydroxo group can be distinguished easily from the aquo group, since the former lacks the HOH bending mode near 1600 cm^{-1}. However, the M—O—H group exhibits the MOH bending mode below 1200 cm^{-1}. For example, Scargill[92] assigned them at 1000–970 cm^{-1} in hydroxo complexes of ruthenium, and Dupuis et al.[93] assigned it at 1150 cm^{-1} for the $[Sn(OH)_6]^{2-}$ ion. Lorenzelli et al.[94] concluded that this ion takes the $\mathbf{S_6}$ symmetry in the solid state.

The OH group also forms a bridge between two metals. Figure III-9 shows the infrared spectrum of a Co(III) complex having bridging OH groups.[95] The band near 1100 cm^{-1} may be assigned to the Co—O—H bending, since it disappears upon deuteration. Ferraro and Walker[96] assigned this mode at 955 cm^{-1} in $[(bipy)Cu(OH)_2Cu(bipy)]\ SO_4 \cdot 5H_2O$. For other hydroxo bridging complexes, see the following references: Cu(II) complexes (96, 97), Cr(III) and Fe(III) complexes (98), Co(III) complexes (99), Pb(II) complexes (100). Hewkin and Griffith[101] have obtained the infrared spectra of metal complexes containing bridges such as O, OH, N, NH, and NH_2.

III-4. CARBONATO, NITRATO, SULFATO, AND OTHER ACIDO COMPLEXES

As has been shown in Sec. III-1, the magnitude of the band shifts caused by coordination can be used as a measure of the strength of the coordinate bond. Where the symmetry of the ligand is lowered by coordination, marked changes in the spectrum are anticipated because of changes in the selection rule. A typical example of this effect is shown by the spectra of two complexes in which the same ligand coordinates to the metal as a unidentate and as a bidentate ligand.

(1) Carbonato (CO_3) Complexes

The carbonate ion coordinates to the metal in one of two ways:

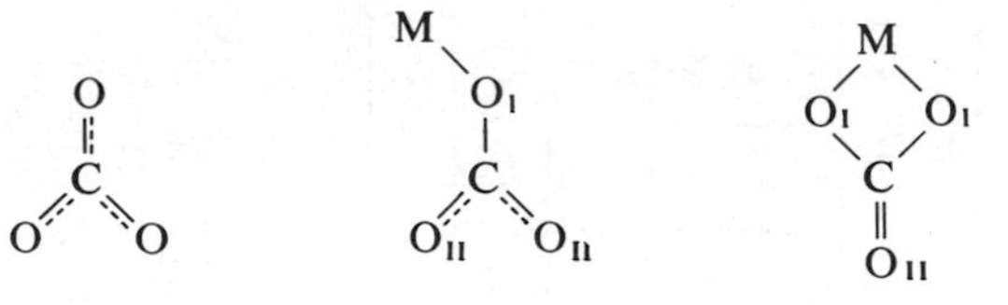

Free ion ($\mathbf{D_{3h}}$) Unidentate ($\mathbf{C_s}$) Bidentate ($\mathbf{C_{2v}}$)

The selection rule changes as shown in Table I-11. In $\mathbf{C_{2v}}$ and $\mathbf{C_s}$*, the ν_1 vibration, which is forbidden in the free ion, becomes infrared active and each of the doubly degenerate vibrations, ν_3 and ν_4, splits into two bands. Although the number of infrared active fundamentals is the same for $\mathbf{C_{2v}}$ and

*Both unidentate and bidentate carbonato groups have the same $\mathbf{C_{2v}}$ symmetry if the metal atom is ignored.

TABLE III-13. CALCULATED AND OBSERVED FREQUENCIES OF UNIDENTATE AND BIDENTATE Co(III) CARBONATO COMPLEXES (cm^{-1})[104]

Species (C_{2v})[a]	$\nu_1(A_1)$	$\nu_2(A_1)$	$\nu_3(A_1)$	$\nu_4(A_1)$	$\nu_5(B_2)$	$\nu_6(B_2)$	$\nu_7(B_2)$	$\nu_8(B_1)$
Calc. Frequency	1376	1069	772	303	1482	676	92	–
Assignment	$\nu(C{-}O_{II}) + \nu(C{-}O_I)$	$\nu(C{-}O_I) + \nu(C{-}O_{II})$	$\delta(O_{II}CO_{II})$	$\nu(Co{-}O_I)$	$\nu(C{-}O_{II})$	$\rho_r(O_{II}CO_{II})$	$\delta(CoO_IC)$	π
$[Co(NH_3)_5CO_3]Br$	1373	1070	756	362	1453	678	–	850
$[Co(ND_3)_5CO_3]Br$	1369	1072	751	351	1471	687	–	854
$[Co(NH_3)_5CO_3]I$	1366	1065	776	360	1449	679	–	850
$[Co(ND_3)_5CO_3]I$	1360	1063	742	341	1467	687	–	853

Species (C_{2v})	$\nu_1(A_1)$	$\nu_2(A_1)$	$\nu_3(A_1)$	$\nu_4(A_1)$	$\nu_5(B_2)$	$\nu_6(B_2)$	$\nu_7(B_2)$	$\nu_8(B_1)$
Calc. Frequency	1595	1038	771	370	1282	669	429	–
Assignment	$\nu(C{-}O_{II})$	$\nu(C{-}O_I)$	Ring def. $+ \nu(Co{-}O_I)$	$\nu(Co{-}O_I)$ + ring def.	$\nu(C{-}O_I) + \delta(O_ICO_{II})$	$\delta(O_ICO_{II}) + \nu(Co{-}O_I) + \nu(Co{-}O_I)$	$\nu(Co{-}O_I)$	π
$[Co(NH_3)_4CO_3]Cl$	1593	1030	760	395	1265	673	430	834
$[Co(ND_3)_4CO_3]Cl$	1645, 1607	(1031)[b]	753	378	1268	672	418	832
$[Co(NH_3)_4CO_3]ClO_4$	1602	–[c]	762	392	1284	672	428	836
$[Co(ND_3)_4CO_3]ClO_4$	1603	–[c]	765	374	1292	676	415	835

[a] Symmetry assuming a linear Co—O—C bond (See Ref. 104)
[b] Overlapped with $\delta_s(ND_3)$.
[c] Hidden by $[ClO_4]^-$ absorption.

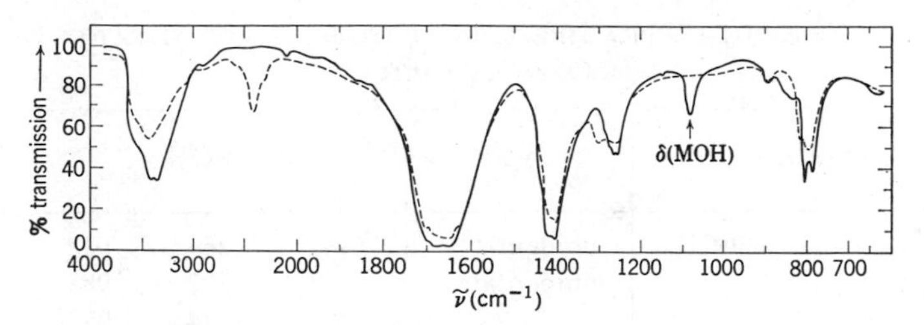

Fig. III-9. Infrared spectra of $K_4[(OX)_2Co(OH)_2Co(OX)_2]$ (solid line) and its deuterated compound (broken line).[95]

$\mathbf{C}_s$, the splitting of the degenerate vibrations is larger in the bidentate than in the unidentate complex.[102] For example, $[Co(NH_3)_5CO_3]Br$ exhibits two CO stretchings at 1453 and 1373 cm^{-1}, whereas $[Co(NH_3)_4CO_3]Cl$ shows them at 1593 and 1265 cm^{-1}. In organic carbonates such as dimethyl carbonate, $(CH_3O_I)_2CO_{II}$, this effect is more striking because the CH_3—O_I bond is strongly covalent. Thus the CO_{II} stretching is observed at 1870 cm^{-1}, whereas the CO_I stretching is at 1260 cm^{-1}. Gatehouse and co-workers[103] have shown that the separation of the CO stretching bands increases along the series,

basic salt $<$ carbonato complex $<$ acid $<$ organic carbonate

Fujita et al.[104] have carried out normal coordinate analyses on unidentate and bidentate carbonato complexes of Co(III). According to their results, the CO stretching force constant, which is 5.46 for the free ion, becomes 6.0 for the CO_{II} bonds and 5.0 for the CO_I bond of the unidentate complex, whereas it becomes 8.5 for the CO_{II} bond and 4.1 for the CO_I bonds of the bidentate complex (all are UBF force constants in units of mdyn/A). The observed and calculated frequencies and theoretical band assignments are shown in Table III-13. Hester and Grossman[105] also carried out normal coordinate analyses on bidentate carbonato and nitrato complexes.

As is shown in Table III-13, normal coordinate analysis predicts that the highest frequency CO stretching band belongs to the B_2 species in the unidentate and the A_1 species in the bidentate complex. Elliott and Hathaway[106] studied the polarized infrared spectra of single crystals of $[Co(NH_3)_4CO_3]Br$, and confirmed these symmetry properties. As will be shown later for nitrato complexes, Raman polarization studies are also useful for this purpose.

(2) Nitrato (NO_3) Complexes

The results obtained for carbonato complexes are also applicable to nitrato complexes. Gatehouse et al.[107] interpreted the infrared spectra of unidentate nitrato complexes based on $\mathbf{C}_{2v}$ symmetry. They show NO stretchings at 1530–1480(B_2), 1290–1250(A_1), and 1035–970(A_1) cm^{-1}. Table III-14 lists the NO stretching frequencies of some unidentate nitrato complexes.

TABLE III-14. NO STRETCHING FREQUENCIES OF NITRATO COMPLEXES (cm^{-1})

Complex	Type	$\nu(NO_3)$			References
$[Co(NH_3)_5(NO_3)]PtCl_4$	unidentate	1481	1269	1012	107
$[Pd(NO_3)_2(bipy)]$	unidentate	1517, 1502	1292, 1274, 1250	989, 979, 975	107
$[Ni(en)_2(NO_3)_2]$	unidentate	1420	1305	818	108
$[Ni(en)_2(NO_3)]ClO_4$	bidentate	1476	1290	809	108
$[Ni(dien)(NO_3)_2]$	unidentate	1440	1315	816	108
	bidentate	1480	1300	808	
$Sn(NO_3)_4 \cdot py$	unidentate	1522	1304	1008	111
$Sn(NO_3)_4$	bidentate	1630	1255	983	111
$UO_2(NO_3)_2 \cdot 2H_2O$	bidentate	1544, 1510	1272	1044, 1021	112
$UO_2(NO_3)_2(py)_2$	bidentate	1500	1282	1028	112

dien : diethylenetriamine

Similar to carbonato complexes, the separation of the first two NO stretching bands increases if the nitrato group acts as a bidentate ligand. For example, $[Ni(en)_2(NO_3)_2]$ (unidentate) exhibits NO stretchings at 1420 and 1305 cm^{-1}, whereas $[Ni(en)_2(NO_3)]ClO_4$ (bidentate) shows them at 1476 and 1290 cm^{-1}.[108] In $[Ni(dien)(NO_3)_2]$, both types of coordination are mixed, since this compound exhibits bands due to unidentate (1440 and 1315 cm^{-1}) and bidentate (1480 and 1300 cm^{-1}) nitrato groups.[108] Lever[109] has concluded that nitrato groups in $[M(A)_2(NO_3)_2]$-type complexes (A, isoquinoline, 2-picoline, etc.; M, Co(II), Ni(II), Cu(II), and Zn(II)) act as bidentate ligands, since they show NO stretching bands at 1517–1484 and 1305–1258 cm^{-1}. Infrared and other studies show that both types of coordination are mixed in $[M(py)_3(NO_3)_2]$ and $[M(py)_3(NO_3)_2] \cdot 3py$ where M is Co(II) or Ni(II).[110]

As stated previously, the highest frequency NO stretching band belongs to the A_1 species in the bidentate and the B_2 species in the unidentate complex. Ferraro et al.[113] have shown that all the nitrato groups in $Th(NO_3)_4 \cdot 2TBP$ (TBP, tributyl phosphate) coordinate to the metal as bidentate ligands, since the Raman line at about 1550 cm^{-1} is polarized (A_1 species).

If the coordinate bond is primarily ionic, it is rather difficult to distinguish unidentate and bidentate coordination from the magnitude of separation of two NO stretching bands. For example, the unidentate nitrato groups in

$Cs_2[U(NO_3)_6]$ give almost the same separation as the bidentate nitrato groups in $Rb[UO_2(NO_3)_3]$; 1531 and 1274 cm^{-1} for the former, and 1536 and 1276 cm^{-1} for the latter. In this case, the metal-oxygen stretching bands may be useful in distinguishing them. According to Topping,[114] the UO stretching bands are at 224 cm^{-1} in the former and at 262 and 223 cm^{-1} in the latter. Ferraro and Walker[115] have shown that anhydrous metal nitrates such as $Cu(NO_3)_2$ and $Co(NO_3)_2$ exhibit strong absorptions between 350 and 250 cm^{-1}. Since ionic nitrates such as KNO_3 and $NaNO_3$ have no absorption in this region, they suggested that these bands are due to metal-oxygen stretching vibrations. Walker and Ferraro[116] have also shown that anhydrous rare earth nitrates such as Pr $(NO_3)_3$ and $Nd(NO_3)_3$ exhibit metal-oxygen stretching bands in the region from 270 to 180 cm^{-1}, and that the nitrato groups in these compounds act as bidentate ligands (chelating or bridging). Ferraro and Walker[117] studied the changes in symmetry of the nitrato groups during the process of dehydration of $UO_2(NO_3)_2 \cdot 6H_2O$ and $Th(NO_3)_4 \cdot 5H_2O$ using infrared spectroscopy. Janz and co-workers[118] studied the Raman spectra of molten nitrates to elucidate the nature of ionic interactions.

(3) Sulfato (SO_4) Complexes

The free sulfate ion belongs to the high symmetry point group $\mathbf{T}_d$. Of the four fundamentals, only ν_3 and ν_4 are infrared active. If the ion is coordinated to a metal, the symmetry is lowered and splitting of the degenerate modes occurs, together with the appearance of new bands in the infrared spectrum corresponding to Raman active bands in the free ion. The lowering of symmetry caused by coordination is different for the unidentate and the bidentate complexes, as shown by the accompanying structures. The change in the

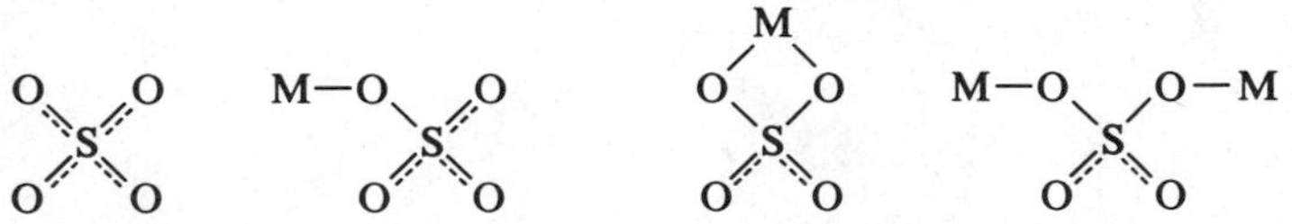

Free ion (T_d) | Unidentate complex (C_{3v}) | Bidentate complex (C_{2v}) | Bridged bidentate complex (C_{2v})

selection rules caused by the lowering of symmetry has already been indicated in Table II-26. Table III-15 and Fig. III-10 give the frequencies and the spectra obtained by Nakamoto et al.[102] for typical sulfato complexes. In $[Co(NH_3)_6]_2(SO_4)_3 \cdot 5H_2O$, ν_3 and ν_4 do not split and ν_2 does not appear; although ν_1 is observed, it is very weak. It is therefore concluded that $\mathbf{T}_d$ symmetry still holds, since the appearance of ν_1 may be attributed to a perturbation caused by the crystal field. In $[Co(NH_3)_5SO_4]Br$, however, both ν_1 and ν_2 appear with medium intensity; moreover, ν_3 and ν_4 each splits into

TABLE III-15. VIBRATIONAL FREQUENCIES OF SULFATO COMPLEXES (CM^{-1})[102]

Compound	Symmetry	ν_1	ν_2	ν_3	ν_4
Free $[SO_4]^{2-}$ ion	T_d	–	–	1104 (vs)[a]	613 (s)
$[Co(NH_3)_6]_2[SO_4]_3 \cdot 5H_2O$	T_d	973 (vw)	–	1130–1140 (vs)	617 (s)
$[Co(NH_3)_5SO_4]Br$	C_{3v}	970 (m)	438 (m)	1032–1044 (s) 1117–1143 (s)	645 (s) 604 (s)
$[(NH_3)_4Co(NH_2)(SO_4)Co(NH_3)_4][NO_3]_3$	C_{2v}	995 (m)	462 (m)	1050–1060 (s) 1170 (s) 1105 (s)	641 (s) 610 (s) 571 (m)

[a] vs, very strong; s, strong; m, medium; w, weak; vw, very weak

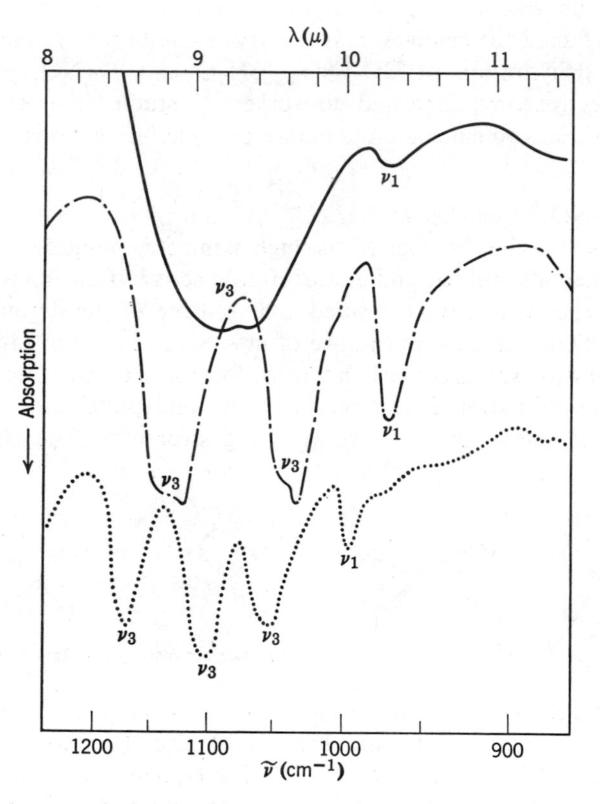

Fig. III-10. Infrared spectra of $[Co(NH_3)_6]_2(SO_4)_3 \cdot 5H_2O$ (solid line); $[Co(NH_3)_5SO_4]Br$ (dot-dash line); $[(NH_3)_4Co(NH_2)(SO_4)Co(NH_3)_4](NO_3)_3$ (dotted line).[102]

two bands. Table II-26 suggests that this result can be explained by assuming lowering of the symmetry from $\mathbf{T}_d$ to $\mathbf{C}_{3v}$. In

$$\left[(NH_3)_4Co\begin{matrix} NH_2 \\ SO_4 \end{matrix}Co(NH_3)_4\right](NO_3)_3$$

both ν_1 and ν_2 appear with medium intensity, and ν_3 and ν_4 each splits into three bands. These results suggest that the symmetry is once again lowered and probably reduced to $\mathbf{C}_{2v}$, as indicated in Table II-26. This conclusion was confirmed by Barraclough and Tobe,[119] who obtained the infrared spectrum of a bidentate sulfato complex, $[Co(en)_2SO_4]Br$. The sulfato group in $[Cr(H_2O)_5(SO_4)]Cl \cdot \frac{1}{2}H_2O$ is a unidentate ligand, since it gives three SO stretchings at 1118, 1068, and 1002 cm^{-1},[120] whereas the sulfato group in $[Cu(bipy)SO_4] \cdot 2H_2O$ is probably a bridging bidentate ligand because it shows four stretchings at 1166, 1096, 1050–1035, and 971 cm^{-1}.[121]

Since both bidentate chelating and bridging sulfato groups belong to $\mathbf{C}_{2v}$, it is not possible to distinguish them from the number of SO stretching bands. It should be noted, however, that the SO stretching frequencies of $[Co(en)_2SO_4]Br$ (chelating)[119] are higher (1211, 1176, 1075, and 993 cm^{-1}) than those of the bridging complex listed in Table III-15. As is seen in Table III-16, the same trend is also seen in the Pd(II) sulfato complexes.[122]

TABLE III-16. SO STRETCHING FREQUENCIES OF BRIDGING AND CHELATING SULFATO COMPLEXES (CM^{-1})[122]

A. Bridging sulfato complexes	
$Pd(NH_3)_2SO_4$	1195, 1110, 1030, 960
$PdSO_4$	1160, 1105, 1035, 996
B. Chelating sulfato complexes	
$[Pd(phen)SO_4]$	1240, 1125, 1040–1015, 955
$[Pd(py)_2(SO_4)] \cdot H_2O$	1235, 1125, 1020, 930
$[Pd(SO_4)(H_2O)_2]$	1230, 1089, 995, 960

The symmetries of the sulfate and nitrate anions of metal salts at various stages of hydration have been discussed from their infrared spectra.[123] Normal coordinate analyses on unidentate and bidentate (bridging) sulfate ions have been carried out by Tanaka et al.[124]

(4) Perchlorato (ClO_4) Complexes

The same symmetry arguments used for sulfato complexes are also applicable to perchlorato complexes. Table III-17 summarizes the results obtained by several workers.

TABLE III-17. ClO STRETCHING FREQUENCIES OF PERCHLORATO COMPLEXES (CM^{-1})

Complex	Structure	ν_3	ν_4	References
$K[ClO_4]$	ionic	1170–1050	(935)[a]	
$Cu(ClO_4)_2{\cdot}6H_2O$	ionic	1160–1085	(947)[a]	125
$Cu(ClO_4)_2{\cdot}2H_2O$	unidentate	{1158, 1030	920	125
$Cu(ClO_4)_2$	bidentate	{1270–1245, 1130, 948–920	1030	125
$Mn(ClO_4)_2{\cdot}2H_2O$	bidentate	{1210, 1138, 945	1030	126
$Co(ClO_4)_2{\cdot}2H_2O$	bidentate	{1208, 1125, 935	1025	126
$[Ni(en)_2(ClO_4)_2]$[b]	bidentate	{1130, 1093, 1058	962	127
$Ni(CH_3CN)_4(ClO_4)_2$	unidentate	{1135, 1012	912	128
$Ni(CH_3CN)_2(ClO_4)_2$	bidentate	{1195, 1106, 1000	920	128
$[Ni(4\text{-}Me\text{-}py)_4](ClO_4)_2$	ionic	1040–1130	(931)[a]	129
$Ni(3\text{-}Br\text{-}py)_4(ClO_4)_2$	unidentate	{1165–1140, 1025	920	129

[a] Weak
[b] Blue form

(5) Complexes of Other Tetrahedral Anions

Siebert[130] reports four PO stretchings (1109, 1043, 918, and 895 cm^{-1}, bidentate coordination) in the infrared spectrum of $[Co(NH_3)_4PO_4] \cdot 2H_2O$. Peters and Fraser[131] concluded that $[Co(NH_3)_5(S_2O_3)]Cl$ is a mixture of the O-bonded (predominant) and S-bonded complexes. They assigned the bands at 1167 and 1137 cm^{-1} to the SO stretchings of the O- and S-bonded complexes, respectively. Costamagna and Levitus,[132] on the other hand, attribute both bands to the S-bonded complex. They also obtained the infrared spectra of other thiosulfato complexes in which the thiosulfato groups act as unidentate (S-bonded) ligands.

(6) Sulfito (SO_3) Complexes

The pyramidal sulfite (SO_3^{2-}) ion may coordinate to a metal as a unidentate, bidentate, or bridging ligand. The following two structures are probable for unidentate coordination.

$$\text{M—S(=O)}_2\text{O} \quad (\mathbf{C}_{3v}) \qquad \text{M—O—S(O)}_2 \quad (\mathbf{C}_s)$$

If coordination occurs through sulfur, the $\mathbf{C}_{3v}$ symmetry of the free ion will be preserved. If coordination occurs through oxygen, the symmetry may be lowered to $\mathbf{C}_s$. In this case, the doubly degenerate vibrations of the free ion will split into two bands. It is anticipated[133] that coordination through sulfur will shift the SO stretching bands to higher frequencies, whereas coordination through oxygen will shift them to lower frequencies relative to those of the free ion. Based on these criteria, Newman and Powell[134] have shown that the sulfito groups in $K_6[Pt(SO_3)_4] \cdot 2H_2O$ and $[Co(NH_3)_5(SO_3)]Cl$ are S-bonded and those in $Tl_2[Cu(SO_3)_2]$ are O-bonded. Baldwin[135] suggested that the sulfito groups in *cis*- and *trans*-$Na[Co(en)_2(SO_3)_2]$ and $[Co(en)_2(SO_3)X]$ (X; Cl or OH) are S-bonded, since they show only two SO stretchings between 1120 and 930 cm^{-1}. Table III-18 shows some results obtained by these investigators.

TABLE III-18. VIBRATIONAL FREQUENCIES OF UNIDENTATE SULFITO COMPLEXES (CM^{-1}).

Complex	Structure	$\nu_3(E)$	$\nu_1(A_1)$	$\nu_2(A_1)$	$\nu_4(E)$	References
Free SO_3^{2-}	–	1010	961	633	496	
$K_6[Pt(SO_3)_4] \cdot 2H_2O$	S-bonded	1082–1057	964	660	540	134
$[Co(NH_3)_5(SO_3)]Cl$	S-bonded	1110	985	633	519	134
Trans-$Na[Co(en)_2(SO_3)_2]$	S-bonded	1068	939	630	–	135
$[Co(en)_2(SO_3)Cl]$	S-bonded	1117–1075	984	625	–	135
$Tl_2[Cu(SO_3)_2]$	O-bonded	902, 862	989	673	506, 460	134

The structure of complexes containing bidentate sulfito groups is rather difficult to deduce from their infrared spectra. Bidentate sulfito groups may be chelating or bridging either through oxygen or sulfur or both, all resulting in $\mathbf{C}_s$ symmetry. Baldwin[135] prepared a series of complexes of the type, $[Co(en)_2(SO_3)]X$ (X; Cl, I, or SCN), which are monomeric in aqueous solution. They show four strong bands in the SO stretching region (one of them may be an overtone or a combination band). She suggests a chelating structure in which two oxygens of the sulfito group coordinate to the Co(III) atom. Newman and Powell[134] obtained the infrared spectra of

$$K_2[Pt(SO_3)_2]\cdot 2H_2O, \quad K_3[Rh(SO_3)_3]\cdot 2H_2O,$$

and other complexes for which bidentate coordination of the sulfito group is expected. It was not possible, however, to determine their structures from infrared spectra alone.

III-5. CYANO AND NITRILE COMPLEXES

(1) Cyano (CN) Complexes

(a) C≡N Stretching Bands

Cyano complexes exhibit sharp C≡N stretching bands at 2200–2000 cm^{-1}. Table III-19 lists the observed C≡N stretching frequencies for a number of cyano complexes. Some complexes exhibit band splitting in the crystalline state. Polynuclear complexes give several bands. Hydrates exhibit spectra different from those of the anhydrates. The C≡N stretching frequencies of cyano complexes are generally higher than that of the free CN^- ion (2080 cm^{-1} for KCN). In terms of the following resonance structures,[142] this result suggests that structure IIA contributes

Free ion: $:\bar{C}{\equiv}N: \longleftrightarrow :C{=}\ddot{N}\!:^{-}$ (IA ⟷ IIA)

Complex: $M{-}C{\equiv}N: \longleftrightarrow \overset{+}{M}{=}C{=}\underset{\cdot\cdot}{\bar{N}}:$ (IB ⟷ IIB)

more to the total structure of the free ion than does structure IIB to the total structure of the complex. In metal cyano complexes, the larger the contribution of IIB, the lower the C≡N stretching frequency.

According to El-Sayed and Sheline,[142] the C≡N stretching frequencies of cyano complexes depend upon: (1) the electronegativity, (2) the oxidation state, and (3) the coordination number of the metal. The effect of electronegativity is seen in the frequency order, $[Ni(CN)_4]^{2-} < [Pd(CN)_4]^{2-} < [Pt(CN)_4]^{2-}$. The electronegativity of Ni(II) is smallest and the contribution of structure IIB is largest. Therefore the C≡N stretching frequency of the Ni(II) complex is the lowest in this series. The effect of oxidation state is seen in the frequency order, $[Ni(CN)_4]^{4-} < [Ni(CN)_4]^{2-}$. Here the larger negative charge of Ni(O) increases the contribution of structure IIB, resulting in a lower frequency shift in the Ni(O) complex. The effect of coordination number[155–157] is seen in the frequency order, $[Ag(CN)_4]^{3-} < [Ag(CN)_3]^{2-} < [Ag(CN)_2]^{-}$. Here an increase in the coordination number results in an

TABLE III-19. C≡N STRETCHING FREQUENCIES IN VARIOUS CYANO COMPLEXES IN THE SOLID STATE (cm^{-1})

Compound	Frequency[a] (cm^{-1})	References
$K_3[Cr(CN)_6]$	2135	136
$K_3[Mn(CN)_6]$	2125	136
$K_4[Mn(CN)_6]$	2060	136
$K_5[Mn(CN)_6]$	2048	136
$K_4[Fe(CN)_6]$	2094, 2073, 2062, 2044, 2031, 2026, 2006	137, 138
	2041 (aqueous)	
$Na_3[Fe(CN)_5NH_3]\cdot 6H_2O$	2036	139
$Na_3[Fe(CN)_5H_2O]\cdot H_2O$	2043	139
$K_3[Fe(CN)_6]$	2125	139
$Na_2[Fe(CN)_5NH_3]\cdot H_2O$	2126	139
$Na_2[Fe(CN)_5H_2O]$	2120	139
$Na_2[Fe(CN)_5NO]\cdot 2H_2O$	2152	139, 140
$K_6[Co_2(CN)_{10}]$	2133, 2090, 2079	141
$K_3[Co(CN)_5H_2O]$	2095	141
$K_3[Co(CN)_6]$	2143, 2129, 2126	141, 137
$K_2[Co(CN)_5H_2O]$	2140	141
$K_3[Co_2(CN)_8]$	2120, 2062	141
$K_2[Ni(CN)_4]\cdot H_2O$	2128	142
$K_4[Ni(CN)_4]$	1985	142, 143
$K_2[Ni(CN)_4]$	2135	144
$K_4[Ni_2(CN)_6]$	2128, 2079, 2055	144
$K_2[Ni(CN)_3NO]$	2133, 2118	141
$K_3[Cu(CN)_4]$	2094, 2081, 2075	145, 147
$K_4[Mo(CN)_8]\cdot 2H_2O$	2119, 2096, 2049	148, 149
$K_4[Mo(CN)_8]$	2128, 2105	148
$K_3[Mo(CN)_8]\cdot 2H_2O$	2119, 2096, 2045	148
$K_3[Mo(CN)_8]$	2105, 2128	148
$K_3[Rh(CN)_6]$	2163	141
$K_3[Rh_2(CN)_8]$	2130, 2070	141
$K[Ag(CN)_2]$	2140	150
$K_2[Pd(CN)_4]\cdot H_2O$	2143	142
$K_3[W(CN)_8]\cdot 2H_2O$	2119, 2088, 2041	148
$K_4[W(CN)_8]$	2110	148
$K_3[W(CN)_8]$	2128, 2105	148
$K_3[Ir(CN)_6]$	2185	141
$K_2[Pt(CN)_4]\cdot 3H_2O$	2150	151
$K[Au(CN)_2]$	2141	153
$K_2[Hg(CN)_4]$	2152	154
$K_2[Hg_2(CN)_6]$	2148, 2158	154

[a] More bands may be observed for some of these compounds under high resolution. For example, Bor[140], using a LiF prism, observed four C≡N stretching bands for $[Fe(CN)_5NO]^{2-}$.

increase in the negative charge on the metal which, in turn, increases the contribution of structure IIB, thus decreasing the frequency (Table III-20).

TABLE III-20. FREQUENCIES AND MOLECULAR EXTINCTION COEFFICIENTS OF CYANO COMPLEXES IN AQUEOUS SOLUTIONS

Ion	Frequency (cm^{-1})	Mol. Extinction Coefficient	References
Free $[CN]^-$ ion	2080 ± 1	29 ± 1	155
$[Ag(CN)_2]^-$	2135 ± 1	264 ± 12	155
$[Ag(CN)_3]^{2-}$	2105 ± 1	379 ± 23	155
$[Ag(CN)_4]^{3-}$	2092 ± 1	556 ± 83	155
$[Cu(CN)_2]^-$	2125 ± 3	165 ± 25	156
$[Cu(CN)_3]^{2-}$	2094 ± 1	1090 ± 10	156
$[Cu(CN)_4]^{3-}$	2076 ± 1	1657 ± 15	156
$[Zn(CN)_4]^{2-}$	2149	113	157
$[Cd(CN)_4]^{2-}$	2140	75	157
$Hg(CN)_2$	2194	3	157
$[Hg(CN)_3]^-$	2161	26	157
$[Hg(CN)_4]^{2-}$	2143	113	157
$[Ni(CN)_4]^{2-}$	2124 ± 1	1068 ± 95	158
$[Ni(CN)_5]^{3-}$	2102 ± 2	1730 ± 230	158

Jones and co-workers[155–157] have made an extensive infrared study of the equilibria of cyano complexes in aqueous solution. (For aqueous infrared spectroscopy, see Sec. III-21.) Figure III-11 shows the infrared spectra of aqueous silver cyano complexes obtained by changing the ratio of Ag^+ to CN^- ions. Table III-20 lists the frequencies and extinction coefficients from which equilibrium constants can be calculated. Chantry and Plane[159] have studied the same equilibria using Raman spectroscopy.

(b) Lower Frequency Bands

In addition to the C≡N stretching bands, the cyano complexes exhibit M—C stretching, and M—C≡N and C—M—C bending bands in the lower frequency region. Figure III-12 shows the infrared spectra of $K_3[Co(CN)_6]$ and $K_2[Pt(CN)_4]\cdot 3H_2O$. Normal coordinate analyses have been carried out on various cyano complexes to assign these low frequency bands. As Table III-21 shows, the results of these calculations indicate that the M—C stretching, M—C≡N bending, and C—M—C bending vibrations appear in the regions 600–350, 500–350, and 130–60 cm^{-1}, respectively. The M—C and C≡N stretching force constants obtained are also given in the table.

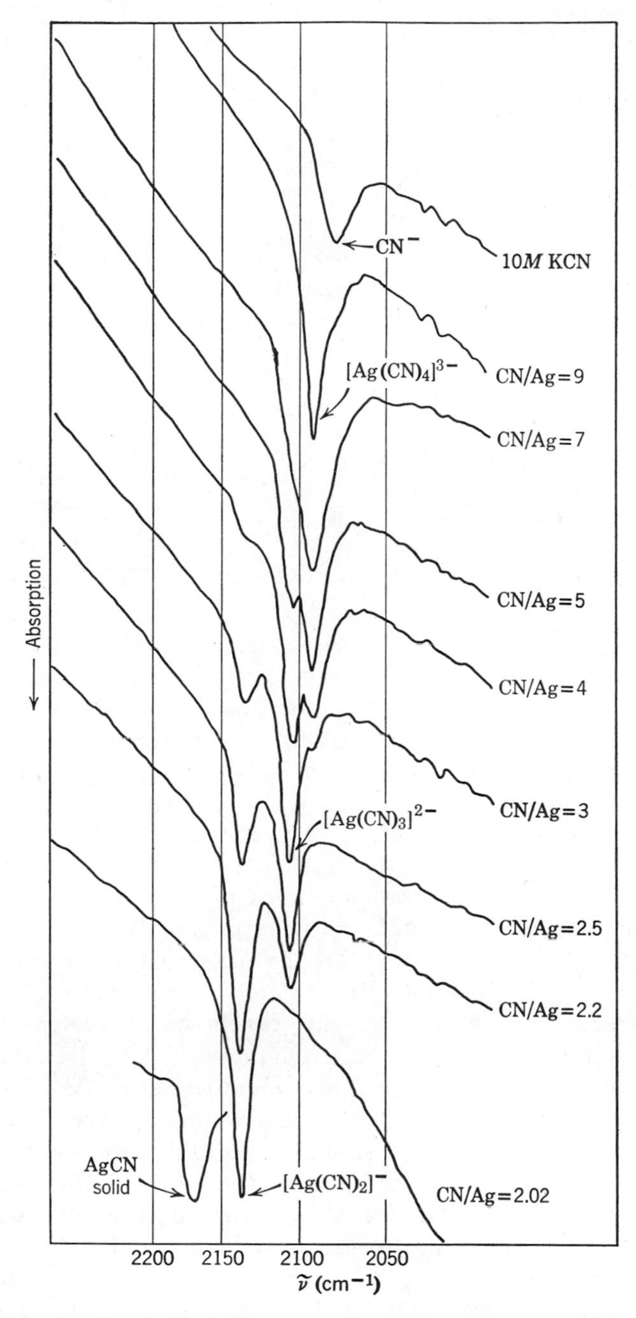

Fig. III-11. Infrared spectra of silver cyano complexes in aqueous solution.[155]

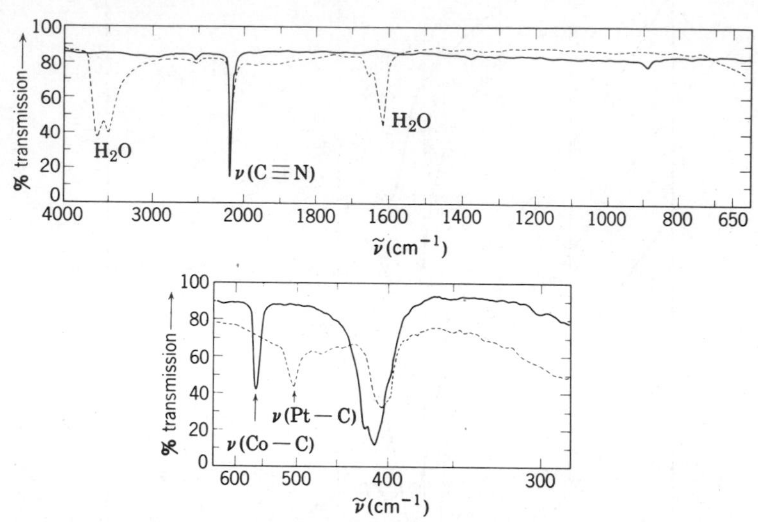

Fig. III-12. Infrared spectra of $K_3[Co(CN)_6]$(solid line) and $K_2[Pt(CN)_4]\cdot 3H_2O$ (broken line).

Nakagawa and Shimanouchi[164] have noted that the M—C stretching force constant increases in the order Fe(III) < Co(III) < Fe(II) < Ru(II) < Os(II) and the C≡N stretching force constant decreases in the same order of metals. An increase in the M—C π-bonding may increase the M—C stretching force constant with a concomitant decrease in the C≡N stretching force constant. Therefore the M—C π-bonding is probably increasing in the above order of metals. On the other hand, the degree of the M—C π-bonding may be proportional to the number of *d*-electrons in the F_{2g} electronic level. According to Jones,[165] the integrated absorption coefficient of the C≡N stretching band (F_{1u}) becomes larger as the number of *d*-electrons in the F_{2g} level increases. Therefore the results shown in Table III-22 suggest that the M—C π-bonding increases Cr(III) < Mn(III) < Fe(III) < Co(III). The order of the M—C stretching frequency also confirms this conclusion. However, the C≡N stretching frequency is relatively constant in the series. Since the masses of the four metals are similar, these results may indicate that the M—C stretching force constant increases in the above series without appreciable changes in the C≡N stretching force constant. Jones[162] noted the same trend in another series in which the M—C stretching force constant increases in the order Co(III) < Rh(III) < Ir(III) but the C≡N stretching force constant stays constant. These results obtained for trivalent metals can be accounted for if we assume that (1) both the M—C π- and σ-bondings are increasing along these series, and (2) a decrease in the C≡N stretching force constant due to π-bonding is largely compensated for by an increase due to σ-bonding. A more rigorous discussion requires a separation of each force constant into σ- and π-components.[166]

TABLE III-21. VIBRATIONAL FREQUENCIES AND BAND ASSIGNMENTS OF HEXACYANO COMPLEXES (cm^{-1})

	$[Cr(CN)_6]^{3-}$	$[Co(CN)_6]^{3-}$	$[Ir(CN)_6]^{3-}$	$[Rh(CN)_6]^{3-}$	$[Co(CN)_6]^{3-}$	$[Fe(CN)_6]^{4-}$	$[Fe(CN)_6]^{3-}$	$[Ru(CN)_6]^{4-}$	$[Os(CN)_6]^{4-}$
A_{1g}, ν(MC)	374	408	469	445	406	(410)	(390)	(460)	(480)
E_g, ν(MC)	336	391	450	435	(375)	(390)	–	(410)	(450)
F_{1g}, δ(MCN)	536	358	415	380	(380)	(350)	–	(340)	(360)
F_{1u}, ν(MC)	457	564	520	520	565	585	511	550	554
δ(MCN)	694	416	398	386	414	414	387	376	392
δ(CMC)	124	92	82	88	–	–	89	–	–
F_{2g}, δ(MCN)	536	480	483	475	–	(420)	–	(400)	(430)
δ(CMC)	106	103	95	94	98	–	99	–	–
F_{2u}, δ(MCN)	496	440	445	–	(395)	–	–	(365)	(390)
δ(CMC)	95	72	69	–	–	–	70	–	–
Force Field	GVF	GVF	GVF	GVF	UBF	UBF	UBF	UBF	UBF
K(M—C), mdyn/A	1.928	2.063	2.704	2.366	2.308	2.428	1.728	2.793	3.343
K(C≡N), mdyn/A	16.422	16.767	16.678	16.831	16.5	15.1	17.0	15.3	14.9
References	160	161, 162, 163	162	162	164	164	164, 163	164	164

TABLE III-22. RELATION BETWEEN INFRARED SPECTRUM AND ELECTRONIC STRUCTURE IN HEXACYANO COMPLEXES[165]

Compound	No. of *d*-electrons in F_{2g} level	ν(CN) in cm^{-1}	ν(MC) in cm^{-1}	Integrated Abs. coefficient in mole^{-1} cm^{-2}
$K_3[Cr(CN)_6]$	3	2128	339	2,100
$K_3[Mn(CN)_6]$	4	2112	361	8,200
$K_3[Fe(CN)_6]$	5	2118	389	12,300
$K_3[Co(CN)_6]$	6	2129	416	18,300

The infrared spectra of solid $K_2[Pt(CN)_4X_2]$($X = Cl^-$, Br^-, and I^-) have been studied by Jones and Smith.[167] Tosi and Danon[168] studied the infrared spectra of the $[Fe(CN)_5X]$-type ions ($X = H_2O$, NH_3, NO_2^-, NO^-, and SO_3^{2-}). The $[Fe(CN)_5(NO)]^{2-}$ ion gives unusually high C≡N stretching frequencies (2148, 2160, and 2170 cm^{-1}) in the series because the M—C π-bonding in this compound is much less than in other compounds due to extensive M—NO π-bonding.

The structure of the $[Mo(CN)_8]^{4-}$ ion has been controversial. Among the four probable structures (Part II-11), a dodecahedron ($\mathbf{D}_{2d}$) or an Archimedean antiprism ($\mathbf{D}_{4h}$) is preferred to others. According to x-ray analysis,[169] the anion in the potassium salt is approximately dodecahedral, the essential symmetry being $\mathbf{C}_s$. However, Stammreich and Sala[170] proposed the $\mathbf{D}_{4d}$ symmetry both in aqueous solution and in the crystalline state. Kettle and Parish,[171] on the other hand, suggest that the infrared spectrum is consistent with the $\mathbf{D}_{2d}$ symmetry in the solid state and that vibrational spectroscopy cannot be used to elucidate the structure in solution. Parish[172] studied the vibrational spectrum of the $[W(CN)_8]^{4-}$ ion in various salts.

Normal coordinate analyses have been made on tetrahedral, square-planar, and linear cyano complexes of various metals. Table III-23 gives the results of these studies. Far-infrared spectra of various cyano complexes have been measured.[177] Jones and co-workers[178] have carried out an extensive study on mixed cyano-halide complexes such as $[Au(CN)_2X_2]^-$ where X is Cl^-, Br^-, or I^-. A UV and IR study[179] shows that the $[Ni(CN)_4]^{2-}$ and $[Ni(CN)_5]^{3-}$ ions are in equilibrium in a solution containing $Na_2[Ni(CN)_4]$, KCN, and KF. The integrated absorption coefficient of the C≡N stretching band increases in the order Hg(II) < Ag(I) < Au(I) in linear dicyano complexes.[165] This may indicate that the M—C π-bonding increases in the above order of metals. From the measurement of infrared dichroism, Jones has determined the orientation of the $[Ag(CN)_2]^-$ and $[Au(CN)_2]^-$ ions in the potassium salts.[150,153] His results are in good agreement with those of x-ray analysis.

TABLE III-23. FREQUENCIES AND BAND ASSIGNMENTS OF THE LOWER FREQUENCY BANDS OF CYANO COMPLEXES (CM^{-1})

Ion	Symmetry	ν(MC)	δ(MCN)	δ(CMC)	Force Const.[a] K(MC)	Force Const.[a] K(CN)	References
$[Cu(CN)_4]^{3-}$	T_d	364(IR)	324(R)	(74)	1.25–	16.10–	145
		288(R)	306(IR)	(63)	1.30	16.31	147
$[Zn(CN)_4]^{2-}$	T_d	359(IR)[b]	315(IR)[b]	71(R)	1.30	17.22	146
		342(R)	230(R)				
$[Cd(CN)_4]^{2-}$	T_d	316(IR)[b]	250(R)[b]	61(R)	1.28	17.13	146
		324(R)	194(R)				
$[Hg(CN)_4]^{2-}$	T_d	330(IR)[b]	235(R)[b]	54(R)	1.53	17.08	146
		335(R)	180(R)				
$[Pt(CN)_4]^{2-}$	D_{4h}	505(IR)	318(R)	95(R)	3.425	16.823	151
		465(R)	300(IR)				173
		455(R)					
$[Ni(CN)_4]^{2-}$	D_{4h}	543(IR)	433(IR)		2.6	16.67	152
		(419)	421(IR)	(54)			
		(405)	488(IR)				
			(325)				
$[Au(CN)_4]^-$	D_{4h}	462(IR)	415(IR)	110(R)	3.28–	17.40–	175
		459(R)			3.42	17.44	
		450(R)					
$[Hg(CN)_2]$	$D_{\infty h}$	442(IR)	341(IR)	(100)	2.607	17.62	174
		412(R)	276(R)				176
$[Ag(CN)_2]^-$	$D_{\infty h}$	390(IR)	(310)	(107)	1.826	17.04	150
		(360)	250(R)				
$[Au(CN)_2]^-$	$D_{\infty h}$	427(IR)	(368)	(100)	2.745	17.17	153
		445(R)	305(R)				

[a] Force constants (mdyn/$Å$) were obtained by using the GVF field for all ions except for the $[Pt(CN)_4]^{2-}$ ion, for which the UBF field was used.
[b] Coupled vibrations between ν(MC) and δ(MCN).

(c) Bridged Cyano Complexes

As is seen in Table III-19, polynuclear cyano complexes absorb in two frequency ranges, one at 2130 and the other at 2090–2050 cm^{-1}. El-Sayed and Sheline[144] proposed the bridged structure I for $K_4[Ni_2(CN)_6]$, whereas

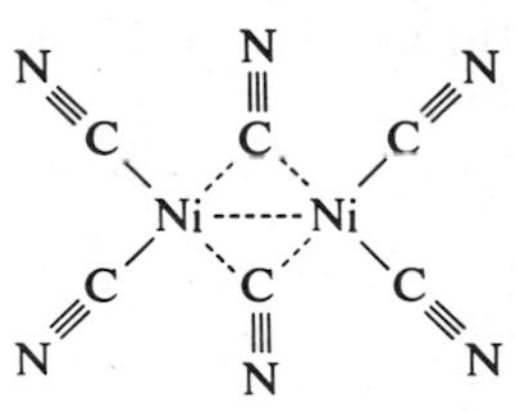

I

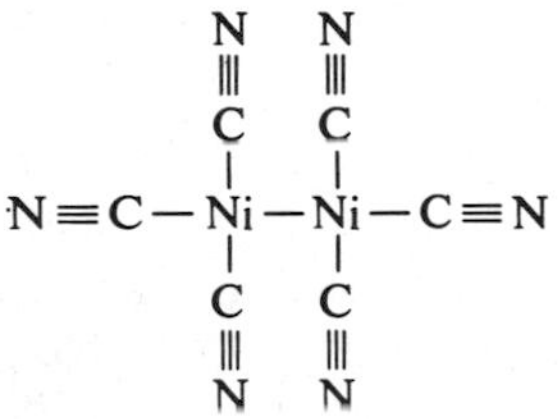

II

Griffith and Wilkinson[141] preferred II which has no bridging cyano groups. Nast and co-workers[180] concluded from infrared spectra that two metal atoms in the $[Co_2(CN)_{10}]^{6-}$ ion are bonded directly as in $[Mn_2(CO)_{10}]$ (see Sec. III-7). Wilmarth and co-workers[181,182] have found that the M—C≡N—M′ bridging cyano groups absorb at higher frequencies than the terminal cyano groups. For example, $[Na_2Co(CN)_5H_2O]_x$ exhibits the bridging and terminal C≡N stretchings at 2202 and 2130 cm^{-1}, respectively. This result is noteworthy, since the bridging group usually absorbs at a lower frequency than the terminal group.

Shriver[183] has noted that the C≡N stretching band at 2130 cm^{-1} of $K_2[Ni(CN)_4]$ is shifted to 2250 cm^{-1} in $K_2[Ni(CN)_4]\cdot 4BF_3$. More generally, the C≡N stretching bands are shifted by 90–120 cm^{-1} to higher frequencies by forming a M—C≡N—BF_3-type bridge. Shriver and co-workers[184] also studied the infrared spectra of cyano complexes containing M—C≡N—M′-type bridges. For $KFeCr(CN)_6$, they have demonstrated that one isomer containing the Fe(II)—C≡N—Cr(III) bridges exhbits the C≡N stretching at 2092 cm^{-1}, whereas the other isomer containing the Cr(III)—C≡N—Fe(II) bridges exhibits them at 2168 and 2114 cm^{-1}.

(2) Nitrile and Isonitrile Complexes

Nitriles (R—C≡N, R=alkyl, or phenyl) form a number of metal complexes by coordination through their nitrogen atoms. Again, the C≡N stretching frequency becomes higher upon coordination. Walton[185] obtained the infrared spectra of $MX_2(RCN)_2$-type compounds (M=Pt(II), Pd(II); $X=Cl^-$, Br^-). The spectra indicate that $PdCl_2(RCN)_2$ is trans-planar, whereas $PtCl_2(RCN)_2$ is *cis*-planar. Evans and Lo[186] suggest that a strong band at 174 cm^{-1} of $ZnCl_2(CH_3CN)_2$ may be due to the Zn—N stretching mode. Kubota and Johnston[187] observed the C≡N stretching bands of mono- and dinitrile complexes of Cu(I) at 2200–2300 cm^{-1}. Reedijk and Groeneveld[188] have carried out an extensive infrared study on acetonitrile complexes. They assigned the M—N stretching bands in the region from 450 to 160 cm^{-1}.

Cotton and Zingales[189] have studied the N≡C stretching bands of isonitrile complexes. When isonitriles are coordinated to zero-valent metals such as Cr(O), back donation of electrons from the metal to the ligand is extensive and the N≡C stretching band is shifted to a lower frequency. For mono- and di-positive metal ions, little or no back donation occurs and the N≡C stretching band is shifted to a higher frequency as a result of the inductive effect of the metal ion. Sacco and Cotton[190] obtained the infrared spectra of $Co(CH_3NC)_4X_2$ and $[Co(CH_3NC)_4][CoX_4]$-type compounds (X=Cl, Br, etc.).

III-6. THIOCYANATO AND OTHER PSEUDOHALOGENO COMPLEXES

The CN^-, OCN^-, SCN^-, $SeCN^-$, CNO^-, and N_3^- ions are called "pseudohalide ions," since they resemble halide ions in their chemical properties. These ions may coordinate to a metal through either one of the end atoms. As a result, the following linkage isomers are possible:

M—CN, cyano complex	M—NC, isocyano complex
M—OCN, cyanato complex	M—NCO, isocyanato complex
M—SCN, thiocyanato complex	M—NCS, isothiocyanato complex
M—SeCN, selenocyanato complex	M—NCSe, isoselenocyanato complex
M—CNO, fulminato complex	M—ONC, isofulminato complex

Two compounds are called true linkage isomers if they have exactly the same composition and different linkages mentioned above. A well-known example is nitro (and nitrito) pentammine Co(III) chloride discussed in Sec. III-2. A pair of true linkage isomers is difficult to obtain since, in general, one form is much more stable than the other. As will be shown later, several pairs of new linkage isomers have been isolated, and infrared spectroscopy has proved to be very useful in distinguishing them.

(1) Thiocyanato (SCN) Complexes

The SCN group may coordinate to a metal through the nitrogen or the sulfur or both (M—NCS—M′). In general, Class A metals (first transition series such as Cr, Mn, Fe, Co, Ni, Cu, and Zn) form the M—N bonds, whereas Class B metals (second half of the second and third transition series such as Rh, Pd, Ag, Cd, Ir, Pt, Au, and Hg) form the M—S Bonds.[191] However, other factors such as the oxidation state of the metal, the nature of other ligands in a complex, and steric consideration also influence the mode of coordination. Mitchell and Williams[192] have shown that the CN stretching frequencies are generally lower in the M—NCS complexes than in the M—SCN complexes. The C—S stretching frequency is more useful in distinguishing these two isomers: 780–860 cm^{-1} for the M—NCS and 690–720 cm^{-1} for the M—SCN group.[193-195] The NCS bending frequency is also different between two isomers; 450–490 cm^{-1} for the M—NCS and 400–440 cm^{-1} (often accompanied by weaker bands at higher frequency side) for the M—SCN group.[194,195] Table III-24 lists some results obtained by Sabatini and Bertini.[195] The M—NCS group is linear or bent, while the M—SCN group is always bent. This seems to suggest that the following resonance structures are predominant in each case:

M—NCS Group

$M-\overset{+}{N}\equiv C-S^{-}$

$\ddot{N}=C=S$ (bonded to M)

M—SCN Group

$\ddot{\underset{\cdot\cdot}{S}}-C\equiv N$ (bonded to M)

Pecile[196] has noted that the integrated intensity of the CN stretching band can also be used to distinguish two linkage isomers.

Clark and Williams[197] have obtained the infrared spectra of tetrahedral $[M(NCS)_2L_2]$ and octahedral $[M(NCS)_2L_4]$-type complexes (M = Fe, Co, Ni, etc.; L = pyridine, etc.) and discussed stereochemistry from infrared spectra. A similar study has been made by Nelson and Shepherd.[198] Bennett et al.[199] have obtained the infrared spectra of $[Cr(NCS)_4L_2]^-$, where L is an amine or phosphine, and have assigned the Cr—N and Cr—L stretching bands in addition to the NCS group vibrations. In $[M(NCS)_2L_2]$-type complexes in which M is Pt(II) or Pd(II), the NCS group is S-bonded if L is an amine and N-bonded if L is a phosphine.[193] This is because a strong π-acceptor such as a tertiary phosphine makes the *d*-orbitals of the metal less available for bonding with the π-orbitals of sulfur. Jennings and Wojcicki[200] obtained the infrared spectra of square-planar $[Rh(CO)(NCS)L_2]$, $[Rh(NCS)L_3]$, $[Rh(CO)_2(NCS)_2]^-$, and $[Rh_2(NCS)_2L_4]$-type complexes in

TABLE III-24. VIBRATIONAL FREQUENCIES OF ISOTHIOCYANATO AND THIOCYANATO COMPLEXES (cm^{-1})[195]

Compound[a]	$\nu(CN)$	$\nu(CS)$[b]	$\delta(NCS)$
K[NCS]	2053	748	486, 471
$(Et_4N)_3[Cr(-NCS)_6]$	2078	–	483
$(Et_4N)_2[Co(-NCS)_4]$	2065	844, 838	481
$(Et_4N)_4[Ni(-NCS)_6]$	2112, 2103	828	470
$(Et_4N)_2[Zn(-NCS)_4]$	2072	837	482
$(Et_4N)_2[Pd(-SCN)_4]$	2112, 2109	698, 694	465, 433, 429, 418
$(Et_4N)[Au(-SCN)_4]$	2127	695	454, 415
$(Et_4N)_2[Pt(-SCN)_6]$	2120	692	461, 457, 418, 415

[a] Et, C_2H_5.

[b] The S-bonded complex may exhibit a band at 880–810 cm^{-1} which is the first overtone of $\delta(NCS)$. Care must be taken, therefore, in distinguishing this band from $\nu(CS)$.[205]

which L is a phosphine, an arsine, a stibine, or a phosphite. Infrared spectra show that all these Rh(I) complexes are N-bonded. The *cis* and *trans* isomers of Co(III)-NCS complexes can be distinguished from infrared spectra.[201,202]

As stated before, the distinction of true linkage isomers by infrared spectroscopy has been made for $[Co(NH_3)_5(-NO_2$ or $-ONO)]Cl_2$[10] and $[KFeCr(CN)_6]$[184]. Basolo et al.[203] succeeded in isolating true linkage isomers of

$$trans\text{-}[Pd(As(Ph)_3)_2(NCS)_2]:$$

The bright-yellow, N-bonded complex absorbs at 2089 (CN stretching) and 854 cm^{-1} (CS stretching); the yellow-orange, S-bonded complex absorbs at 2119 cm^{-1} (CN stretching), although the phenyl absorption near 700 cm^{-1} obscures its C—S stretching band. Infrared spectra of linkage isomers have also been reported for $[Pd(bipy)(NCS/SCN)_2]$[204,205] and $[Pd(As(n\text{-}C_4H_9)_3)_2(NCS/SCN)_2]$.[205] A complex of the composition $[Cu(tren)(NCS)_2]$ (tren, triaminotriethylamine) exhibits two CN stretching (2094 and 2060 cm^{-1}) and two CS stretching (818 and 745 cm^{-1}) bands. Based on this and other data, Raymond and Basolo[206] suggested that this complex may contain both Cu—NCS and Cu—SCN linkages. Later, Jain and Lingafelter[207] have shown from x-ray analysis that one NCS group is bonded to the metal through nitrogen and the other group is not coordinated with it. Bertini and Sabatini[208] suggest that

$$[Pd(4,4'\text{-dimethyl-bipy})(NCS)_2]$$

may contain both types of coordination, since it exhibits two CN stretchings at 2120 and 2090 cm^{-1}.

The NCS group also forms a bridge between two metal atoms. The CN stretching frequency of a bridging group is generally higher than that of a terminal group. For example, $HgCo(NCS)_4$(Co—NCS—Hg) absorbs at 2137 cm^{-1}, whereas $(Et_4N)_2$ $[Co(—NCS)_4]$ absorbs at 2065 cm^{-1}. According to Chatt and Duncanson,[209] the CN stretching frequencies of Pt(II) complexes are 2182–2150 cm^{-1} for the bridging and 2120–2100 cm^{-1} for the terminal NCS group. $[(P(n\text{-}Pr)_3)_2Pt_2(SCN)_2Cl_2]$ (compound I) exhibits one bridging CN stretching, whereas $[(P(n\text{-}Pr)_3)_2Pt_2(SCN)_4]$ (compound II) exhibits both bridging and terminal CN stretching bands. Thus the infrared spectra suggest the structure of each compound to be:

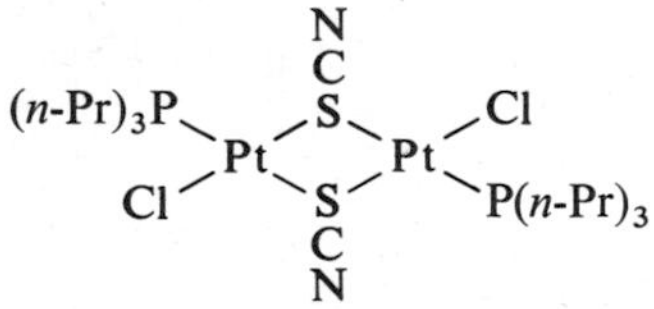

```
                 N                                  N
                 C                                  C
(n-Pr)3P        S        Cl          (n-Pr)3P       S        SCN
        \     /   \     /                    \    /   \    /
          Pt        Pt                         Pt       Pt
        /     \   /     \                    /    \   /    \
     Cl           S        P(n-Pr)3      NCS        S        P(n-Pr)3
                 C                                  C
                 N                                  N
                 I                                  II
```

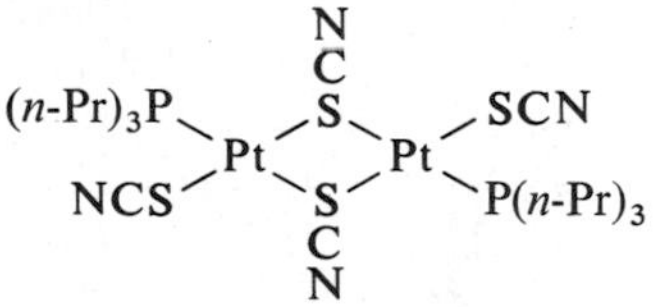

(*n*-Pr, *n*-Propyl)

Compound I, however, exists as two isomers, α and β, which absorb at 2162 and 2169 cm^{-1}, respectively. Chatt and Duncanson[209] originally suggested a geometrical isomerism in which two SCN groups were in a *cis* or *trans* position with respect to the central ring. Later,[210–212] "bridge isomerism" such as shown below was demonstrated by x-ray analysis.

$$\begin{array}{ccc} (n\text{-Pr})_3P & & SCN & & Cl \\ & Pt & & Pt & \\ Cl & & NCS & & P(n\text{-Pr})_3 \end{array} \qquad \begin{array}{ccc} (n\text{-Pr})_3P & & NCS & & Cl \\ & Pt & & Pt & \\ Cl & & SCN & & P(n\text{-Pr})_3 \end{array}$$

α $\qquad\qquad$ β

(2) Selenocyanato (SeCN) Complexes

The SeCN group also coordinates to a metal through the nitrogen (M—NCSe) or the selenium (M—SeCN) or both (M—NCSe—M′). Again, Class A metals tend to form the M—N bonds, while Class B metals prefer the M—Se bonds. The CN stretching frequencies of the Se-bonded complexes are, in general, higher than those of the N-bonded complexes. The distinction of these two types can be made more easily from the C—Se stretching frequency; 690—620 cm^{-1} for the N-bonded and 540–510 cm^{-1} for the Se-bonded complexes.[213] The CN stretching frequencies of N-bonded (terminal) Co(II) complexes are below 2080 cm^{-1}, but that of a bridging complex, $HgCo(NCSe)_4$, (Co—NCSe—Hg) is 2146 cm^{-1}.[214] Table III-25 lists the observed frequencies of N- and Se-bonded complexes.

TABLE III-25. VIBRATIONAL FREQUENCIES OF ISOSELENOCYANATO AND SELENOCYANATO COMPLEXES (cm^{-1})

Compound	ν(CN)	ν(CSe)	δ(NCSe)	References
K[NCSe]	2070	558	424, 416	128 (Part II)
$R_4[(Mn(—NCSe)_6]$	2079, 2082 2070	640 617	424	215
$R_2[Fe(—NCSe)_4]$	2067, 2055	673, 666	432	215
$R_4[Ni(—NCSe)_6]$	2118, 2102	625	430	215
$[Ni(pn)_2(—NCSe)_2]$	2096, 2083	692	–	216
$R'_2[Co(—NCSe)_4]$	2053	672	433, 417	217
$[Co(NH_3)_5(—NCSe)](NO_3)_2$	2116	624	–	218
$R_2[Zn(—NCSe)_4]$	2087	661	429	215
$[Cu(pn)_2(—SeCN)_2]$	2053, 2028	–	–	216
$R_3[Rh(—SeCN)_6]$	2104, 2071	515	–	215
$R''_2[Pd(—SeCN)_4]$	2114, 2105	521	410, 374	219
$R_2[Pt(—SeCN)_4]$	2105, 2060	516	–	215
$[Pt(bipy)(—SeCN)_2]$	2135, 2125	532, 527	–	218
$K_2[Pt(—SeCN)_6]$	2130	519	390, 379 367	217

R, $[N(n\text{-}C_4H_9)_4]^+$; R′, $[N(C_2H_5)_4]^+$; R″, $[N(CH_3)_4]^+$
pn, propylenediamine; bipy, 2,2′-bipyridine

The integrated intensity of the CN stretching band is also useful in distinguishing the N- and Se-bonded complexes.[215] Burmeister and Gysling[220] prepared a series of complexes of the type $[Pd(SeCN)_2L_2]$ to study the effect of L on the Pd-SeCN bond. They also prepared a pair of true linkage isomers for $[Pd(Et_4dien)$*(—SeCN/—NCSe)](BPh_4).[221] The CN and CSe stretching frequencies of the Se-bonded complex are 2123 and 532 cm^{-1}, respectively, whereas those of the N-bonded complex are 2090 and 618 cm^{-1}.

(3) Cyanato (OCN) Complexes

The OCN group may coordinate to a metal through the oxygen (M—OCN) or the nitrogen (M—NCO), or both. So far, only the N-bonded complexes have been reported. Table III-26 lists the observed frequencies of the N-bonded complexes.

TABLE III-26. VIBRATIONAL FREQUENCIES OF ISOCYANATO COMPLEXES (CM^{-1})

Compound	ν_a(NCO)	ν_s(NCO)	δ(NCO)	References
K[NCO]	2165	1207	637, 628	129, 130 (Part II)
$Si(NCO)_4$	2284	1482	608, 546	146 (Part II)
$Ge(NCO)_4$	2247	1426	608, 528	146 (Part II)
$[Zn(NCO)_4]^{2-}$	2208	1326	624	223
$[Mn(NCO)_4]^{2-}$	2222	1335	623	224
$[Fe(NCO)_4]^{2-}$	2182	1337	619	224
$[Co(NCO)_4]^{2-}$	2217, 2179	1325	620, 617	224
$[Ni(NCO)_4]^{2-}$	2237, 2186	1330	619, 617	224
$[Fe(NCO)_4]^{-}$	2208, 2171	1370	626, 619	224
$[Pd(NCO)_4]^{2-}$	2200–2190	1319	613, 604, 594	222
$[Sn(NCO)_6]^{2-}$	2270, 2188	1307	667, 622	222

The M—NCO bond is linear or almost linear in these complexes. Forster and Horrocks[223] carried out a normal coordinate analysis on the $[Zn(NCX)_4]^{2-}$-type anions where X is O, S, or Se. The Zn—N stretching force constants were 1.80, 1.68, and 1.50 mdyn/A for X = O, S, and Se, respectively.

* Et_4dien: N,N,N′,N′-tetraethyldiethylenetriamine.

(4) Azido (N_3) Complexes

Beck et al.[225] have carried out an extensive infrared study on azido complexes. Table III-27 lists the observed frequencies of some azido complexes. Forster and Horrocks[226] have made complete band assignments on the $[Co(N_3)_4]^{2-}$ and $[Zn(N_3)_4]^{2-}$ ($\mathbf{D}_{2d}$ symmetry) and $[Sn(N_3)_6]^{2-}$ ($\mathbf{D}_{3d}$ symmetry) ions. The spectra suggest that the M—NNN bonds in these anions are not linear.

TABLE III-27. VIBRATIONAL FREQUENCIES OF AZIDO COMPLEXES (cm^{-1})

Compound	ν_a(NNN)	ν_s(NNN)	δ(NNN)	ν(M—N)	References
$K[N_3]$	2041	1344	645	–	71, 72 (Part II)
$R_2[Pt(N_3)_4]$	2075, 2060 2024, 2029	1276	582	394	225
$R[Au(N_3)_4]$	2030, 2034	1261 1251	578	432	225
$R''_2[Zn(N_3)_4]$	2097, 2058	1330 1282	–	–	225
$R_2[VO(N_3)_4]$	2088, 2051 2092, 2060 2005	1340	652	442 405	225
$R_2[Pb(N_3)_6]$	2045, 2056 2037	1262 1253	640 597	327 313	225
$R_2[Pt(N_3)_6]$	2022, 2028	1275 1262 1253	578	402 397 320	225
$R'_2[Co(N_3)_4]$	2089, 2050	1338 1280	642 610	368	226
$R'_2[Zn(N_3)_4]$	2098, 2055	1342 1290	649 615	351 295	226
$R'_2[Sn(N_3)_4]$	2115, 2080	1340	659 601	390 330	226

R, $[As(Ph)_4]^+$; R′, $[N(C_2H_5)_4]^+$; R″, $[P(Ph)_4]^+$.

III-7. CARBONYL, NITROSYL, AND HYDRIDO COMPLEXES

(1) Carbonyl (CO) Complexes

Since metal carbonyls exhibit a variety of structures, their elucidation by means of vibrational spectra has been a subject of considerable interest. The method of determining molecular structure using group theory has already been explained in Sec. I-10.

As a matter of convenience, metal carbonyls may be classified into the following four groups.

(a) Mononuclear Carbonyls

Since the structure of a mononuclear carbonyl is relatively simple, it can be determined from its vibrational spectrum without much difficulty. Normal coordinate analyses have been carried out by a number of investigators, and complete band assignments are available for most of the compounds. Table III-28 lists the symmetry, the observed CO and MC stretching frequencies and

TABLE III-28. STRUCTURES AND VIBRATIONAL FREQUENCIES OF MONONUCLEAR METAL CARBONYLS (cm^{-1})

Compound	Symmetry	IR or Raman	Obs. $\nu(CO)$	Obs. $\nu(MC)$	K(M—C) mdyn/A	References
BH_3CO	C_{3v}	IR	2165	691	2.629(GVF)	227–230
$Ni(CO)_4$	T_d	IR	2057	422	2.12(GVF)	231–235
		R	2121 2039	422 381		236, 237
$[Co(CO)_4]^-$	T_d	R	1918 1883	532 439	3.55(GVF)	238–240
$[Fe(CO)_4]^{2-}$	T_d	R	1788[b]	550 464	4.06(GVF)	238, 240
$Fe(CO)_5$	D_{3h}	IR	2028 1994	472 430	3.09– 3.27(GVF)	241
		R	2114 2031 1984	492 414 377	2.47(GVF)	242, 243 244–246
$Fe(CO)_5$	C_{4v}	IR	2027 2045	377 364	1.625(GVF)	248
$[Mn(CO)_5]^-$	D_{3h}	IR	1898 1863			240
$Cr(CO)_6$	O_h	IR	2000	441	2.034(πIVF)[a]	249–150 251–254, 160
		R	2118 2026	390 363		
$Mo(CO)_6$	O_h	IR	2004	368	1.806(πIVF)[a]	249, 250 251–254
		R	2124 2027	392 344		
$W(CO)_6$	O_h	IR	1998	374	2.148(πIVF)[a]	249, 250 253
		R	2124 2019	420 363		

[a] πIVF, π interaction valence force field.[256]

[b] Two bands are too close to be observed separately (Ref. 238).

the MC stretching force constants. For a tetrahedral $M(CO)_4$ molecule, group theory predicts only one infrared active (F_2) and two Raman active CO stretching vibrations (A_1 and F_2). As Table III-28 shows, the number of bands observed agrees with this prediction. The CO stretching frequency decreases remarkably in going from $Ni(CO)_4$ to $[Fe(CO)_4]^{2-}$. This may be due to an increase in the back donation of electrons from the metal to the CO group in the same order of metals. As a result, the MC stretching frequency and the corresponding force constant increase in the same order.

Figure III-13 shows the structures of various metal carbonyls. If the $Fe(CO)_5$ molecule is trigonal bipyramidal (structure I), group theory predicts two infrared active (A_2'' and E') and three Raman active ($2A_1'$ and E') fundamentals for both the CO and FeC stretching vibrations. If the molecule is tetragonal pyramidal (structure II), it must exhibit three infrared active ($2A_1$ and E) and four Raman active ($2A_1$, B_1, and E) CO stretching bands. Although O'Dwyer[248] has proposed structure II, the infrared and Raman spectra fit structure I better. The $\mathbf{D}_{3h}$ structure of $Fe(CO)_5$ was also proved by x-ray analysis.[255]

For an octahedral $M(CO)_6$ molecule, the theory predicts one infrared active (F_{1u}) and two Raman active (A_{1g} and E_g) fundamentals, both for the CO and MC stretching modes. Since only these fundamentals are observed, $Cr(CO)_6$, $Mo(CO)_6$, and $W(CO)_6$ must be octahedral. It is interesting to note that the order of the MC stretching force constant is $W > Cr > Mo$ in these hexacarbonyls. Jones[256] has proposed the "π interaction valence force (πIVF) field" on the assumption that the interaction force constants in metal carbonyls arise mainly from interatomic interactions of the π-electrons and π-orbitals.

Solvent effects on infrared spectra of metal carbonyls have been studied.[257,258] Edgell and co-workers[259] have attributed the band at 407 cm^{-1} of $Li^+[Co(CO)_4]^-$ in a THF solution to ion pair formation or formation of higher aggregates involving both ions. The corresponding bands are observed at 190 and 150 cm^{-1} for the sodium and potassium salts, respectively.

(b) Hydrocarbonyls

The infrared spectrum of $HCo(CO)_4$ exhibits one CO stretching band at 2049 cm^{-1} with a shoulder at 2066 cm^{-1}, in agreement with structure IV (Fig. III-13). However, the OH stretching band expected from structure IV was not observed. Thus Friedel and co-workers[239,260] concluded that the hydrogen may be close enough to one or more of the oxygens to give two infrared active CO stretching modes, but not bound strongly enough to produce an observable OH stretching band (an intermediate structure between III and IV). Edgell and colleagues[261] proposed structure V, in which the

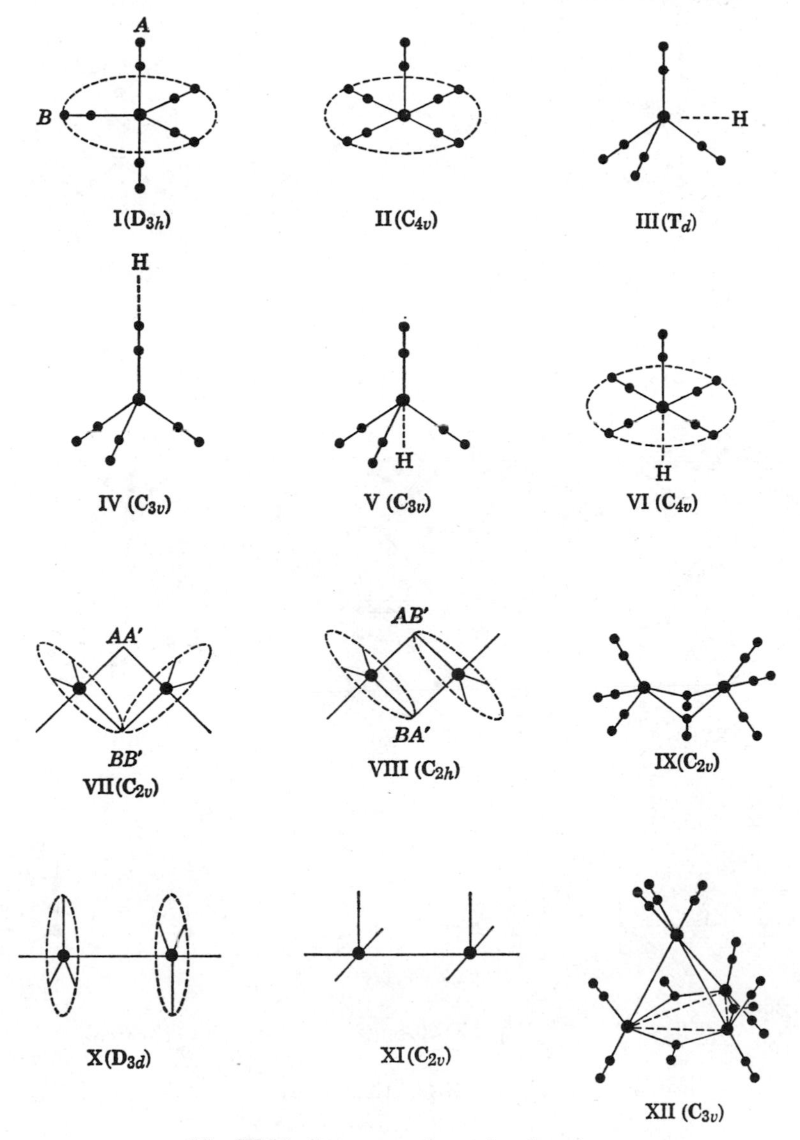

Fig. III-13. Structures of metal carbonyls.

hydrogen atom is on the three-fold axis and about 2A away from the Co atom. Cotton[262] has shown, however, that the maximum overlap between the 1*s* orbital of hydrogen and the 2*pπ* orbitals of carbon and oxygen occurs at a distance of about 1.2A from the Co atom. Later, Farrar et al.[263] estimated the Co—H distance to be 1.2±0.1A from its broad line PMR spectrum.

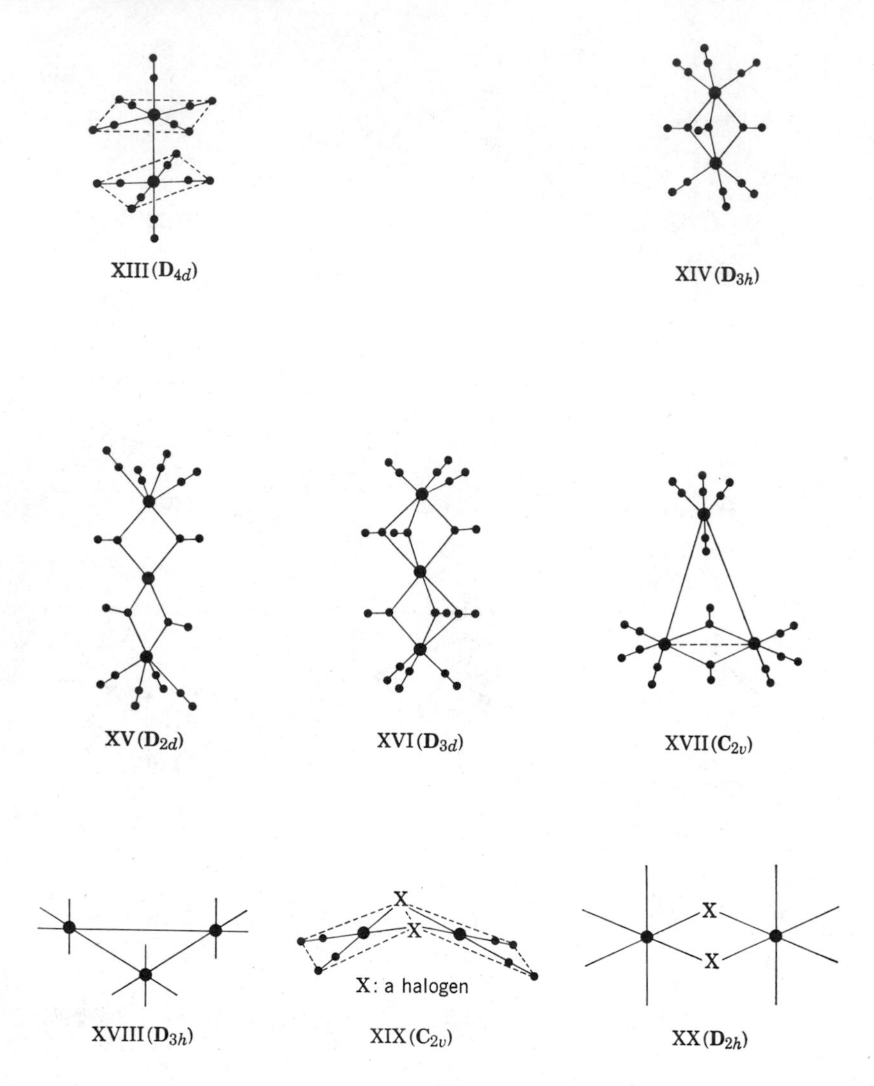

Fig. III-13. (Continued).

Although the band at 704 cm^{-1} has been suggested as a Co—C stretching mode,[239] Edgell and co-workers concluded that this band is associated with a motion involving the hydrogen atom. Their conclusion has been confirmed by Cotton and Wilkinson.[264] Edgell and Summitt[265] assigned the band at 1934 cm^{-1} to the Co—H stretching (corresponding force constant, ca. 2.22 mdyn/A), and the band at 704 cm^{-1} to the Co—H bending mode. Zingales et al.[266] report the Co—H stretching frequency of $[(Ph)_2P(CH_2)_2P(Ph)_2]_2CoH$ to be 1884 cm^{-1}.

A structure similar to $HCo(CO)_4$ may be expected for $[HFe(CO)_4]^-$, the Raman spectrum of which was obtained by Stammreich et al.[238] From PMR studies, Bishop et al.[267] have estimated the Fe—H distance in $H_2Fe(CO)_4$ to be about 1.1A.

The infrared spectrum of $HMn(CO)_5$ has been studied by Wilson,[268] who concluded that a structure of $\mathbf{C}_{4v}$ symmetry (structure VI of Fig. III-13) can be ruled out. Cotton and co-workers[269] reached the same conclusion, but none of these investigators could determine the structure. However, LaPlaca et al.[270] have shown from x-ray analysis that the $Mn(CO)_5$ portion of the molecule has $\mathbf{C}_{4v}$ symmetry with the carbon atoms at five of the six vertices of an octahedron. Huggins and Kaesz[271] reinterpreted the infrared spectrum based on this structure. Braterman et al.[272] have carried out a more detailed study on the infrared spectra of $HMn(CO)_5$ and $DMn(CO)_5$. Farrar et al.[273] estimated the Mn—H distance to be 1.28 ± 0.01A from its broad line PMR spectrum.

The infrared spectra of $HRe(CO)_5$ and $DRe(CO)_5$ have been obtained by Beck and co-workers[274] who assigned the bands at 1832 and 1318 cm^{-1} to the Re—H and Re—D stretching modes, respectively. However, no definite structures were deduced from their results. Smith et al.[275,276] studied the infrared and Raman spectra of trimeric $[ReH(CO)_4]_3$ and suggested that the band near 1100 cm^{-1} may be due to the Re—H stretching mode. The unusually low Re—H stretching frequency may indicate that the hydrogens act as bridging groups between two Re atoms. Hayter[277] studied the infrared spectra of the $[MM'H(CO)_{10}]^-$-type anions (M, M′: Cr, Mo, W) in which M and M′ are presumably bonded through a hydrogen bridge. There are many other hydrocarbonyls for which the CO and MH stretching frequencies have been reported. Table III-29 lists some of these frequencies.

TABLE III-29. VIBRATIONAL FREQUENCIES OF METAL HYDROCARBONYL COMPOUNDS (CM^{-1})*

Compound	ν(CO)	ν(MH)	δ(MH)	References
$RhH(CO)(PPh_3)_3$	1926	2004	784	278
$IrH(CO)(PPh_3)_3$	1930	2068	822	278
$IrHCl_2(CO)(PEt_2Ph)_2$	2101	2008	–	279
$IrHBr_2(CO)(PEt_2Ph)_2$	2035	2232	–	279
$IrHCl_2(CO)(PPh_3)_2$	2027	2240	–	280
$OsHCl(CO)(PPh_3)_3$	1912	2097	–	280

* For the configuration of these molecules, see original references.

(c) Polynuclear Carbonyls

In general, the terminal CO group absorbs at 2100–2000 cm^{-1}, whereas the bridging CO group absorbs at 1900–1800 cm^{-1}. Using an NaCl prism, Cable et al.[281] observed three terminal (2070, 2043, and 2025 cm^{-1}) and one bridging (1858 cm^{-1}) CO stretching bands for $Co_2(CO)_8$. Among several possible structures, structures VII and VIII of Fig. III-13 were considered to be more probable than others. According to group theory, structure VII should show five terminal and two bridging CO stretching bands, while structure VIII should show three terminal and one bridging CO stretching bands in the infrared spectrum. Since the spectrum obtained by Cable and co-workers[281] and Friedel et al.[239,282] agrees with the prediction based on structure VIII, it was concluded that VIII is more probable than VII. However, Cotton and Monchamp[283] and Bor and Markó,[284] using a CaF_2 and a LiF prism, found five terminal (2075, 2064, 2047, 2035, and 2028 cm^{-1}) and two bridging (1867 and 1859 cm^{-1}) CO stretching bands (Fig. III-14). This result agrees rather well with structure VII. It was found, however, that $Co_2(CO)_8$ assumed structure IX in the crystalline state (x-ray analysis).[285] Since this structure belongs to $\mathbf{C}_{2v}$, it also accounts for the infrared spectrum observed by Cotton and Monchamp.[283] According to Noack[286] and Bor,[287] this compound consists of two isomers in solution: one isomer has structure IX, and the other has structure X according to Noack and structure XI according to Bor.[288]

The infrared spectrum of $Co_4(CO)_{12}$ was first obtained by Friedel et al.[239] who found two terminal (2058 and 2030 cm^{-1}) and one bridging (1873 cm^{-1})

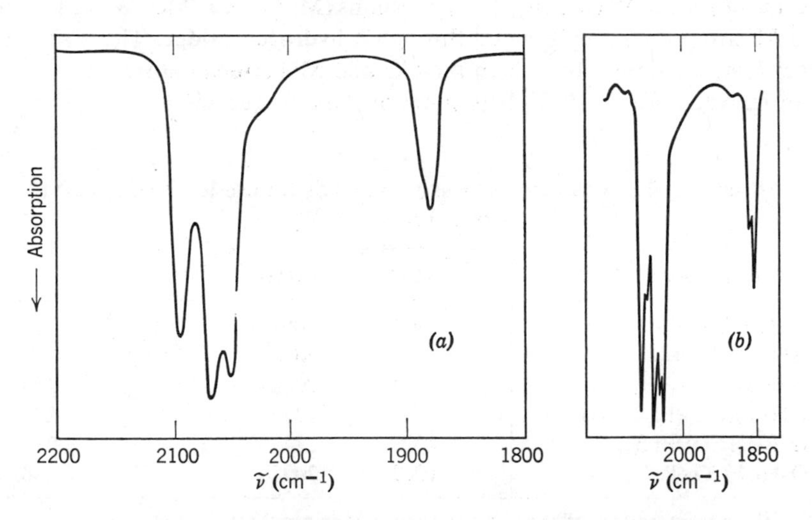

Fig. III-14. Infrared spectra of $Co_2(CO)_8$ in CO stretching region: (*a*) NaCl prism (hexane solution); (*b*) CaF_2 prism (pentane solution).

CO stretching bands. Using a CaF_2 prism, however, Cotton and Monchamp[283] have found four terminal (2070, 2062, 2045, and 2033 cm^{-1}) and one bridging (1869 cm^{-1}) CO stretching bands. The structure of this compound may be too complex to allow its determination from vibrational spectra alone. Corradini[289] deduced structure XII (Fig. III-13) from x-ray analysis. According to group theory, this structure should exhibit six terminal and two bridging CO stretching bands. Then the observed spectrum[283] is simpler than this prediction. A careful study by Bor[287] revealed, however, that the observed spectrum is in perfect agreement with the prediction from Corradini's structure.

The infrared spectra of $Mn_2(CO)_{10}$ and $Re_2(CO)_{10}$ were first reported by Brimm et al.[290] Since no bridging CO stretching bands were found, they suggested that dimerization occurs through formation of an M—M bond. Cotton and co-workers[291] also studied the infrared spectra of these compounds and found three CO stretching bands between 2070 and 1985 cm^{-1}. They concluded that structures in which two $M(CO)_5$ pentagonal pyramids are joined by an M—M bond can be ruled out because they are expected to show only two infrared active CO stretching bands. Later, Dahl et al.,[292] using x-ray analysis, proved that the compound actually has structure XIII.

The infrared spectrum of $Fe_2(CO)_9$ was first obtained by Sheline and Pitzer,[246] who found two terminal (2080 and 2034 cm^{-1}) and one bridging (1828 cm^{-1}) CO stretching bands. This result agrees with that predicted for structure XIV, obtained by x-ray analysis.[293] The infrared spectrum of $Fe_3(CO)_{12}$ was first reported by Sheline,[294] who observed two terminal (2043 and 2020 cm^{-1}) and one bridging (1833 cm^{-1}) CO stretching bands. On the basis of this result, Sheline proposed structure XV (4, 2, 2, 4 structure). Cotton and Wilkinson,[295] however, observed three terminal (2043, 2020, and 1997 cm^{-1}) and two bridging (1858 and 1826 cm^{-1}) CO stretching bands. Mills[296] rejected structure XV, which lacks a center of symmetry, from his x-ray study. Dahl and Rundle,[297,298] however, stated that the symmetry argument fails due to a disorder in the crystal lattice; they presented evidence that the three Fe atoms are located at the corners of an essentially equilateral triangle. On the other hand, Brown[299] suggested structure XVI (3, 3, 3, 3 structure) to account for its magnetic property. This structure was also supported by its Mossbauer[300] and infrared spectra.[301] Finally, Dahl and Blount[302] proposed structure XVII which accounts for x-ray, Mossbauer,[303] and solid-state infrared spectra.

According to x-ray analysis,[304] $Os_3(CO)_{12}$, takes structure XVIII in the solid state. Huggins et al.[305] interpreted its infrared spectrum based on this structure. Stammreich and co-workers[306] have found from Raman spectra that $M[Co(CO)_4]_2$ (M = Cd or Hg) has a linear OC—Co—M—Co—CO bond and takes a staggered conformation of $\mathbf{D}_{3d}$ symmetry. Flitcroft et al.[307] compared the infrared spectrum of $(OC)_5Mn—Re(CO)_5$ with those of $Mn_2(CO)_{10}$ and $Re_2(CO)_{10}$.

(d) Substituted Metal Carbonyls

Orgel[308] has discussed the CO stretching frequencies of $M(CO)_nL_{6-n}$-type compounds where L is an axially symmetric ligand such as Cl and CF_3. From an approximate normal coordinate analysis, Cotton and Kraihanzel[309] derived simple secular equations to calculate the CO stretching frequencies of the $M(CO)_nL_{6-n}$-type compounds. The force constants used in these approximate calculations were compared with those obtained from more rigorous calculations.[310]

Irving and Magnusson[311] have reported the CO stretetching frequencies of a number of Pt(II) carbonyl complexes such as $Pt(CO)_2X_2$, $Pt_2(CO)_2X_4$ and $[Pt(CO)X_3]^-$ (X = Cl, Br, or I). The CO stretching frequency of the chloro complex was the highest and that of the iodo complex was the lowest in all series of halogeno compounds. El-Sayed and Kaesz[312] studied the CO stretching bands of the $[M(CO)_4X]_2$-type compounds (M = Mn, Tc, or Re, and X = Cl, Br, or I), and proposed the halogen-bridged structure (structure XX of Fig. III-13). Garland and Wilt[313] have shown that the infrared spectra of $Rh_2(CO)_4X_2$ (X = Cl or Br) can be interpreted on the basis of structure XIX (Fig. III-13), which was found from x-ray analysis.[314] The infrared spectra of the $M(CO)_5X$-type compounds (M = Mn or Re and X = Cl, Br, or I) have been studied extensively.[315–317] and the relative intensity of CO stretching bands in these compounds has been discussed.[318,319] Hieber and Beck[320] obtained the infrared spectra of the $Fe_2(CO)_6X_2$-type compounds, where X is S, Se, SC_2H_5, or SeC_2H_5. Apparently the two Fe atoms are bridged by two S or Se atoms in these compounds.

It should be noted that the terminal CO stretching frequency can be as low as 1700 cm^{-1}. For example, Abel and co-workers[321] have found that the CO stretching frequencies of the $M(CO)_3L_3$-type compounds (M = Cr, Mo, or W; L = pyridine, PPh_3, $AsPh_3$, etc.) are in the region from 2100 to 1700 cm^{-1}. Only the references are cited for other substituted metal carbonyls: arsines (Fe, Ni, W),[322–325] nitriles (Cr, Mo, W),[323,326,327] isonitriles (Fe),[328] $Ni(CO)_3(PPh_3)$,[329] $Co(CO)_3(NO)$,[330,331] $RCo(CO)_4$ (R = H, CH_3, etc.),[332,333] $Fe(CO)_x(PF_3)_{5-x}$,[334] $M(CO)_3XL_2$ (M = Mn or Re, L = PPh_3, $AsPh_3$, etc., and X = Cl, Br, or I).[335]

Winkhaus and Wilkinson[336] have obtained the infrared spectra of binuclear olefin-substituted Co carbonyl compounds of the type $Co_2(CO)_6L$ and $Co_2(CO)_4L_2$ where L is a diene. Appearance of bridging CO stretching bands suggests that the two bridging CO groups originally present in $Co_2(CO)_8$ are retained in these compounds.

(2) Nitrosyl (NO) Complexes

As stated in Sec. II-1, the nitrosonium ion ($[NO]^+$) absorbs at about 2220 cm^{-1}, whereas nitric oxide (NO) absorbs at 1876 cm^{-1}. The NO group in metal nitrosyl complexes may be cationic ([NO)]$^+$), anionic ([NO]$^-$), or nearly neutral, depending on the nature and valence state of metal, and on the ligands present in the complex. From infrared data on a wide range of nitrosyl complexes, Lewis et al.[337,338] have suggested that the cationic stretching frequency falls in the range between 1940 and 1575 cm^{-1}, whereas the anionic stretching frequency is in the range between 1200 and 1040 cm^{-1}. This criterion may need revision, since no monomeric nitrosyl complexes that exhibit NO stretchings in the latter region are known. Gans[339] suggests the 1700-1500 cm^{-1} region for the anionic NO group. For example, [NO]$^-$ is present in the ions such as $[V(NO)(CN)_5]^{5-}$ (1575 cm^{-1}), $[Cr(NO)(CN)_5]^{4-}$ (1515 cm^{-1}), and $[Mn(NO)(CN)_5]^{3-}$ (1700 cm^{-1}).[337,338]

It has been known for many years that the complexes of the composition, $[Co(NH_3)_5NO]X_2$, exist in two isomeric forms: the black salt ($X = Cl^-$) and the red salt ($X = Br^-$ or NO_3^-). The structures of these two isomers have been a subject of controversy. Griffith et al.[340] proposed a dimeric structure having an N—N bond for the black salt and a monomeric structure for the red salt. They have assigned the bands near 1100 cm^{-1} of both salts to the anionic NO stretching modes. Later, it was shown by x-ray analysis[341,342] that the black salt, $[Co(NH_3)_5NO]Cl_2$, is rather monomeric. This complex shows the NO stretching at 1610 cm^{-1}.[343,344] Although the red salt is dimeric,[345] its detailed structure is not known yet. Mercer et al.[344] have assigned the bands at 1030 and 920 cm^{-1} of the red salt to the antisymmetric and symmetric NO stretching modes, respectively.

The infrared spectra of nitrosyl complexes of the types, $[MX_5NO]^{n-}$ (M = Ru, Os, Ir; $X = Cl^-, Br^-, I^-$) and $[M(CN)_5NO]^{n-}$ (M = Cr, Mn, Fe) have been studied by Gans et al.[346] and Durig et al.[347] The NO stretching and M—N stretching bands were assigned at 1950–1600 and 660–520 cm^{-1}, respectively, in these complexes. Mercer et al.[348] noted that, in a series of complexes of the type, *trans*-$[Ru(NH_3)_4(NO)X]^{n+}$, the Ru—N stretching band is shifted to a lower frequency as X is changed in the order, OH > Cl > NH_3 > Br > I. By comparing the infrared spectra of $K_2[Ru(NO)X_5]$ ($X = Cl^-$ or Br^-) with its N^{15} analog, Miki et al.[349] have shown that the Ru atom in this complex is bonded to the nitrogen atom of the NO group. They assigned two bands between 610 and 550 cm^{-1} to the Ru—N stretching (higher frequency) and Ru—NO bending (lower frequency). Jahn[350] and Beck and Lottes[351] studied the infrared spectra of nitrosyl complexes of Co, Fe, Ni, and Mn and dimeric complexes of the type:[350,352]

```
ON\   /X\   /L
   M     M
L/   \X/   \NO
```

M = Co or Ni
X = a halogen
L = NO, a phosphine or an arsine

Waddington and Klanberg[353] have measured the NO stretching frequencies of nitrosyl compounds such as $NOBF_3Cl$(2335 cm^{-1}), $NOSbF_5Cl$(2300 cm^{-1}), $NOSbCl_4$(1900 cm^{-1}), and $NOAsCl_4$(1860 cm^{-1}). The first two frequencies may represent the free $[NO]^+$ ion, whereas the latter two are close to those of coordinated cationic $[NO]^+$.

(3) Hydrido Complexes

The hydrogen atom directly bonded to a metal generally exhibits a strong and sharp metal-hydrogen stretching band between 2250 and 1700 cm^{-1}.[354] Chatt et al.[355] studied the infrared spectra of a number of complexes of the type, *trans*-$[PtHXL_2]$, where L is a phosphine or an arsine and X is a unidentate anion, and found that the Pt—H stretching frequency decreases with increasing *trans*-effect of X. For example, the order of the Pt—H stretching frequency in *trans* $[PtHX(PEt_3)_2]$ is

$$X{=}NO_3^- < Cl^- < Br^- < I^- < NO_2^- < SCN^- < CN^-$$

$$\nu(Pt{-}H) \quad 2242 > 2183 > 2178 > 2156 > 2150 > 2112 > 2041 \ cm^{-1}$$

Bailar and Itatani[356] report that the Pt—H stretching frequencies of $[Pt(H)(Cl)(PPh_3)_2]$ are 2220 cm^{-1} for the *trans* isomer and 2260 and 2225 cm^{-1} for the *cis* isomer in the solid state. Kruck[357] has found the M—H stretching bands of the $HM(PF_3)_4$-type compounds (M = Co, Rh, or Ir) at 2080–1970 cm^{-1}. Table III-30 lists the M—H stretching and bending frequencies of other hydrido complexes.

TABLE III-30. VIBRATIONAL FREQUENCIES OF METAL HYDRIDO COMPLEXES (cm^{-1})

Compound	ν(MH)	δ(MH)	References
trans-$[Fe(H)(Cl)(C_2H_4(PEt_2)_2)_2]$	1849	656	358
trans-$[Fe(H)_2(o\text{-}C_6H_4(PEt_2)_2)_2]$	1726	716	358
$[Ir(H)(Cl)_2(PPh_3)_3]$, isomer I	2197	840, 804	359
isomer II	2049	835, 820	
isomer III	2243	806	
trans-$[Ru(H)(Cl)(C_2H_4(PEt_2)_2)_2]$	1938	–	360
trans-$[Os(H)_2(C_2H_4(PEt_2)_2)_2]$	1721	–	360

III-8. COMPLEXES OF MOLECULAR NITROGEN AND OTHER GASES

Allen and Senoff[361] first prepared a series of molecular nitrogen complexes of the type, $[Ru(N_2)(NH_3)_5]X_2$, where X is Br^-, I^-, BF_4^-, or PF_6^-. They were prepared by reacting hydrazine hydrate with Ru(III) or Ru(IV) salts or by adding the azide ion to the $[Ru(NH_3)_5H_2O]^{3+}$ ion. These compounds exhibit a strong and sharp NN stretching band between 2170 and 2115 cm^{-1}. Figure III-15 shows the infrared spectrum of $[Ru(N_2)(NH_3)_5]Br_2$ obtained by Allen et al.[362] The NN stretching frequency (Raman active) of free molecular nitrogen is 2331 cm^{-1}. Thus, coordination to the Ru atom lowered the NN stretching frequency by about 200 cm^{-1}. The Ru—N—N bond is linear

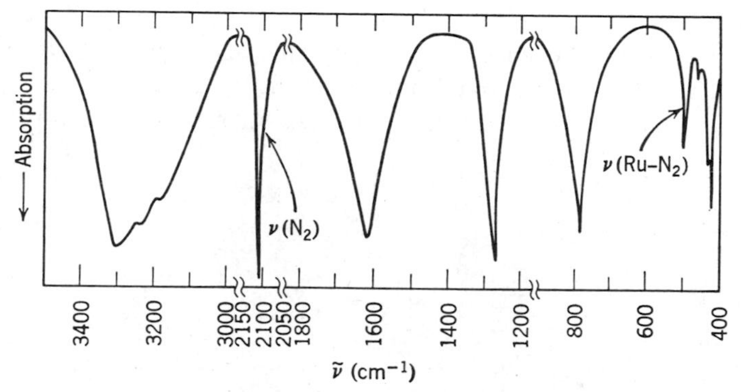

Fig. III-15. Infrared spectrum of $[Ru(NH_3)_5N_2]Br_2$.[362]

according to x-ray analysis of the chloride.[363] This structure may suggest that the Ru—N_2 bonding is similar to the metal-CO bonding. Since N_2 is a weaker Lewis base than CO, the $d\pi$—$p\pi$ back-bonding may be more important in nitrogen complexes than in metal carbonyls.[362] The Ru—N_2 stretching bands of these compounds were assigned at 510–470 cm^{-1}.

Collman and Kang[364] prepared $Ir(N_2)Cl(PPh_3)_2$, which exhibits the NN stretching band at 2095 cm^{-1}. They tentatively proposed a structure in which the N_2 molecule is coordinated to the metal perpendicularly. Yamamoto et al.[365] prepared $Co(N_2)(PPh_3)_3$, which exhibits the NN stretching band at 2088 cm^{-1}. Although the complexes of the type $Co(N_2)HL_3$ ($L = PPh_3$ or PPh_2Et) exhibit the NN stretchings between 2084 and 2080 cm^{-1}, they do not show Co—H stretching bands between 2000 and 1900 cm^{-1}.[366] Enemark et al.[367] have found from x-ray analysis that the Co—N—N linkage in $Co(N_2)H(PPh_3)_3$ is linear, and suggested that the bands at 2105 and 2085 cm^{-1} may be due to the NN and Co—H stretchings, respectively. The Co—H stretching frequency of this compound is much higher than that of $CoH(CO)_4$(1934 cm^{-1}). This may be due to the fact that the Co—H bond in the former is *trans* to N_2, which exerts more $d\pi$—$p\pi$ back-bonding than CO.

In 1963, Vaska[368] found that complexes of the type $Ir(CO)X(PPh_3)_2$ (X, a halogen) react with a variety of gases such as H_2, O_2, X_2, HX, CO, and SO_2 to form octahedral complexes:

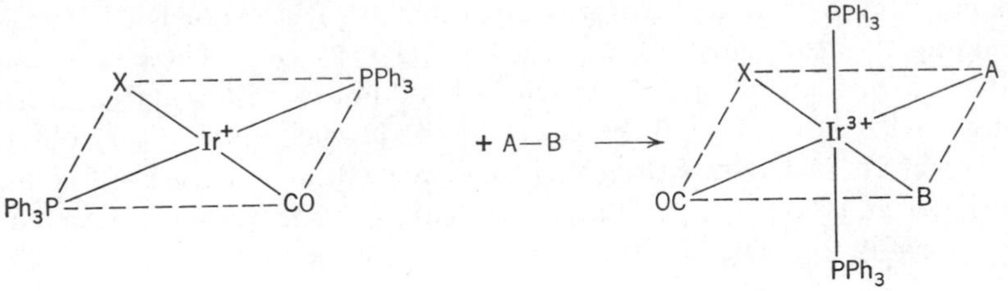

La Placa and Ibers[369] carried out an x-ray analysis on Vaska's oxygen complex, $Ir(CO)Cl(PPh_3)_2O_2$, and found that the Ir atom may be described either five- or six-coordinated with a trigonal-bipyramidal structure:

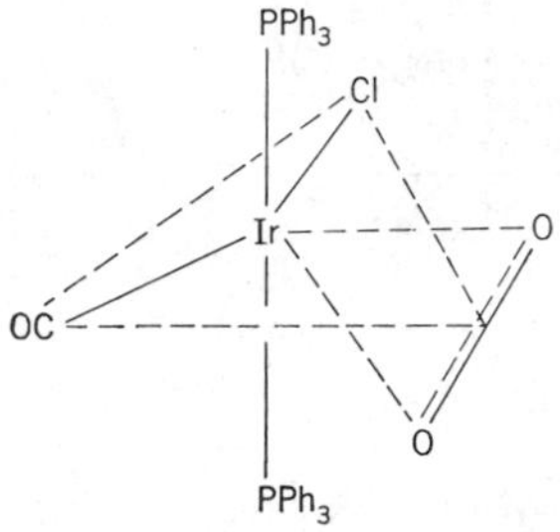

The O—O distance of this compound (1.30A) is between those of free O_2 molecule (1.21A) and the $O_2{}^{2-}$ ion (1.49A) and close to that of the O_2^- ion (1.28A). The O—O stretching frequency of this compound is reported to be 857 cm^{-1},[370] which is much lower than that of free O_2 molecule (1555 cm^{-1} for $^{16}O_2$). Vaska and Catone[371] reports the O—O stretching frequency of $[Ir(O_2)(Ph_2P(CH_2)_2PPh_2)_2]$ Cl at 845 cm^{-1}.

Hirota and Yamamoto[372] obtained the infrared spectra of $Ni(O_2)$ $(t\text{-}BuNC)_2$ and its Pd analog, which contain 25.5% O_2^{18}. The 898 cm^{-1} band of the Ni complex splits into three bands (898, 873, and 848 cm^{-1}) upon introducing O_2^{18}. If these three bands are assigned to the O—O stretchings of the O^{16}—O^{16}, O^{16}—O^{18}, and O^{18}—O^{18} species, respectively, the intensity ratio is in good agreement with that predicted from the isotopic composition of the oxygen gas used. Although the bands near 520 cm^{-1} are also sensitive to isotopic substitution, the nature of these bands is not clear at present.

Vaska and Bath[373] isolated the SO_2 complex of the type $M(SO_2)Cl(CO)$ $(PPh_3)_2$ where M is Ir or Rh. They exhibit the antisymmetric and symmetric SO_2 stretchings at 1220–1185 and 1057–1048 cm^{-1}, respectively. These frequencies are lower than those of free SO_2 gas (1362 and 1151 cm^{-1}).

Different from the gases discussed above, molecular hydrogen, when coordinated to a metal, seems to convert into atomic hydrogens to give metal hydrido complexes. Thus complexes containing H_2 exhibit strong and sharp M—H stretching bands between 2250 and 1700 cm^{-1}, but do not show H—H stretching bands in the high frequency region. For example, Chock and Halpern[370] observed two Ir—H stretchings at 2190 and 2100 cm^{-1} for $Ir(H_2)(CO)(Cl)(PPh_3)_2$. Vaska and Catone[371] suggested a *cis*-dihydrido *trans*-diphosphine structure from the IR and NMR spectra of this complex. The Ir—H stretching frequencies of $[Ir(H_2)(Ph_2P(CH_2)_2PPh_2)_2]BPh_4$ are reported to be 2091 and 2080 cm^{-1}.[371] Vaska[374] reports the Ir—H stretching frequencies of a number of other Ir complexes. The Ir—H stretching bands of $Ir(H_2)(CO)(PPh)_3$ at 2160–2107 cm^{-1} are shifted to the region of 1620–1548 cm^{-1} in its deutero analog.[375] Misono et al.[376] assigned the band at 1755 cm^{-1} of $Co(H_2)(PPh_3)_3$ to the Co—H stretching, and Sacco and Rossi[366] assigned two bands at 1933 and 1745 cm^{-1} of $CoH_3(PPh_3)_3$ to the Co—H stretchings.

III-9. COMPLEXES OF PHOSPHINES AND ARSINES

Ligands such as phosphines (PR_3) and arsines (AsR_3) (R = alkyl, aryl, or halogen) form metal complexes with Group VIII and its neighboring elements. Among them, triphenylphosphine (PPh_3) is most common. These triphenyl compounds exhibit a number of phenyl group absorptions in the high frequency region. In addition, they exhibit phenyl ring deformations, M′—Ph stretching and Ph—M′—Ph bending vibrations (M′ = P or As) below 550 cm^{-1}.[377] Therefore, the far-infrared spectra of metal complexes containing these ligands must be interpreted with caution.

Table III-31 summarizes the M—P stretching frequencies of PPh_3 complexes reported thus far. It is seen that these frequencies scatter over a wide frequency region from 460 to 170 cm^{-1}. Several investigators[382,383] have suggested that this scattering of frequencies may be due to the differences in the nature of the metals (oxidation state, electronic configuration, etc.) and in the structure of the complex (stereochemistry, coordination number, etc.). Complexes of Pt and Pd give exceptionally high M—P stretching frequencies, and this has been attributed to the ability of these metals to form strong π-back bonds with PPh_3.[382]

The infrared spectra of triethylphosphine and its metal complexes have been studied by Green.[384] He noted that the P—C stretching bands of triethylphosphine are shifted by about 25 cm^{-1} to higher frequencies upon complex formation. The M—P stretching frequencies of $P(C_2H_5)_3$ complexes are summarized in Table III-32a.

TABLE III-31. M—P STRETCHING FREQUENCIES OF TRIPHENYLPHOSPHINE COMPLEXES (cm^{-1})

Metal	Compound	ν(M—P)	References
Pt(O)	$Pt(PPh_3)_3$	422	378
Pt(II)	$Pt_2Cl_4(PPh_3)_2$	452, 435, 398	379
Pd(II)	$Pd_2Cl_4(PPh_3)_2$	450, 428	379
Ni(O)	$Ni(CO)_3(PPh_3)$	192	380
Ni(II)	$NiCl_2(PPh_3)_2$	189, 166	381
Co(II)	$CoCl_2(PPh_3)_2$	187, 151	381
Zn(II)	$ZnCl_2(PPh_3)_2$	166, 140	381, 382
Cd(II)	$CdCl_2(PPh_3)_2$	136	382
Hg(II)	$HgCl_2(PPh_3)_2$	137, 108	382
Cr(O), Mo(O), W(O)	$M(CO)_x(PPh_3)_{6-x}$, $x = 3$–5	230–173	383

TABLE III-32a. M—P STRETCHING FREQUENCIES OF TRIETHYLPHOSPHINE COMPLEXES (cm^{-1})

Metal	Compound	ν(M—P)	References
Pt(II)	*trans*-$[PtCl_2(PEt_3)_2]$	415	385, 386
	cis-$[PtCl_2(PEt_3)_2]$	442, 427	385, 386
Pd(II)	*trans*-$[PdCl_2(PEt_3)_2]$	410	385
Ni(II)	*trans*-$[NiCl_2(PEt_3)_2]$	415	385
Au(I)	$[AuX_3(PEt_3)]^{2-}$(X: a halogen)	388–347	387
Cr(III)	$[Cr(NCS)_4(PEt_3)_2]^-$	304	388

In these previous investigations, the metal-ligand vibrations have been assigned by using either one or a combination of the following methods: (1) A comparison of infrared spectra between a free ligand and its metal complex; the metal-ligand vibrations should be absent in the spectrum of the free ligand. (2) The metal-ligand vibration should show a relatively large shift by changing the metal or its oxidation state. Method (1) often fails to give a clear-cut assignment since some ligand vibrations, activated by complex formation, may appear in the same region as the metal-ligand vibrations. Method (2) is applicable when a series of metal complexes have exactly the same structure, only the central metal being different.

Recently, Nakamoto et al.[389] developed the *metal isotope technique* to assign metal-ligand vibrations of coordination compounds. In this method, a pair of metal complexes in which only the metals are isotopically substituted are prepared on a milligram scale, and their spectra are compared in the far-infrared region. Only the vibrations involving the motion of the metal atom are shifted by metal isotope substitution. It is, therefore, possible to provide a unique band assignment of a metal-ligand vibration from such an experiment.* The magnitude of an isotopic shift is small because a relative mass difference between metal isotopes is small. Even so, the observed isotopic shifts (2–10 cm^{-1}) are often beyond possible experimental errors (± 0.5 cm^{-1}).

As an example, Fig. III-16 shows an actual tracing of the far-infrared

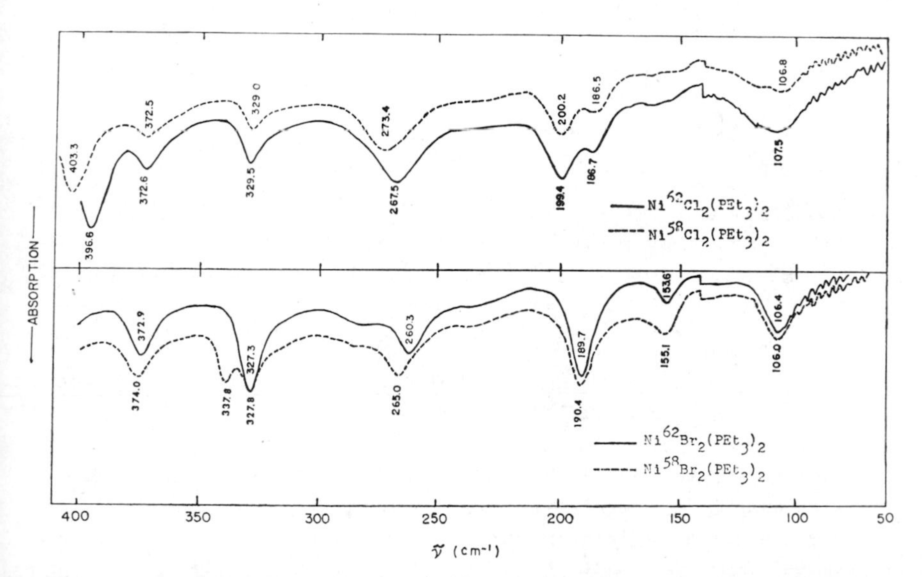

Fig. III-16. Far-infrared spectra of $Ni^{58}X_2(PEt_3)_2$ and $Ni^{62}X_2(PEt_3)_2$ (X═Cl and Br).[390]

spectra of trans-$Ni^{58}X_2(PEt_3)_2$ (X = Cl and Br) and their Ni^{62} analogs. Table III-32b lists the observed frequencies, isotopic shifts, and band assignments.[390] One Ni—X stretching and one Ni—P stretching vibration is expected to be infrared active for trans-planar complexes of this type. In agreement with this prediction, the results show that each compound exhibits two bands which give large isotopic shifts relative to other bands. The band at 403.3 cm^{-1} of $Ni^{58}Cl_2(PEt_3)_2$ is clearly due to the Ni—Cl stretching mode since this band is strong and not present in its bromo analog.[385, 387] Then, the other band at 273.4 cm^{-1} must be assigned to the Ni—P stretching mode.

* Metal isotope shifts of bending modes are generally smaller than those of stretching modes. Thus, the metal isotope method is not so effective in assigning the former.

TABLE III-32b. INFRARED FREQUENCIES, ISOTOPIC SHIFTS, AND BAND ASSIGNMENTS OF $NiX_2(PEt_3)_2$ (X = Cl AND Br)(cm^{-1})[390]

PEt_3	$Ni^{58}Cl_2(PEt_3)_2$		$Ni^{58}Br_2(PEt_3)^2$		Assignment[b]
$\tilde{\nu}$	$\tilde{\nu}$	$\Delta\tilde{\nu}$[a]	$\tilde{\nu}$	$\Delta\tilde{\nu}$[a]	
408	416.7	0.0	413.6	1.2	δ(CCP)
—	*403.3*	6.7	*337.8*	10.5[c]	ν(Ni–X)
365	372.5	−0.1	374.0	1.1	δ(CCP)
330	329.0	−0.5	327.8	0.5[c]	δ(CCP)
—	*273.4*	*5.9*	*265.0*	*4.7*	ν(Ni–P)
245	(hidden)		(hidden)		δ(CCP)
	200.2	0.8	190.4	0.7	δ(CPC)
	186.5	−0.2	155.1	1.5	δ(Ni–X)
	161.5	−0.5	(hidden)		δ(Ni–P)

[a] $\Delta\tilde{\nu}$ indicates metal isotope shift, $\tilde{\nu}(Ni^{58}) - \tilde{\nu}(Ni^{62})$.
[b] Ligand vibrations were assigned according to Ref. 384.
[c] These two bands are overlapped (Fig. III-16). Therefore, $\Delta\tilde{\nu}$ values are only approximate.

$Ni^{58}Br_2(PEt_3)_2$ exhibits the Ni—P stretching band at 265.0 cm^{-1}. Previous investigators[385, 387] assigned the Ni—P stretching bands of these compounds at 426–410 cm^{-1}. They are probably due to a ligand vibration.

A tetrahedral complex such as $NiCl_2(PPh_3)_2$ is expected to show two Ni—Cl stretching and two Ni—P stretching bands in the infrared. A comparison of the infrared spectrum of $Ni^{58}Cl_2(PPh_3)_2$ with that of its Ni^{62} analog reveals that four bands at 341.2, 305.0, 189.6, and 164.0 cm^{-1} of the former give large isotopic shifts relative to those of other bands.[390] The former two bands are assigned to the Ni—Cl stretchings since they are strong and absent in the bromo analog. Then, the latter two bands must be assigned to the Ni—P stretching modes. These assignments are in complete agreement with those of previous workers.[381] It should be noted that both Ni—Cl and Ni—P stretching frequencies are markedly different between trans-planar and tetrahedral complexes.

Trifluorophosphine (PF_3) forms a number of complexes with transition metals.[391] The P—F stretching bands of PF_3 (982 and 848 cm^{-1}) are shifted to higher frequencies (960–850 cm^{-1}) in complexes of the types, $M(PF_3)_n$ (n = 4, 5, or 6) and HM $(PF_3)_4$, and to lower frequencies (850–750 cm^{-1}) in complexes of the types, $[M(PF_3)_4]^-$ (M = Co, Rh, or Ir) and $Ni(PF_3)_n(PPh_3)_{4-n}$ (n = 2 or 3). These observations have been explained by assuming that the P—F bond possesses double bond character, which is influenced by the nature of the M—P bond.[391]. The M—P stretching bands of several PF_3

complexes have been assigned: $Ni(PF_3)_4$ (218 cm^{-1}(IR) and 195 cm^{-1} (Raman));[392] $Ni(PF_3)_n(CO)_{4-n}$ ($n = 1, 2$, or 3) (270–200 cm^{-1});[393] $M(PF_3)_5$ (M = Fe, Ru, or Os) (250–200 cm^{-1}).[394]

Tertiary phosphine oxides and arsine oxides also form coordination compounds with a variety of first-row transition metals. Cotton et al.[395] noted that the P=O stretching frequency of $OPPh_3$(1195 cm^{-1}) is lowered by ca. 50 cm^{-1} upon coordination to a metal. A similar observation has been made for the As=O stretching frequency of $OAsPh_3$ (880 cm^{-1}).[396] Exceptions to this rule are found in $Mn(OAsPh_3)_2X_2$ (X = Cl or Br) which shift to higher frequencies by about 20–30 cm^{-1} upon coordination. Rodley et al.[397] have studied the far-infrared spectra of metal complexes of tertiary arsine oxides and assigned the M—O stretching bands between 440 and 370 cm^{-1}.

III-10. COMPLEXES OF UREA, SULFOXIDES, AND RELATED COMPOUNDS

(1) Complexes of Urea and Related Compounds

Penland et al.[398] first studied the infrared spectra of urea complexes to determine whether coordination occurs through nitrogen or oxygen. The electronic structure of urea may be represented by a resonance hybrid of structures I, II, and III, with each contributing roughly an equal amount.

$$O{=}C(NH_2)_2 \quad (I) \qquad {}^{-}O{-}C({=}NH_2^+)(NH_2) \quad (II) \qquad O{-}C(NH_2)({=}NH_2^+) \quad (III)$$

If coordination occurs through nitrogen, contributions of structures II and III will decrease. This results in an increase of the CO stretching frequency with a decrease of the CN stretching frequency. The NH stretching frequency in this case may fall in the same range as those of the amido complexes (Sec. III-1). If coordination occurs through oxygen, the contribution of structure I will decrease. This may result in a drecrease of the CO stretching frequency but no appreciable change in the NH stretching frequency. Since the spectrum of urea itself has been analyzed completely,[399] band shifts caused by coordination can be checked immediately. The results shown in Table III-33 indicate that coordination occurs through nitrogen in the Pt(II) complex, and through oxygen in the Cr(III) complex. It was also found that Pd(II) coordinates to the nitrogen, whereas Fe(III), Zn(II), and Cu(II) coordinate to the oxygen of urea.

From infrared studies on thiourea ($(NH_2)_2CS$) complexes, Yamaguchi et al.[400] found that all the metals studied (Pt, Pd, Zn, and Ni) form M—S

TABLE III-33. SOME VIBRATIONAL FREQUENCIES OF UREA AND ITS METAL COMPLEXES (cm^{-1})[398]

$[Pt(urea)_2Cl_2]$	Urea	$[Cr(urea)_6]Cl_3$	Predominant Mode
3390	350	3440	$\nu(NH_2)$, free
3290	3350	3330	
3130, 3030		3190	$\nu(NH_2)$, bonded
1725	1683	1505[a]	$\nu(C{=}O)$
1395	1471	1505[a]	$\nu(C{-}N)$

[a] $\nu(C{-}O)$ and $\nu(C{-}N)$ couple in the Cr complex.

bonds, since the CN stretching frequency increases and the CS stretching frequency decreases upon coordination without an appreciable change of the NH stretching frequency. Using the same criterion, thiourea complexes of Fe(II),[401] Mn(II), Co(II), Cu(I), Hg(II), Cd(II), and Pb(II) were shown to be S-bonded.[402] Several investigators[403–405] studied the far-infrared spectra of thiourea complexes and assigned the M—S stretching bands between 300 and 200 cm^{-1}. Thus far, the only metal that is reported to be N-bonded is Ti(IV).[406] Infrared spectra of alkylthiourea complexes have also been studied. Lane and colleagues[407] studied the infrared spectra of methylthiourea complexes and concluded that methylthiourea forms M—S bonds with Zn(II) and Cd(II) and M—N bonds with Pd(II), Pt(II), and Cu(I). For other alkylthiourea complexes, see Refs. 408 and 409.

(2) Complexes of Sulfoxides and Related Compounds

Cotton et al.[410] studied the infrared spectra of sulfoxide complexes to see whether coordination occurs through oxygen or sulfur. The electronic structure of sulfoxides may be represented by a resonance hybrid of the structures:

$$\underset{\text{I}}{:\ddot{\underset{..}{O}}^{-}-\overset{+}{S}R_2:} \longleftrightarrow \underset{\text{II}}{\ddot{O}{=}SR_2:}$$

If coordination occurs through oxygen, contribution of structure II will decrease and result in a decrease of the SO stretching frequency. If coordination occurs through sulfur, contribution of structure I will decrease and may result in an increase of the SO stretching frequency. It has been concluded that coordination occurs through oxygen in the $Co(DMSO)_6^{2+}$ ion, since the SO stretching frequency of this ion (950 cm^{-1}) is lower than that of free DMSO (dimethylsulfoxide), which absorbs at 1100–1055 cm^{-1}. On the other hand, coordination may occur through sulfur in $PdCl_2(DMSO)_2$ and $PtCl_2(DMSO)_2$, since the SO stretching frequencies of these compounds (1157–1116 cm^{-1}) are higher than that of the free ligand. Other ions such as Mn(II), Fe(II, III), Ni(II), Cu(II), Zn(II), and Cd(II) are all coordinated through oxygen, since the DMSO complexes of these metals exhibit the SO stretching bands between 960 and 910 cm^{-1}. Drago and Meek,[411] however, assigned the SO stretching bands of O-bonded complexes in the region of 1025–985 cm^{-1}, since they are metal-sensitive. The bands between 960 and 930 cm^{-1}, which were previously assigned to the SO stretchings, are not metal-sensitive and are now assigned to the CH_3 rocking modes.

Ligands such as DPSO(diphenylsulfoxide) and TMSO (tetramethylenesulfoxide) do not exhibit the CH_3 rocking bands near 950 cm^{-1}. Thus the SO stretching bands of metal complexes containing these ligands can be assigned without difficulty. Table III-34 lists the SO stretching frequencies and the magnitude of band shifts in DPSO complexes.[412] Van Leeuwen and Groeneveld[412] have noted that the shift becomes larger as the electronegativity of the metal increases. In Table III-34, the metals are listed in the order of increasing electronegativity.

TABLE III-34. SHIFTS OF SO STRETCHING BANDS IN DPSO AND DMSO COMPLEXES (CM^{-1})[412]

	DPSO Complex		DMSO Complex
Metal	ν(SO)	Shift	Shift
Ca(II)	1012–1035	0 – (–23)	–
Mg(II)	1012	–23	–
Mn(II)	983–991	–45	–41
Zn(II)	987–988	–47	–
Fe(II)	987	–48	–
Ni(II)	979–982	–55	–45
Co(II)	978–980	–56	–51
Cu(II)	1012, 948	–23, –87	–58
Al(III)	942	–93	–
Fe(III)	931	–104	–

The far-infrared spectra of DMSO complexes have been studied by Johnson and Walton,[413] who assigned the M—O stretching bands at 500–400 cm^{-1}. The S-bonded DMSO complexes (Pt(II) and Pd(II)) also exhibit a strong band in the same region. Hence it is difficult to differentiate the O-

and S-bonded complexes from far-infrared spectra. Tanaka[414] assigned the Sn(IV)—O stretching bands of DMSO complexes at 500–400 cm^{-1}. Francis and Cotton[415] studied the infrared spectra of metal complexes of tetrahydrothiophene oxide. Using the same arguments developed for DMSO complexes, they concluded that Co(II), Ni(II), and Cu(II) are O-bonded while Pt(II) and Pd(II) are S-bonded.

III-11. COMPLEXES OF PYRIDINE AND RELATED LIGANDS

In the high frequency region (above 650 cm^{-1}), the pyridine (py) vibrations show very little shift upon complex formation.[416] However, those at 604 (in-plane ring deformation and 405 cm^{-1} (out-of-plane ring deformation) are shifted to higher frequencies upon coordination to a metal. Clark and Williams[417] have carried out an extensive far-infrared study on metal pyridine complexes. Table III-35 lists the observed frequencies of these metal-sensitive

TABLE III-35. VIBRATIONAL FREQUENCIES OF PYRIDINE COMPLEXES (CM^{-1})[417]

Complex	Structure	py	py	ν(M—py)
$Co(py)_2Cl_2$[a]	monomeric, tetrahedral	642	422	253[b]
$Ni(py)_2I_2$	monomeric, tetrahedral	643	428	240
$Cr(py)_2Cl_2$	polymeric, octahedral	640	440	219
$Cu(py)_2Cl_2$	polymeric, octahedral	644	441	268
$Co(py)_2Cl_2$[a]	polymeric, octahedral	631	429	243, 235[b]
trans-$[Rh(py)_3Cl_3]$	monomeric, octahedral	650	468	265, 245, 230
cis-$[Rh(py)_3Cl_3]$	monomeric, octahedral	643	464	266, 245
trans-$[Ni(py)_4Cl_2]$	monomeric, octahedral	626	426	236
trans-$[Ir(py)_4Cl_2]Cl$	monomeric, octahedral	650	469	260, (255)
cis-$[Ir(py)_4Cl_2]Cl$	monomeric, octahedral	656	468	287, 273
trans-$[Pt(py)_2Br_2]$	monomeric, square-planar	656	476	297
cis-$[Pt(py)_2Br_2]$	monomeric, square-planar	659 644	448	260, 234

[a] $Co(py)_2Cl_2$ exists in two forms: blue complex (monomeric, tetrahedral) and violet complex (polymeric, octahedral). The far-infrared spectra of these two forms are shown in Fig. III-19 (Sec. III-12).
[b] Assignments made by Postmus et al. (Ref. 418).

pyridine vibrations and metal-py stretching vibrations. They have demonstrated that the metal-py stretching and metal-halogen stretching (Sec. III-12) vibrations are very useful in elucidating the stereochemistry of these complexes. For example, *cis*-$[Rh(py)_3Cl_3]$ exhibits two Rh—py stretchings

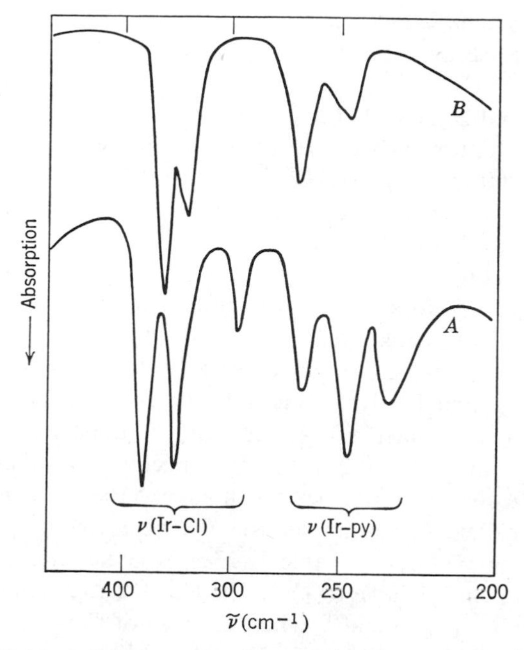

Fig. III-17. Far-infrared spectra of *trans* (A) and *cis* (B) $Rh(py)_3Cl_3$.[417]

($\mathbf{C}_{3v}$ symmetry), whereas *trans*-$[Rh(py)_3Cl_3]$ exhibits three Rh—py stretchings ($\mathbf{C}_{2v}$ symmetry) near 250 cm^{-1}. Figure III-17 shows the infrared spectra of these two compounds.

The infrared spectra of metal complexes with substituted pyridines have been studied: picolines (419, 420, 421), halogenopyridines (420, 422), cyano-pyridines (423, 424) and pyridones (425). Goldstein et al.[426] assigned the Cu-py stretching bands of CuX_2L_2-type compounds (X = Cl or Br, and L = an alkyl-substituted pyridine) between 270 and 240 cm^{-1}.

Metal complexes of pyridine N-oxide exhibit the NO stretchings at about 1200 cm^{-1} and the M—O stretchings below 450 cm^{-1}.[427,428] The C—H out-of-plane bendings (1000–700 cm^{-1}) of the pyridine ring were assigned from the analysis of combination bands at 2000–1600 cm^{-1}.[428] Whyman and Hatfield[429] have assigned the M—O stretching bands of the CuL_2X_2-type compounds (L = a substituted pyridine N-oxide) between 460 and 330 cm^{-1}.

III-12. HALOGENO COMPLEXES

Halogens are the most common ligands in coordination chemistry. Several reviews[430,431] summarize the results of extensive infrared studies on halogeno complexes. Part II of this book lists the vibrational frequencies of a number of simple halogeno compounds. Section III-1 gives the M—X(metal-halogen) stretching frequencies of halogenoammine complexes.

(1) Terminal Metal-Halogen Bond

Terminal M—X stretching bands appear in the regions at 750–500 cm^{-1} for M—F, 400–200 cm^{-1} for M—Cl, 300–200 cm^{-1} for M—Br and 200–100 cm^{-1} for M—I. For the same metal, the ratios of the M—X stretching frequencies are 0.77–0.74 for ν(M—Br)/ν(M—Cl) and 0.65 for ν(M–I)/ν(M–Cl).[417] According to Clark,[432] the M—X stretching frequency is governed by several factors. If other conditions are equal, the M—X stretching frequency is higher as the *oxidation state* of the metal is higher. For example, Table II-23*b*(Part II) shows that the Fe-Cl stretching frequencies (ν_3) of tetrahedral $FeCl_4^-$ and $FeCl_4^{2-}$ ions are 385 and 286 cm^{-1}, respectively. Table II-32*a* also shows that the Pt—F stretching frequencies (ν_3) of octahedral PtF_6 and PtF_6^{2-} are 705 and 571 cm^{-1}, respectively. If other conditions are equal, the M—X stretching frequency is higher as the *coordination number* of the metal is smaller. For example, the Ge—Cl stretching frequencies of $GeCl_4$ are 451 and 397 cm^{-1} (Table II-23*a*), whereas those of $GeCl_6^{2-}$ are 318, 293, and 213 cm^{-1} (Table II-32*b*). The Co—Cl stretching band of $CoCl_2$ (triatomic, linear) is at 493 cm^{-1}, whereas those of $Co(py)_2Cl_2$ (monomeric, tetrahedral) are at 344 and 304 cm^{-1}. Wharf and Shriver[433] have shown that the Sn—X stretching force constants of halogenotin compounds are approximately proportional to the oxidation number of the metal divided by the coordination number of the complex.

The M—X stretching vibrations are very useful in distinguishing *stereochemical isomers*. As is shown in Sec. I-10, the number of infrared active M—X stretching bands can be predicted from group theory. It is then possible to determine the structure of a complex by comparing the observed number of the M—X stretching bands with that predicted theoretically for each isomer. Several examples of this method will be given in the following sections.

(a) Square-planar Complexes

The M—X and M—N stretching frequencies of the $M(NH_3)_2X_2$-type compounds (M = Pt(II) or Pd(II)) are listed in Table III-6. The symmetry of the PtN_2X_2 skeleton is $\mathbf{C}_{2v}$ for the *cis* and $\mathbf{C}_{2h}$ for the *trans* isomer. The former is expected to show two infrared active M—X stretchings ($A_1 + B_2$), whereas the latter is expected to give only one M—X stretching (B_u). This is illustrated by the spectra of *cis*- and *trans*-$[Pd(NH_3)_2Cl_2]$ shown in Fig. III-4. The same is true for a pair of *cis*- and *trans*-$[Pt(py)_2Cl_2]$.[434,435] Adams et al.[436] have studied the far-infrared spectra of *cis*- and *trans*-$[PtL_2X_2]$-type compounds (L = a neutral ligand).

Complexes of *cis*-(square-planar) and tetrahedral ML_2X_2-types exhibit two M—X stretching bands in the infrared. In general, the antisymmetric M—X stretching band is stronger and at a higher frequency than the symmetric M—X stretching band. These two modes can also be distinguished by studying the effect of pressure on the M—X stretching bands.[437] Figure III-18 shows the effect of pressure on the Pt—Cl stretching bands of

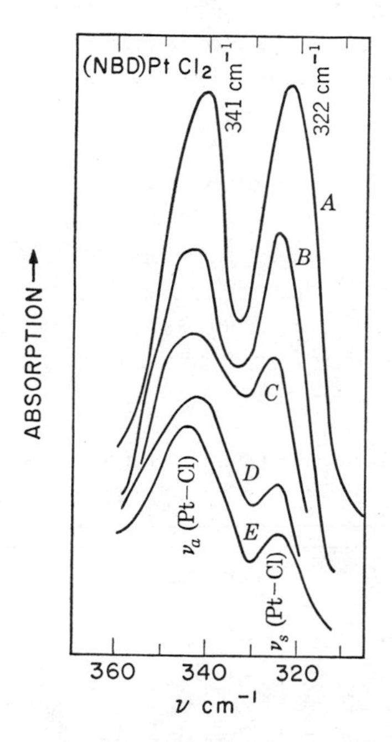

Fig. III-18. Effect of pressure on Pt—Cl stretching bands: *A*, 1 atm; *B*, 6,000 atm; *C*, 12,000 atm; *D*, 18,000 atm; and *E*, 24,000 atm.

$Pt(NBD)Cl_2$(NBD, norbornadiene). It is seen that the intensity of the symmetric mode (322 cm^{-1}) is more suppressed by increasing pressure than that of the antisymmetric mode (341 cm^{-1}). Similar observations have been made for a number of tetrahedral ML_2X_2-type compounds.

(b) Octahedral Complexes

Cis-ML_4X_2-type compounds ($\mathbf{C}_{2v}$) should exhibit two M—X stretchings ($A_1 + B_2$), while *trans*-ML_4X_2-type compounds ($\mathbf{D}_{4h}$) should give only one M—X stretching (A_{2u}) in the infrared. In fact, *cis*-$[Ir(py)_4Cl_2]Cl$ shows two Ir—Cl stretchings at 333 and 327 cm^{-1}, while *trans*-$[Ir(py)_4Cl_2]Cl$ shows one Ir—Cl stretching at 335 cm^{-1}.[417] If ML_3X_3-type compounds are *cis* ($\mathbf{C}_{3v}$), two M—X stretchings ($A_1 + E$) are expected in the infrared. If they are *trans* (C_{2v}), three M—X stretchings ($2A_1 + B_2$) should be infrared active. In accord with these predictions, *cis*-[Rh(py)$_3$Cl$_3$] exhibits two bands at 341 and 325 cm^{-1} and *trans*-[Rh(py)$_3$Cl$_3$] gives three bands at 355, 322, and 295 cm^{-1} (see Fig. III-17). Durig et al.[438] have found that *cis*-[Ru(H$_2$O)$_3$Cl$_3$] exhibits two Ru—Cl stretchings (315, and 311 or 302 cm^{-1}) and its *trans* isomer exhibits two Ru—Cl stretchings (314 and 291 cm^{-1}).

In the ML$_2$X$_4$-type compounds, group theory predicts one M—X stretching (E_u) for the *trans* isomer (D_{4h}) and four M—X stretchings ($2A_1 + B_1 + B_2$) for the *cis* isomer. For example, *trans*-[Pt(NH$_3$)$_2$Cl$_4$] exhibits one Pt—Cl stretching at 352 cm^{-1} (with a shoulder at 346 cm^{-1}), whereas *cis*-[Pt(NH$_3$)$_2$Cl$_4$] exhibits four Pt—Cl stretchings at 353, 344, 330, and 206 cm^{-1}.[439] Beattie et al.[440] have carried out normal coordinate analyses on *cis*- and *trans*-ML$_2$X$_4$-type compounds. Beattie and Rule[441] obtained the infrared spectra of *cis*- and *trans*-SnL$_2$Cl$_4$-type compounds (L, acetone, trimethylamine, etc.). It was found that the complex is *cis* when L is acetone and becomes *trans* if a bulky group such as trimethylamine is introduced. The M—X stretching frequencies are reported by many other investigators. Table III-36 lists the M—X stretching frequencies of other metals.

TABLE III-36. METAL-HALOGEN STRETCHING FREQUENCIES (CM^{-1})

Type of Compound	Bond	Frequency	References
CuCl$_2$L and CuCl$_2$L$_2$ (L, subst.-py N-oxide)	Cu—Cl	350–280	429
CuX$_2$L$_2$ (L, subst.-py, etc.)	Cu—Cl	330–287	426
	Cu-Br	266–230	
ZnX$_2$L$_2$ (L, neutral ligand)	Zn—Cl	350–225	442
	Zn—Br	270–200	
CdX$_2$L$_2$ (L, neutral ligand)	Cd—Cl	270–210	442
	Cd—Br	195	
HgX$_2$L$_2$ (L, neutral ligand)	Hg—Cl	350–270	442
	Hg—Br	220–200	
TiF$_4$L$_2$ (L, neutral ligand)	Ti—F	650–450	443
ZrF$_4$L$_2$ (L, neutral ligand)	Zr—F	570–450	443
MX$_5$O^{2-} (M, Nb, Mo or W)	M—Cl	340–310	444
	M—Br	255–220	
MCl$_2$ (acac)(OR)$_2$	Nb—Cl	340–270	445
(M = Nb or Ta, R = Me or Et)	Ta—Cl	310–270	

(2) Bridging Metal-Halogen Bond

Halogens tend to form bridges between two metal atoms. In general, bridging M—X stretching frequencies (ν_b(M—X)) are lower than terminal M—X stretching frequencies (ν_t(M—X)). Table III-37 lists the terminal and bridging M—X stretching frequencies reported so far. Planar M_2X_6-type compounds ($\mathbf{D}_{2h}$) exhibit two terminal ($B_{2u} + B_{3u}$) and two bridging ($B_{2u} + B_{3u}$) M—X stretchings in the infrared. Table III-37 shows that the ratio, ν_b(M—X)/ν_t(M—X), is about 0.85–0.80 in the M_2X_6- and $M_2L_2X_4$-type complexes. For the $[Pt_2Cl_6]^{2-}$ ion, the terminal and bridging Pt—Cl stretching force constants were calculated to be 2.02 and 1.42 mdyn/A, respectively.[446] Adams et al.[447] studied the far-infrared spectra of anyhydrous metal halides of the type MX_2 (M is Pt, Pd, Cu, Cd, or Hg; X is Cl or Br)

TABLE III-37. TERMINAL AND BRIDGING METAL-HALOGEN STRETCHING FREQUENCIES (CM^{-1})

Compound	ν_t(M—X)	ν_b(M—X)	ν_b(M—X)[a] / ν_t(M—X)	References
$[Cr_2Cl_9]^{3-}$	341	322	0.94	432
$[V_2Cl_9]^{3-}$	335	296	0.88	432
$[Cu_2Cl_6]^{2-}$	301	278, 236	0.85	450
$[Cu_2Br_6]^{2-}$	237	224, 168	0.83	450
$[Cu(py)_2Cl_2]$[b]	287	229	0.80	426, 451
$[Cu(py)_2Br_2]$[b]	256	204	0.80	426, 451
$[Pt_2Cl_6]^{2-}$	350, 346	315, 302	0.89	446
$[Pt_2Br_6]^{2-}$	243, 237	211, 196	0.85	446
$[Pt_2I_6]^{2-}$	194, 179	157, 144	0.80	446
$Pt_2L_2Cl_4$[c]	365–340	335–310 295–250	0.85	386
$Pt_2L_2Br_4$[c]	260–235	230–210 190–175	0.91	386
$Pt_2L_2I_4$[c]	200–170	190–150 150–135	0.84	386
$Pd_2L_2Cl_4$[c]	370–345	310–300 280–250	0.80	386
$Pd_2L_2Br_4$[c]	285–265	220–185 200–165	0.70	386
$Co(py)_2Cl_2$				
monomeric	347, 306	–	0.55	418
polymeric	–	186, 174		

[a] The ratio was calculated by using average frequencies for terminal and bridging vibrations.

[b] In these compounds, the Cu atoms are in distorted octahedral environments; the Cu-Cl bonds of the $CuCl_2(py)_2$ skeleton (*trans*-square-planar) are shorter (2.28A) than those perpendicular to it (3.05A). The latter serve as bridges between two $CuCl_2(py)_2$ units.

[c] L is a neutral ligand such as PEt_3. *Cis* compounds are expected to show two terminal M—X stretching bands.

which contain only bridging M—X bonds. These compounds exhibit strong bands between 375 and 250 cm^{-1}.

Table III-37 indicates that the ratio, ν_b(M—X)/ν_t(M—X), is 0.88—0.94 for the $[M_2X_9]^{3-}$-type compounds and 0.55 for $Co(py)_2Cl_2$. In the latter compound, the bridging frequencies are exceptionally low because the polymeric form of $Co(py)_2Cl_2$ assumes an octahedral structure:

py py
Cl Cl Cl
Co Co
Cl Cl Cl
py py

Thus an increase in coordination number as well as polymeric structure is responsible for the lowering of the bridging frequencies. Figure III-19 compares the far-infrared spectra of both forms of $Co(py)_2Cl_2$.

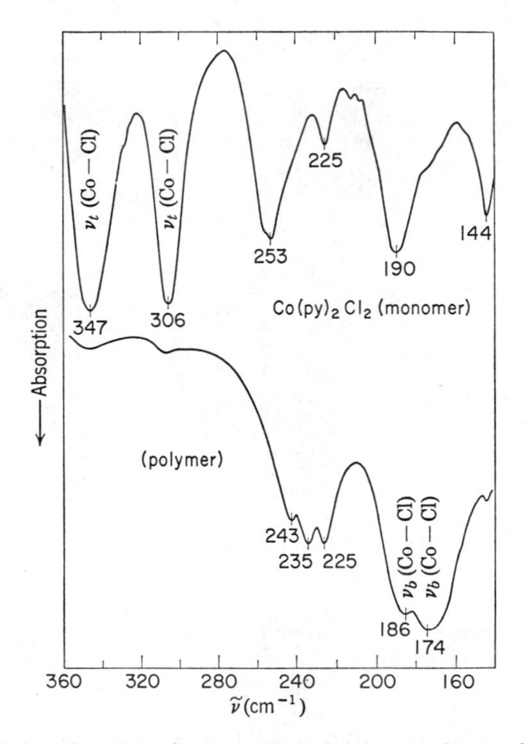

Fig. III-19. Infrared spectra of monomeric and polymeric forms of $Co(py)_2Cl_2$.

Far-infrared spectra of metal cluster ions such as $[Nb_6X_{12}]^{2-}$ and $[Ta_6X_{12}]^{2-}$ (X=Cl or Br) have been studied by Boorman and Straughan.[448] Clark and co-workers[449] also studied the far-infrared spectra of the $[Mo_6Cl_8]^{4+}$ and $[W_6Cl_8]^{4+}$ ions. Although the detailed band assignments are not available, the bands listed in Table III-38 are probably due to the bridging M—X stretching modes. Cotton and colleagues[452] have assigned the far-infrared spectra of the $[M_6X_8Y_6]^{2-}$-type compounds where M is Mo or W and X and Y are halogens. Since these ions belong to the $\mathbf{O}_h$ point group, only five F_{1u} vibrations are infrared active. Only the M—X and M—Y stretching frequencies are listed in Table III-38.

TABLE III-38. METAL-HALOGEN STRETCHING FREQUENCIES OF METAL CLUSTER COMPOUNDS (cm^{-1})

Ion	Bands	References
$[Nb_6Cl_{12}]^{2-}$	345–340, 290–276, 250–220, 150–144	448
$[Nb_6Br_{12}]^{2-}$	280–270, 210–200, 180–170, 145	448
$[Ta_6Cl_{12}]^{2-}$	330–320, 235–230, 190–175, 145–140	448
$[Ta_6Br_{12}]^{2-}$	230–225, 182–175, 145–130, 107–96	448
$[Mo_6Cl_8]^{4+}$	360–320, 310–240	449
$[W_6Cl_8]^{4+}$	320–290, 260–220	449
$[Mo_6Cl_{14}]^{2-}$	350–310, 246	452
$[Mo_6Cl_8Br_6]^{2-}$	330–290, 168	452
$[Mo_6Cl_8I_6]^{2-}$	325–280, 132	452
$[Mo_6Br_8Cl_6]^{2-}$	345–330, 300–260	452
$[Mo_6Br_{14}]^{2-}$	270, 250–220, 170–130	452
$[Mo_6Br_8I_6]^{2-}$	274, 250–210, 120	452
$[W_6Cl_{14}]^{2-}$	320–290, 224	452

III-13. COMPLEXES CONTAINING METAL-METAL BONDS

X-ray analyses have shown the presence of metal-metal bonds in compounds such as $Cu_2(CH_3COO)_4(H_2O)_2$,[453] $Mn_2(CO)_{10}$,[292] and $K_3[W_2Cl_9]$.[454] The metal-metal bonds are also formed by reacting anions such as $SnCl_3^-$ with a variety of metal complexes. Shriver and Johnson[455] have noted that the Sn—Cl stretching bands of the free $SnCl_3^-$ ion (289(A_1) and 252 cm^{-1}(E)) are shifted to higher frequencies upon coordination to a metal. For example, the Sn—Cl stretching frequencies of $[(Ph)_3P]_3CuSnSCl_3$ are 315 and 288 cm^{-1} and those of $[(Me)_4N]_4[Rh_2Cl_2(SnCl_3)_4]$ are 339 and 323 cm^{-1}. The following general rule is applicable to the coordinate bonds of predominantly σ-character:[455] The L—X force constant of the ligand LX_n will increase upon coordination to a metal (electron acceptor) if X is significantly more electronegative than L, and vice versa. In the above example, the

Sn—Cl stretching bands shift to higher frequencies, since Cl is more electronegative than Sn. In metal ammine complexes, the N—H stretching bands shift to lower frequencies because N is more electronegative than H.

The metal-metal stretching bands appear in the 250–100 cm^{-1} region, since the masses of metals are relatively large. Cotton and Wing[456] first assigned the Re—Re stretching of $Re_2(CO)_{10}$ at 120 cm^{-1}(Raman). Since then, the metal-metal stretching bands have been assigned for a number of compounds. Table III-39 lists some of these frequencies.

TABLE III-39. METAL-METAL STRETCHING FREQUENCIES (CM^{-1})

Compound	Bond	ν(M—M)	References
$(Me)_3Sn—Ge(Ph)_3$	Sn—Ge	225(IR)	457
$(Me)_3Sn—Mo(CO)_3(C_5H_5)$	Sn—Mo	172(IR)	457
$(Me)_3Sn—Fe(CO)_2(C_5H_5)$	Sn—Fe	185(IR)	457
$(Me)_3Sn—Mn(CO)_5$	Sn—Mn	182(IR)	458
Cl—Hg—Hg—Cl	Hg—Hg	166(R)	459
$(Ph)_3Sn—Sn(Ph)_3$	Sn—Sn	208(R)	459
$(Ph)_3Sn—Mn(CO)_5$	Sn—Mn	174(R)	459
$(CO)_5Mn—Mn(CO)_5$	Mn—Mn	157(R)	459
$(CO)_5Mn—Re(CO)_5$	Mn—Re	182(R)	459
$(CO)_4Co—Cd—Co(CO)_4$	Co—Cd	152(R)	237
$(CO)_4Co—Hg—Co(CO)_4$	Co—Hg	161(R), 196(IR)	460, 237
$(CO)_5Mn—Hg—Mn(CO)_5$	Mn—Hg	167(R), 188(IR)	460
$[Mo_6Cl_{14}]^{2-}$	Mo—Mo	220(IR)	452
$[Mo_6Cl_8Br_6]^{2-}$	Mo—Mo	232(IR)	452
$[W_6Cl_{14}]^{2-}$	W—W	150 ?(IR)	452

In general, the metal-metal stretching vibration of the heteronuclear diatomic system (such as Sn—Ge) appears strongly in the Raman and weakly in the infrared spectrum.[459] However, some complexes show relatively strong absorption in the infrared.[457] The metal-metal stretching bands appear relatively strongly in polyatomic systems such as metal cluster ions. Figure III-20 shows the Sn—Mn stretching bands observed in the far-infrared region.[458]

III-14. METAL ALKOXIDES, AND COMPLEXES OF ALCOHOLS AND CARBOXYLIC ACIDS

In general, metal alkoxides, $M(OR)_n$ (R, alkyl group), exhibit the C—O stretchings at about 1000 cm^{-1} and the M—O stretchings below 650 cm^{-1}. Table III-40 lists the M—O stretching frequencies of various metal alkoxides. References 464 and 465 give the infrared spectra of other alkoxides.

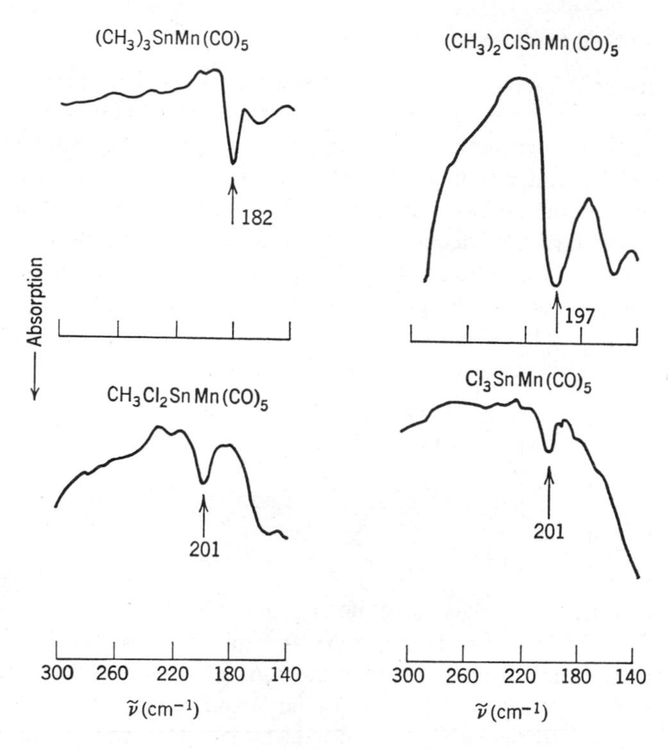

Fig. III-20. Far-infrared spectra of $(Me)_{3-n}Cl_nSnMn(CO)_5$, where $n = 0, 1, 2$, or 3.[458] The arrow indicates the Sn—Mn stretching band.

TABLE III-40. METAL-OXYGEN STRETCHING FREQUENCIES OF METAL ALKOXIDES (CM^{-1})

Compound	ν(M—O)	References
$Ti(OMe)_4$	588, 553	461
$Ti(OEt)_4$	625, 500	462
$Ta(OEt)_5$	556	462
$Nb(OEt)_5$	571	462
$Cr(OMe)_2$	515, 470 ?	463
$Mn(OMe)_2$	360, 307	463
$Fe(OMe)_2$	370, 330, 420 ?	463
$Co(OMe)_2$	491, 412, 340	463
$Ni(OMe)_2$	425, 375	463
$Cu(OMe)_2$	520, 435	463
$Zn(OMe)_2$	465, 325	463

Van Leeuwen[466] studied the infrared spectra of alcohol complexes of the type $[M(EtOH)_6]X_2$, where M is a divalent transition metal and X is ClO_4^-, BF_4^-, or NO_3^-. As expected, the anions have considerable influence on the O—H stretching and bending frequencies. Miyake[467] studied the infrared spectra of metal complexes of ethyleneglycol and its derivatives, and found two types of spectra that depend on the water content of the salts used for preparation of the complexes. The A-type spectra are obtained when the water content of the salts is greater than 15%, and the B-type spectra are obtained when it is less than 15%. He concluded that the structures of the complexes giving rise to the A- and B-type spectra are

(A) (B)

Extensive infrared studies have been made on metal complexes of carboxylic acids.[468,469] Table III-41 gives the infrared frequencies and band assignments for the formate and acetate ions obtained by Itoh and Bernstein.[470] References 471–473 give similar band assignments. Vibrational spectra of metal formates[474–476] and acetates[477,478] can be assigned by using the results obtained for the free ions. Since the antisymmetric COO stretching frequency is most sensitive to a change in the metal, the relationship between this frequency and some physical property of the metal has been discussed by several investigators. For example, Theimer and Theimer[479] claimed that the radius of the metal ion is the main factor in determining the frequency, whereas Kagarise[480] found that the electronegativity of the metal is important. Ellis and Pyszora[481] suggested that the COO stretching frequency is a complicated function of the mass, radius, and electronegativity of the metal.

In examining the effect of coordination on the COO stretching frequency, it is important to interpret the results based on the structures obtained by x-ray analysis. For example, the acetate anion ($(ac)^-$) coordinates to a metal in one of the following schemes:

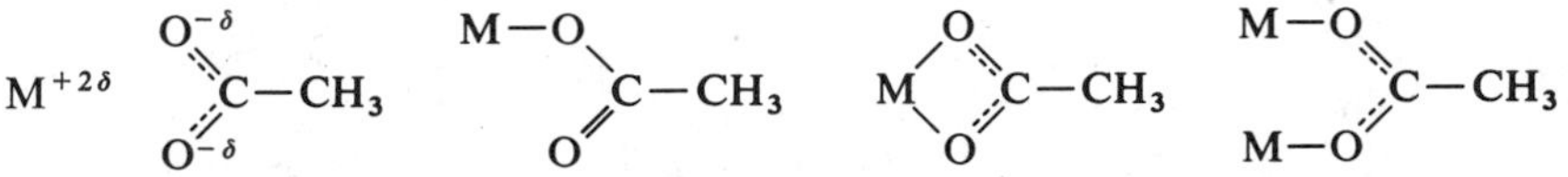

I II III IV

According to x-ray analysis, the structure of sodium formate[482] is I, the two CO bond lengths being equal (1.27A). On the other hand, $Li(ac)\cdot 2H_2O$[483] has structure II, the two CO bond lengths being different (1.33 and 1.22A).

TABLE III-41. INFRARED FREQUENCIES AND BAND ASSIGNMENTS FOR THE FORMATE AND ACETATE IONS (CM^{-1})[470]

$[HCOO]^-$		$[CH_3COO]^-$			
Na Salt	Aq. Sol'n	Na Salt	Aq. Sol'n	C_{3v}	Band Assignment
2841	2803	2936	2935	A_1	$\nu(CH)$
–	–	–	1344		$\delta(CH_3)$
1366	1351	1414	1413		$\nu(COO)$
–	–	924	926		$\nu(CC)$
772	760	646	650		$\delta(OCO)$
–	–	–	–	A_2	$\rho_t(CH_3)$
–	–	2989	3010 or 2981	B_1	$\nu(CH)$
1567	1585	1578	1556		$\nu(COO)$
–	–	1430	1429		$\delta(CH_3)$
–	–	1009	1020		$\rho_r(CH_3)$
1377	1383	460	471		$\delta(CH)$ or $\rho_r(COO)$
–	–	2989	2981 or 3010	B_2	$\nu(CH)$
–	–	1443	1456		$\delta(CH_3)$
–	–	1042	1052		$\rho_r(CH_3)$
1073	1069	615	621		$\pi(CH)$ or $\pi(COO)$

Structure II occurs frequently among the salts of high oxidation state metals. The bidentate structure III is reported for $Zn(ac)_2 \cdot 2H_2O$[484] and $Na[UO_2(ac)_3]$.[485] Although this structure is less common than II, many anhydrous salts such as $Cr(ac)_3$ and $Mn(ac)_3$ may have structure III, since the maximum coordination number of the metal is attainable in this structure. The usual bridging structure of the acetate ion was found in $Cr_2(ac)_4 \cdot 2H_2O$,[486] $Cu_2(ac)_4 \cdot 2H_2O$,[453] $Be_4O(ac)_6$[487] and $Zn_4O(ac)_6$.[488]

The effect of changing the metal on the COO stretching frequencies may be different for each structural type mentioned above. For example, in a series of salts having structure II, the antisymmetric COO stretching frequency will increase and the symmetric COO stretching frequency will decrease, as the M—O bond becomes stronger (see Sec. III-18). This trend is not seen in a series of compounds having structure IV. In fact, both COO stretching bands are shifted to higher frequencies in going from $Cr_2(ac)_4 \cdot 2H_2O$ to $Cu_2(ac)_4 \cdot 2H_2O$.[489] Infrared spectra of the $Cu_2(RCOO)_4X_2$-type compounds (R is phenyl or H, and X is pyridine, etc.) have been reported.[490]

Lappert[491] studied the infrared spectra of metal complexes with carboxylic acid esters. The C=O and C—OEt stretching bands of CH_3COOEt are at 1741 and 1240 cm^{-1}, respectively. Upon coordination to the Sn atom, the former is shifted to 1613 cm^{-1}, whereas the latter is shifted to 1319 cm^{-1}. This result suggests that coordination occurs through the carbonyl oxygen as shown below:

$CH_3-C-\ddot{O}-Et$
$\quad\quad\;\; \| $
$\quad\quad\; {}^{+}O$
$\quad\quad\quad\; \backslash$
$\quad\quad\quad\;\; Sn^{-}$

III-15. COMPLEXES OF ETHYLENEDIAMINE AND RELATED COMPOUNDS

As is shown in Fig. III-21, 1,2-disubstituted ethane may exist in the *cis*,

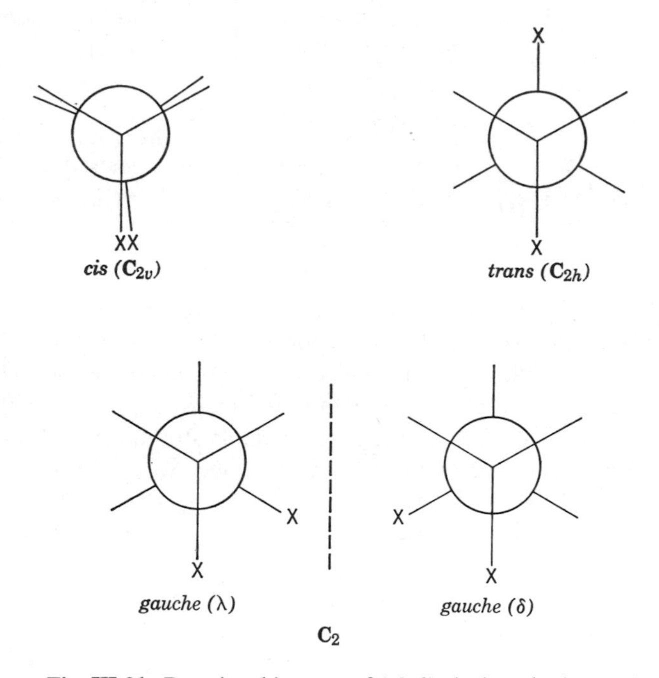

Fig. III-21. Rotational isomers of 1,2-disubstituted ethane.

trans, or *gauche* form, depending upon the angle of internal rotation. The *cis* form may not be stable in the free ligand because of steric repulsion between two X groups. The *trans* form belongs to point group $\mathbf{C}_{2h}$, in which only the *u* vibrations (antisymmetric with respect to the center of symmetry) are infrared active. On the other hand, both *gauche* forms belong to point group $\mathbf{C}_2$ in which all the vibrations are infrared active. Thus the *gauche* form exhibits more bands than the *trans* form. Mizushima and co-workers[492] have shown that 1,2-dithiocyanatoethane (NCS—CH_2—CH_2—SCN) in the crystalline state definitely exists in the *trans* form, because no infrared frequencies coincide with Raman frequencies (mutual exclusion rule). By comparing the spectrum of the crystal with that of a $CHCl_3$ solution, they have concluded that several extra bands observed in solution can be attributed to the *gauche* form. Table III-42 summarizes the infrared frequencies and band assignments obtained by Mizushima et al. It is seen that the CH_2 rocking vibration provides the most clear-cut diagnosis of conformation: one band (A_u) at 749 cm^{-1} for the *trans* form and two bands (A and B) at 918 and 845 cm^{-1} for the *gauche* form.

The compound 1,2-dithiocyanatoethane may take the *cis* or *gauche* form when it coordinates to a metal through the sulfur atoms. The chelate ring formed will be completely planar in the *cis*, and puckered in the *gauche* form. The *cis* and *gauche* forms can be distinguished by comparing the spectrum of a metal chelate with that of the ligand in $CHCl_3$ solution (*gauche* + *trans*). Table III-42 compares the infrared spectrum of 1,2-dithiocyanatoethane-dichloroplatinum (II) with that of the free ligand in a $CHCl_3$ solution. Only the bands characteristic of the *gauche* form are observed in the Pt(II) complex. This result definitely indicates that the chelate ring in the Pt(II) complex is *gauche*. The method described above has also been applied to the metal complexes of 1,2-dimethyl-mercaptoethane (CH_3S—CH_2—CH_2—SCH_3).[493] In this case, the free ligand exhibits one CH_2 rocking at 735 cm^{-1} in the crystalline state (*trans*), whereas the metal complex always exhibits two CH_2 rockings at 920–890 and 855–825 cm^{-1}(*gauche*).

The conformation of ethylenediamine (en) in metal complexes is of particular interest in coordination chemistry. Unfortunately, the CH_2 rocking mode discussed above does not provide a clear-cut diagnosis in this case, since it couples strongly with the NH_2 rocking and C—N stretching modes that appear in the same frequency region. However, x-ray analyses on *trans*-$[Co(en)_2Cl_2]Cl \cdot HCl \cdot H_2O$[4] and other complexes indicate that the chelating ethylenediamine takes the *gauche* conformation without exception. Powell and Sheppard[494] have carried out an extensive infrared study on ethylenediamine and its metal complexes. Other references are: $[M(en)_3]^{2+}$ (M = Zn(II), Cd(II) or Hg(II));[495] $[Ru(en)_3]^{2+}$;[496] $[M(en)_2]^{2+}$ (M = Pt(II) or Pd(II))[497]; *trans* $[Co(en)_2Cl_2]ClO_4$.[498] Infrared spectra of *cis*- and *trans*-$[M(en)_2X_2]^+$ (M = Co(III), Cr(III), Ir(III) or Rh(III) and X = a halogen) have been studied extensively.[499–503] These isomers can be distinguished by the NH_2 bending (1700–1500 cm^{-1}), CH_2 rocking (950–850 cm^{-1}) and M—N stretching (610–500 cm^{-1}) bands.

TABLE III-42. INFRARED SPECTRA OF 1,2-DITHIOCYANATOETHANE AND ITS PT(II) COMPLEX (CM^{-1})[492]

Ligand		Pt Complex (*gauche*)	Assignment
Crystal (*trans*)	$CHCl_3$ Solution (*gauche* + *trans*)		
–	2170 (*g*)	2165 (*g*)	$\nu(C\equiv N)$
2155 (*t*)	2170 (*t*)	–	
1423 (*t*)	1423 (*t*)	–	$\delta(CH_2)$
–	1419 (*g*)	1410 (*g*)	
1291[a] (*t*)	–	–	$\rho_w(CH_2)$
–	1285 (*g*)	1280 (*g*)	
1220 (*t*)	1215 (*t*)	–	
1145 (*t*)	1140 (*t*)	–	$\rho_t(CH_2)$
–	1100 (*g*)	1110 (*g*)	
–	[b] (*g*)	1052 (*g*)	$\nu(C—C)$
1037[a]	–	–	
–	918 (*g*)	929 (*g*)	$\rho_r(CH_2)$
–	845 (*g*)	847 (*g*)	
749 (*t*)	[b]	–	
680 (*t*)	677 (*t*)	–	$\nu(C—S)$
660 (*t*)	660 (*t*)	–	

[a] Raman frequencies in the crystalline state.
[b] Hidden by $CHCl_3$ absoprtion.

The most interesting information obtained from infrared studies is that ethylenediamine takes the *trans* form when it functions as a bridging group between two metal atoms. Powell and Sheppard[504] were the first to suggest that ethylenediamine in $(C_2H_4)Cl_2Pt(en)PtCl_2(C_2H_4)$ is likely to be *trans*, since the infrared spectrum of this compound is simpler than that of other complexes in which ehylenediamine is *gauche*. Similar results have been obtained for $(AgCl)_2en$,[505] $(AgSCN)_2en$,[506] $(AgCN)_2en$,[506] $Hg(en)Cl_2$[507] and $M(en)Cl_2$ (M = Zn or Cd).[505] The structure of these complexes may be depicted as follows:

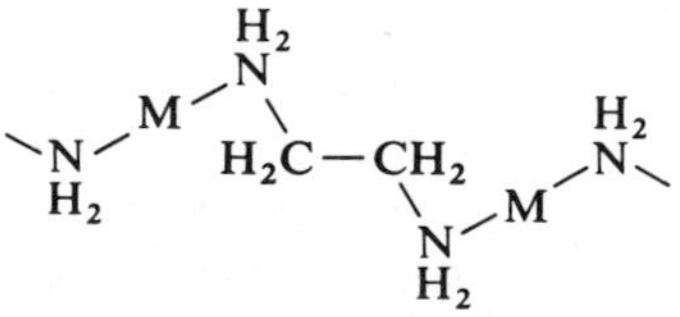

```
                    H2
                    N
              M  /    \                   H2
  \ N /  /            H2C—CH2        /N\
    H2                        \    /M
                               N /
                               H2
```

Succinonitrile (sn), NC—CH_2—CH_2—CN, is another example of 1,2-disubstituted ethanes. According to x-ray analysis,[508] the complex ion in $[Cu(sn)_2]NO_3$ has a polymeric chain structure in which the ligand takes the *gauche* form.

```
\   /NC−CH2−CH2−CN\   /NC−CH2−CH2−CN\   /
 Cu                 Cu                 Cu
/   \NC−CH2−CH2−CN/   \NC−CH2−CH2−CN/   \
```

Infrared and Raman studies of succinonitrille[509,510] indicate that both *trans* and *gauche* forms coexist in the liquid and solid states, and only the latter remains when the solid is cooled to $-50°C$. Again the similarity of the CH_2 rocking frequencies in the free ligand (963 and 820 cm^{-1}) and the Cu(I) complex (966 and 826 cm^{-1}) confirms the *gauche* conformation. On the other hand, the Ag(I) complexes, Ag(sn)X ($X = NO_3^-$ or ClO_4^-) exhibit the CH_2 rocking bands near 760 cm^{-1}, which is characteristic of the *trans* form. It has therefore been suggested that in these Ag(I) complexes succinonitrile acts as a bridge between two Ag atoms and assumes the *trans* configuration.[511]

There are four rotational isomers for glutaronitrile (gn), NC—CH_2—CH_2—CH_2—CN, which are spectroscopically distinguishable. Figure III-22 shows

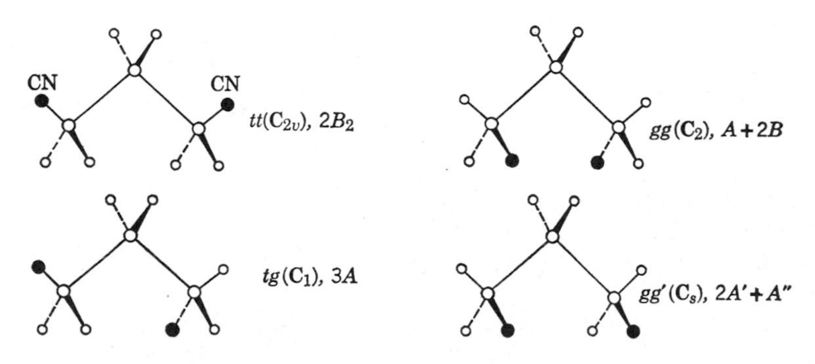

Fig. III-22. Rotational isomers of glutaronitrile.

the conformation, the symmetry, and the number of infrared active CH_2 rocking vibrations for each isomer. According to the x-ray analysis on $Cu(gn)_2NO_3$,[512] the ligand in this complex is in the *gg* conformation. The infrared spectrum of this complex is very similar to that of solid glutaronitrile in the stable form. Matsubara[513] therefore concluded that the latter also takes the *gg* conformation. However, the spectrum of solid glutaronitrile in the metastable form (produced by rapid cooling) is different from that of the *gg* conformation. It could be *tt*, *tg*, or *gg'*. The *tt* conformation was excluded because of the absence of the 730 cm^{-1} band characteristic of the *trans*-planar methylene chain,[514] and the *gg'* conformation was considered to be improbable because of steric repulsion between two CN groups. This left only the *tg* conformation for the metastable solid. The complicated spectrum of liquid glutaronitrile was accounted for by assuming that it is a mixture of the *tg*, *gg*, and *tt* conformations. Kubota and Johnston,[515] using these results, have been able to show that the glutaronitrile molecules in $Ag(gn)_2ClO_4$ and $Cu(gn)_2ClO_4$ are in the *gg* conformation, while those in $TiCl_4gn$ and $SnCl_4gn$ are in the *tt* conformation. Table III-43 summarizes the CH_2 rocking frequencies of glutaronitrile and its metal complexes. An infrared study similar to the above has been extended to adiponitrile ($NC—(CH_2)_4—CN$) and its Cu(I) complex.[516]

TABLE III-43. INFRARED ACTIVE CH_2 ROCKING FREQUENCIES OF GLUTARONITRILE AND ITS METAL COMPLEXES (CM^{-1})

Liquid[c]	945 (*tg*)	904 (*gg*)	835 (*tg, gg*)	757 (*tg, gg*)	737 (*tt*)[a]
Solid[c] (metastable)	943 (*tg*)	–	839 (*tg*)	757 (*tg*)	–
Solid[c] (stable)	–	903 (*gg*)	837 (*gg*)	768 (*gg*)	–
$Cu(gn)_2NO_3$[c]	–	913 (*gg*)	830 (*gg*)[b]	778 (*gg*)	–
$Cu(gn)_2ClO_4$[d]	–	908 (*gg*)	875 (*gg*)	767 (*gg*)	–
$Ag(gn)_2ClO_4$[d]	–	904 (*gg*)	872 (*gg*)	772 (*gg*)	–
$SnCl_4(gn)$[d]	–	–	–	–	733 (*tt*)
$TiCl_4(gn)$[d]	–	–	–	–	730 (*tt*)

[a] The *tt* form should exhibit two IR active CH_2 rocking vibrations. The other one is not known, however.
[b] Overlapped with a NO_3^- absorption.
[c] Reference 513.
[d] Reference 515.

Watt and Klett[517] have assigned the infrared spectra of metal complexes of the type [Pd(dien)X]X where dien is diethylenetriamine ($NH_2—(CH_2)_2—NH—(CH_2)_2—NH_2$) and X is a halogen. Buckingham and Jones[518] have measured the infrared spectra of the $[M(trien)X_2]^+$-type compounds where trien is triethylenetetramine, M is Co(III), Cr(III) or Rh(III), and X is a halogen or an acido anion. Figure III-23 illustrates the structures of three possible isomers. It has been demonstrated that these isomers can be distinguished by comparing the spectra in the regions 3300–3000 and 1100–990 cm^{-1}.

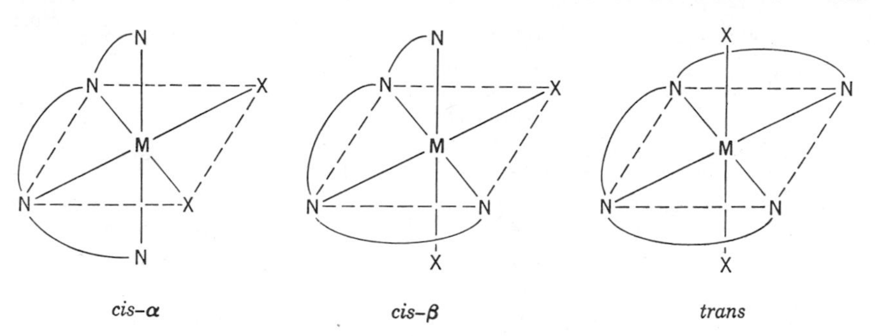

Fig. III-23. Structures of $[M(trien)X_2]^+$-type ions.

III-16. COMPLEXES OF 2,2′-BIPYRIDINE AND RELATED LIGANDS

Infrared spectra of metal complexes of 2,2′-bipyridine (bipy) and 1,10-phenanthroline (phen) have been studied extensively.[519–522] In general, the bands that appear in the high frequency region are not metal-sensitive, since they originate in the heterocyclic or aromatic ring of the ligand. Thus recent work has been concentrated on the low frequency region where the M—N stretching and other skeletal vibrations appear.

Inskeep[523] was the first to assign the M—N stretching bands of these complexes with first row transition metals at 530–250 cm^{-1}. Durig et al.[524] assigned the Pd—N stretchings of the $[Pd(bipy)X_2]$- and $[Pd(py)_2X_2]$-type complexes (X, a halogen) at 310–240 cm^{-1}. Ferraro and co-workers[525] assigned the M—N stretching bands of bipy complexes with rare earth metals at 300–200 cm^{-1}. The Zn—N stretching bands of bipy and trpy (2,2′,2″-terpyridine) complexes were also assigned to the same region.[526] On the other hand, Clark and Williams[527] could find no bands assignable to the M—N stretching modes in the far-infrared spectra (above 200 cm^{-1}) of bipy complexes with Co(II, III), Fe(II), and Ni(II).

As stated in Sec III-9, the metal isotope technique[389] is very useful in assigning the metal-ligand vibrations. Hutchinson et al.[528] applied this technique to the tris-bipy and phen complexes of Fe(II), Ni(II), and Zn(II). Their results indicate that the M—N stretching bands of these compounds are at 375–360, 300–240, and 240–175 cm^{-1} for Fe(II), Ni(II), and Zn(II), respectively. Table III-44 gives the observed frequencies, isotopic shifts, and band assignments for $[Ni(bipy)_3]I_2 \cdot 6H_2O$.

There are many other pyridine derivatives that form metal-chelate rings similar to those of bipy and phen. Infrared studies have been made on metal complexes of biacetyldihydrazone,[529] 2-pyridinaldazine,[530] 2-pyridinaldoxime,[531] 2-pyridylcarbinol,[532] pyridinal-methylamines,[531] and 2,2′-bipyridylamine.[533]

TABLE III-44. METAL ISOTOPE EFFECT ON FAR-INFRARED SPECTRA OF $[Ni(bipy)_3]I_2 \cdot 6H_2O$ (cm^{-1})

bipy	$[Ni(bipy)_3]I_2 \cdot 6H_2O$ Ni^{58}	Ni^{62}	Shift	Probable band assignment
423	437.0	438.4	+1.4	ligand vibration
404	416.6	416.1	−0.5	ligand vibration
–	282.2	278.3	−3.9	Ni—N stretch[b]
–	259.1	254.1	−5.0	Ni—N stretch[b]
(225)[a]	210.5	210.5	0	ligand vibration
–	181.1	181.1	0	ligand vibration
168	167.9	167.9	0	ligand vibration

[a] Observed in Raman spectrum.
[b] Overlapped on ligand vibration.

Infrared studies have also been made on more complicated metal chelate systems such as those of dipyrromethane and prophin derivatives. Infrared spectra have been reported for metal chelates of porphin,[534] tetraphenylporphin,[535] protoporphyrin IX dimethyl ester,[536] hematoporphyrin IX dimethyl ester,[536] chlorophylls,[537] and phthalocyanines.[538–540] Thomas and Martell[535] have made empirical band assignments on the infrared spectra of metal chelates of tetraphenylporphin and its derivatives. They suggest that the bands near 1000 cm^{-1} may be associated with metal-ligand vibrations, since they are metal-sensitive. Boucher and Katz[536] noted that the bands near 970–920, 530–500, and 350 cm^{-1} are metal-sensitive in the infrared spectra of metal complexes they studied. Murakami and Sakata[541] obtained the infrared spectra of metal chelates of dipyrromethane derivatives and found metal-sensitive bands at 400–350 cm^{-1}

III-17. COMPLEXES OF DIMETHYLGLYOXIME

In 1952, Rundle and Parasol[542] noted that the OH stretching band of dimethylglyoxime (DMG—H) near 3100 cm^{-1} is shifted to 1775 cm^{-1} in $Ni(DMG)_2$. This extraordinarily low OH stretching frequency was attributed to strong intramolecular hydrogen bonding, since they observed that, as the OH ··· O distance decreases, the OH stretching frequency becomes lower in a series of compounds in which the OH ··· O distance varies progressively. Later, Nakamoto et al.[543] extended similar "frequency-distance relationships" to other hydrogen bond system such as NH ··· O and OH ··· N. According to x-ray analysis,[544] the OH ··· O bond in $Ni(DMG)_2$ is extremely

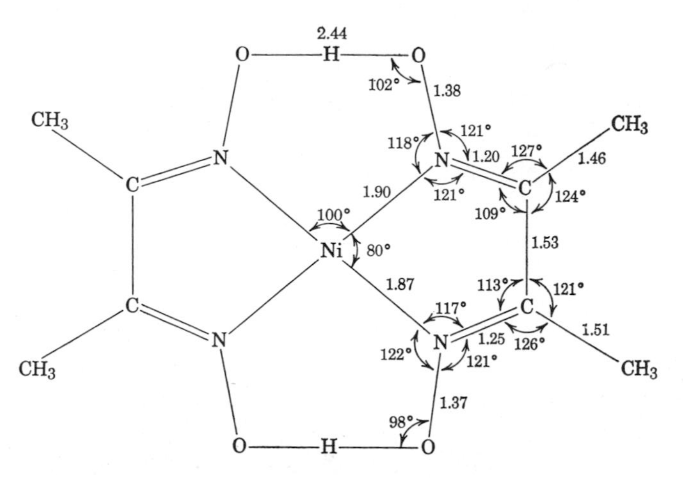

Fig. III-24. The structure of $Ni(DMG)_2$.[544]

short (2.44A) and may be symmetrical, the hydrogen atom being equidistant from the two oxygen atoms. Figure III-24 illustrates the structure of $Ni(DMG)_2$ obtained from x-ray analysis.

Fujita and co-workers[545] found weak OH stretching bands at 1770–1680 cm^{-1} in $[Co(DMG)_2XY]$ (X, Y = Cl^-, Br^-, NO_2^-, or OH_2). Since the presence of a weak OH stretching band in this frequency range is an indication of an OH ··· O bond similar to that of $Ni(DMG)_2$, it was concluded that the structure must be I. (The broken line denotes a hydrogen bond similar to that of $Ni(DMG)_2$.) Nakahara and colleagues[546] have shown that the structure of $Co(DMG)_3$ must be II, since it shows no bands between 1770 and 1680 cm^{-1}. Gillard and Wilkinson[547] proposed structure III for $H[Co(DMG)_2Cl_2]$ (solid state), since it gives the bands characteristic of the free OH(3200 cm^{-1}) and hydrogen-bonded OH(1725 cm^{-1}) groups.

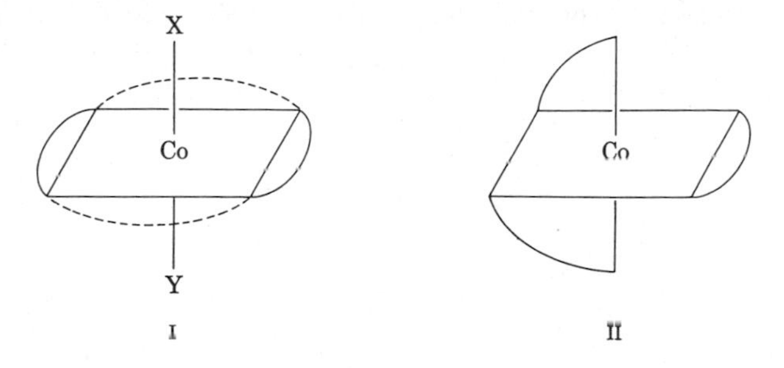

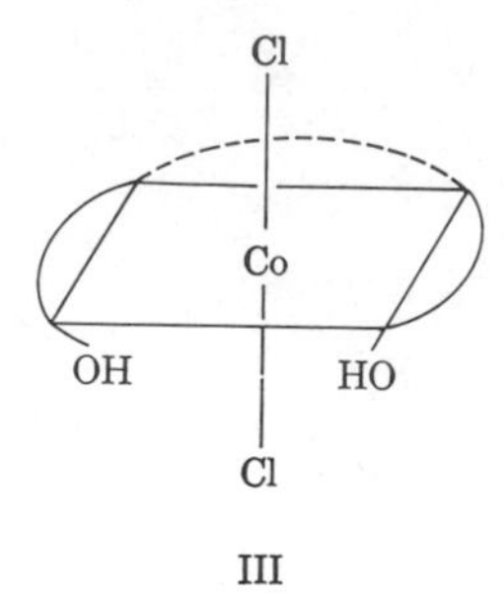

III

Frasson and co-workers[548] made an x-ray analysis on $Pt(DMG)_2$ and obtained an OH ··· O distance of 3.03A. The observed OH stretching frequency (3450 cm^{-1}) of this compound predicts an OH ··· O distance of about 2.9A according to the "frequency-distance relationship."[543]

Blinc and Hadži[549] reinvestigated the infrared spectra of DMG complexes with various metals; they claim that a newly found band at 2350 cm^{-1} in $Ni(DMG)_2$ is the OH stretching band, and that the band at 1775 cm^{-1} is the OH bending band. If this is so, the OH ··· O distance, as calculated from the "frequency-distance relationship," would be 2.56A. Blinc and Hadži, accordingly, suggest that the OH ··· O bond is bent and probably not symmetric.

III-18. COMPLEXES OF AMINO ACIDS, EDTA, AND RELATED COMPOUNDS

Amino acids exist as dipolar (zwitter) ions in the crystalline state. Empirical band assignments have been made for the dipolar ions of glycine,[550] α-alanine[551] and α-aminoisobutyric acid.[552,553] Table III-45 gives the frequencies and band assignments for glycine and α-alanine.

X-ray analysis indicates that, in $[Ni(gly)_2]\cdot 2H_2O$,[554] $[Zn(gly)_2]\cdot H_2O$, and $[Cd(gly)_2]\cdot H_2O$,[555] two glycino anions (gly) coordinate to the metal by forming a *trans* square-planar structure like that shown in Fig. III-25. Furthermore, it has been shown that the oxygens of the carboxyl groups that are not coordinated to the metal are hydrogen-bonded either to the amino group of the neighboring molecule or to water of crystallization, or they are bounded weakly to the metal of the neighboring complex. Thus the COO

Fig. III-25.

TABLE III-45. INFRARED FREQUENCIES AND BAND ASSIGNMENTS OF GLYCINE AND α-ALANINE IN THE CRYSTALLINE STATE (CM^{-1})[550,551]

Glycine	α-Alanine	Band Assignment
1610	1597	$\nu_{as}(COO^-)$
1585	1623	$\delta_d(NH_3^+)$
1492	1534	$\delta_s(NH_3^+)$
–	1455	$\delta_d(CH_3)$
1445	–	$\delta(CH_2)$
1413	1412	$\nu_s(COO^-)$
–	1355	$\delta_s(CH_3)$
1333	–	$\rho_w(CH_2)$
–	1308	$\delta(CH)$
1240(R)	–	$\delta_t(CH_2)$
1131, 1110	1237, 1113	$\rho_r(NH_3^+)$
1033	1148	$\nu_{as}(CCN)$
–	1026, 1015	$\rho_r(CH_3)$
910	–	$\rho_r(CH_2)$
893	918[a], 852[a]	$\nu_s(CCN)$
694	648	$\rho_w(COO^-)$
607	771	$\delta(COO^-)$
516	492	$\rho_t(NH_3^+)$
504	540	$\rho_r(COO^-)$

[a] Both of these bands are coupled vibrations between $\nu_s(CCN)$ and $\nu(C—CH_3)$.

stretching frequencies of complexes of amino acids are affected by coordination as well as by intermolecular interaction.

In order to examine the effect of coordination and hydrogen bonding, Nakamoto et al.[489] have made extensive measurements of the COO stretching frequencies of various metal complexes of amino acids in D_2O solution, in the hydrated crystalline state and in the anhydrous crystalline state. The results show that, in any one physical state, the same frequency order is found for a series of metals, regardless of the nature of the ligand. The antisymmetric frequencies increase, the symmetric frequencies decrease, and the separation between the two frequencies increases in the following order of metals:

$$Ni(II) < Zn(II) < Cu(II) < Co(II) < Pd(II) \approx Pt(II) < Cr(III)$$

An exception is seen in the glycino and α-alanino complexes of Ni(II), Zn(II), and Cu(II) in the hydrated solid state. In general, however, these results indicate that the effect of coordination is still the major factor in determining the frequency order in a given physical state. The frequency order above can best be explained if it is assumed that the covalent character of the M—O bond increases along the series, since an increase of covalent character leads to a more asymmetrical carboxyl group and results in an increase in the frequency separation of the two COO stretching bands. Rosenberg[556] and Quagliano et al.[557] have also discussed the nature of the M—O bond of amino acid complexes from infrared spectra.

Square-planar bis(glycino) complexes can take the *trans* or *cis* configuration. According to x-ray analysis, two glycino anions are in the *cis* configuration in $[Cu(gly)_2]\cdot H_2O$.[558] The *cis* and *trans* isomers of bis(glycino) complexes with Pt(II), Pd(II), and Cu(II) have been prepared and their infrared spectra reported.[559,560] As expected from symmetry consideration, the *cis* isomer exhibits more bands than the *trans* isomer. Octahedral tris(glycino) complexes may take the *cis-cis* and *cis-trans* configurations shown in Fig. III-26. For example, $Co(gly)_3$ exists in two forms: purple crystals, $[Co(gly)_3]\cdot 2H_2O$ (α form), and red crystals, $[Co(gly)_3]\cdot H_2O$ (β form). It has been suggested[561] that the α form is *cis-trans*($\mathbf{C}_1$) and the β form *cis-cis*($\mathbf{C}_3$), since the *d—d** band in the visible region splits into two bands in the α form. In accordance with the electronic spectra, the infrared spectrum of the α form also exhibits more bands than that of the β form, which has a higher symmetry.[562]

In order to give theoretical band assignments on metal glycino complexes, Condrate and Nakamoto[560] have carried out a normal coordinate analysis on the metal-glycino chelate ring. Figure III-27 shows the infrared spectra of bis(glycino) complexes of Pt(II), Pd(II), Cu(II), and Ni(II). Table III-46 lists the observed frequencies and theoretical band assignments. The CH_2 group frequencies are not listed, since they are not metal-sensitive. It is seen that the C=O stretching, NH_2 rocking, and M—N and M—O stretching bands are metal-sensitive and are shifted progressively to higher frequencies as the metal is changed in the order Ni(II) < Cu(II) < Pd(II) < Pt(II). Table III-46

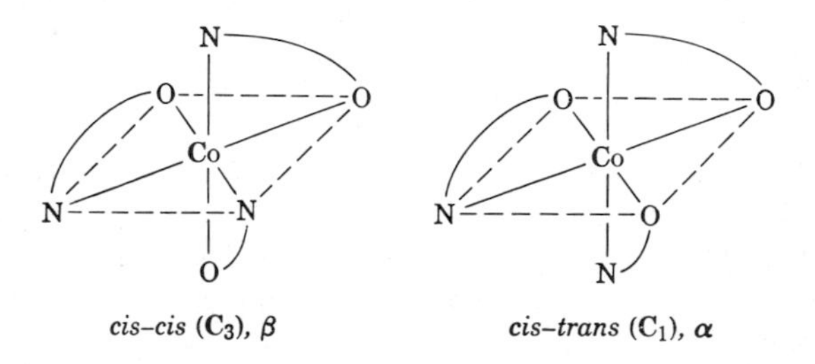

cis–cis ($\mathbf{C}_3$), β   *cis–trans* ($\mathbf{C}_1$), α

Fig. III-26.

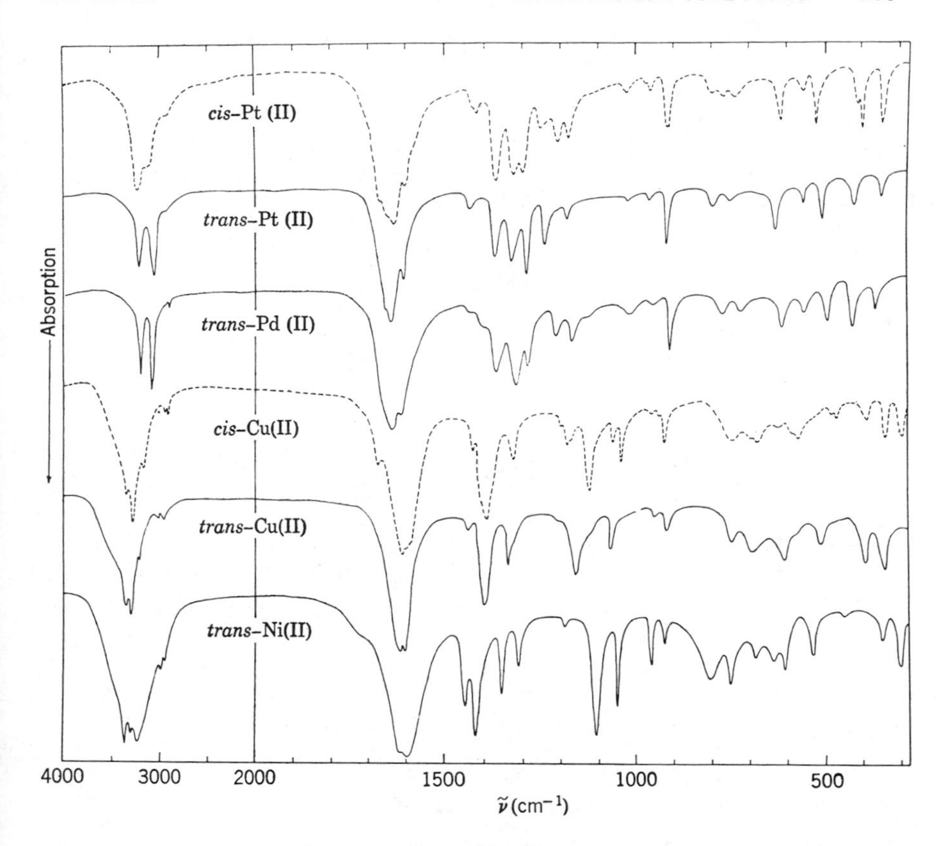

Fig. III-27. Infrared spectra of bis(glycino) complexes of divalent metals.[560]

shows that both the M—N and M—O stretching force constants also increase in the same order of the metals. These results provide further support to the previous discussion on the M—O bonds of glycino complexes.

The infrared spectra of amino acid complexes with these and other metals have been studied by other investigators. Watt and Knifton[563] noted that the Ni—N stretching band of $[Ni(gly)_2]$ at 435 cm^{-1} is shifted to about 480 cm^{-1} in $K_2[Ni(NHCH_2COO)_2]$ in which the coordinated glycino group is deprotonated. This result indicates that the Ni—N bond becomes stronger by deprotonation of the NH_2 group. Other references are: glycino complexes,[559] valine complexes,[564] isoleucine complexes,[565] leucine complexes,[566] and α-alanine complexes.[567] The Raman spectra of glycino complexes with Zn(II), Cd(II), and Be(II) have been reported.[568] Infrared spectra of metal halide complexes with esters of amino acids have been studied.[569]

TABLE III-46. OBSERVED FREQUENCIES AND BAND ASSIGNMENTS OF BIS(GLYCINO) COMPLEXES (cm^{-1})

trans-[Pt(gly)$_2$]	*trans*-[Pd(gly)$_2$]	*trans*-[Cu(gly)$_2$]	*trans*-[Ni(gly)$_2$]	Band Assignment
3230, 3090	3230, 3120	3320, 3260	3340, 3280	$\nu(NH_2)$
1643	1642	1593	1589	ν(C=O)
1610	1616	1608	1610	$\delta(NH_2)$
1374	1374	1392	1411	ν(C—O)
1245	1218	1151	1095	$\rho_t(NH_2)$
1023	1025	1058	1038	$\rho_w(NH_2)$
792	771	644	630	$\rho_r(NH_2)$
745	727	736	737	δ(C=O)
620	610	592	596	π(C=O)
549	550	439	439	ν(M—N)
415	420	360	290	ν(M—O)
2.10	2.00	0.90	0.70	K(M—N), mdyn/A
2.10	2.00	0.90	0.70	K(M—O), mdyn/A

Glycine also coordinates to the Pt(II) atom as a unidentate ligand.

$$-\overset{|}{\underset{|}{Pt}}-NH_2-CH_2-C\begin{matrix}\nearrow O\\ \searrow OH\end{matrix} \qquad -\overset{|}{\underset{|}{Pt}}-NH_2-CH_2-C\begin{matrix}\nearrow O^{-1/2}\\ \searrow O^{-1/2}\end{matrix}$$

The carboxyl group is not ionized in *trans*-[Pt(glyH)$_2$X$_2$](X, a halogen), whereas it is ionized in *trans*-[Pt(gly)$_2$(NH$_3$)$_2$]. The former exhibits the un-ionized COO stretching band near 1710 cm^{-1}, while the latter shows the ionized COO stretching band near 1610 cm^{-1}.[570]

The distinction of uni- and bidentate glycino complexes of Pt(II) can be made readily from their infrared spectra. Figure III-28 illustrates the infrared spectra of *trans*-[Pt(glyH)$_2$Cl$_2$] and K[Pt(gly)Cl$_2$] in the COO stretching and Pt—O stretching regions. The bidentate (chelated) glycino group absorbs at 1643 cm^{-1}, which is different from either the ionized unidentate group (1610 cm^{-1}) or the unionized unidentate group (1710 cm^{-1}). Furthermore, the bidentate glycino group exhibits the Pt—O stretching band at 388 cm^{-1}, whereas the unidentate glycino group has no absorption between 470 and 350 cm^{-1}. Figure III-28 also shows the spectrum of [Pt(gly)(glyH)Cl] in which both the unidentate and bidentate glycino groups are present. It is seen that the spectrum of this compound can be interpreted as a superposition of those of the former two compounds.[570]

The infrared spectra of metal glycolato complexes have been studied by Nakamoto et al.[571]

$$\begin{array}{c} \quad\quad\quad O_1 \\ \quad\quad\quad // \\ H_2C - C \\ / \quad\quad \backslash \\ H{-}O_3 \quad\quad O_2 \\ \backslash \quad / \\ M \end{array}$$

According to normal coordinate analysis, bis(glycolato) cobalt(II) exhibits the bands characteristic of the coordinated OH and COO groups at 3090 (OH stretching), 1595 (C=O stretching), 1482 (OH bending), 1402 ($C—O_2$ stretching), 1062 ($C—O_3$ stretching), 689 (C=O bending), 540 (ring deformation), and 460 and 300 cm^{-1} (Co—O stretchings). The infrared spectra of metal complexes of oxy-acids such as lactic, tartaric, and citric acids have

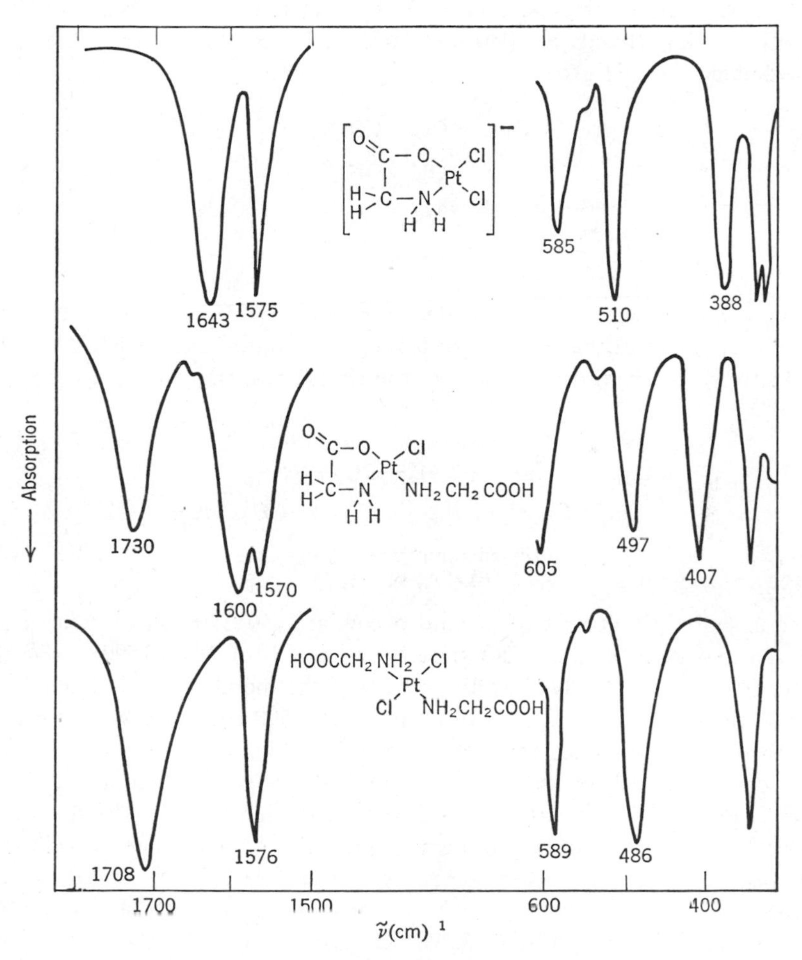

Fig. III-28. Infrared spectra of $K[Pt(gly)Cl_2]$, $[Pt(gly)(glyH)Cl]$ and trans-$[Pt(glyH)_2Cl_2]$.[570]

been studied. Goulden[572] has found that the OH in-plane bending band of the lactate ion at 1275 cm^{-1} is shifted to 1390 cm^{-1} upon chelation to zinc. Girard and Lecomte,[573] from infrared studies, have proposed the following structure for bismuth and antimony tartrates.

```
 [ O                     O ]
 [  \\                  // ]
 [    C—CH——CH—C        ]  K        M = Bi(III) or Sb(III)
 [  //    |     |    \\ ]
 [ O      O     O      O ]
 [         \   /          ]
 [           M            ]
```

Kirschner and Kiesling[574] proposed the following structure for copper tartrate, since it shows the coordinated COO stretching band at 1634 cm^{-1} and C—O stretching bands at 1080 and 1063 cm^{-1}, the latter being due to the coordinated C—OH group.

```
OOC—CH—CH—COO
  \   |   |   /
   \  OH  OH /
    \  \    /
       Cu
```

Busch and co-workers[575] have determined from infrared spectra the coordination number of the metals in metal chelate compounds of EDTA and its derviatives. The method is based on the simple rule that the un-ionized and

```
HOOCH2C                    CH2COOH
        \                  /
         N—CH2—CH2—N
        /                  \
HOOCH2C                    CH2COOH
```

Ethylenediaminetetraacetic acid
(EDTA) or (H_4Y)

uncoordinated COO stretching band occurs at 1750–1700 cm^{-1}, whereas the ionized and coordinated COO stretching band is at 1650–1590 cm^{-1}. The latter frequency depends upon the nature of the metal: 1650–1620 cm^{-1} for metals such as Cr(III) and Co(III), and 1610–1590 cm^{-1} for metals such as Cu(II) and Zn(II). Since the free ionized COO stretching band is at 1630–1575 cm^{-1}, it is also possible to distinguish the coordinated and free COO^- stretching bands if a metal such as Co(III) is chosen for complex formation. Table III-47 shows the results obtained by Busch et al. The coordination number of EDTA in other metal complexes (rare earth metals[577] and Rh(III)[578]) has also been studied by using the same method. Sawyer and co-workers[579] concluded that the M—O bond is covalent if the COO stretching band is between 1660 and 1630 cm^{-1}.

TABLE III-47. ANTISYMMETRIC COO STRETCHING FREQUENCIES AND NUMBER OF FUNCTIONAL GROUPS USED FOR COORDINATION IN EDTA COMPLEXES (cm^{-1})[575]

Compound	Un-ionized COOH	Coordinated $COO^- \cdots M$	Free COO^-	Number of Coordinated Groups
$H_4[Y]$	1697	–	–	[a]
$Na_2[H_2Y]$	1668	–	1637	[a]
$Na_4[Y]$	–	–	1597	[a]
$Ba[Co(Y)]_2 \cdot 4H_2O$	–	1638	–	6
$Na_2[Co(Y)Cl]$	–	1648	1600	5
$Na_2[Co(Y)NO_2]$	–	1650	1604	5
$Na[Co(HY)Cl] \cdot \frac{1}{2}H_2O$	1750	1650	–	5
$Na[Co(HY)NO_2] \cdot H_2O$	1745	1650	–	5
$Ba[Co(HY)Br] \cdot 9H_2O$	1723	1628	–	5
$Na[Co(YOH)Cl] \cdot \frac{3}{2}H_2O$	–	1658	–	5
$Na[Co(YOH)Br] \cdot H_2O$	–	1654	–	5
$Na[Co(YOH)NO_2]$	–	1652	–	5
$[Pd(H_2Y)] \cdot 3H_2O$	1740	1625	–	4
$[Pt(H_2Y)] \cdot 3H_2O$	1730	1635	–	4
$[Pd(H_4Y)Cl_2] \cdot 5H_2O$	1707, 1730	–	–	2
$[Pt(H_4Y)Cl_2] \cdot 5H_2O$	1715, 1730	–	–	2

[Y], tetranegative ion; [HY], trinegative ion; $[H_2Y]$, dinegative ion; $[H_4Y]$, neutral species of EDTA; [YOH], trinegative ion of HEDTA (hydroxyethylenediaminetri-acetic acid).

[a] Reference 576.

Tomita and Ueno[580] have studied the infrared spectra of metal complexes of NTA using the method described above. They concluded that NTA acts

$$N\begin{cases} CH_2COOH \\ CH_2COOH \\ CH_2COOH \end{cases} \qquad \text{Nitrilotriacetic acid (NTA)}$$

as a quadridentate ligand in complexes of Cu(II), Ni(II), Co(II), Zn(II), Cd(II), and Pb(II), and as a tridentate in complexes of Ca(II), Mg(II), Sr(II), and Ba(II).

III-19. INFRARED SPECTRA OF AQUEOUS SOLUTIONS

Infrared studies of aqueous solution provide a valuable tool for elucidating the structure of complex ions in equilibria. In order to observe the infrared spectrum of an aqueous solution, it is necessary to use window materials such as AgCl and BaF_2, which are insoluble in water, and a thin spacer (0.02–0.01 mm) to reduce strong absorption of water. The latter condition necessitates a solution of relatively high concentration. Even if these conditions are met, it is still difficult to measure the spectrum of a solute in the regions at 3700–2800, 1800–1600 cm^{-1} and below 1000 cm^{-1}, where water (H_2O) absorbs strongly. The C≡N stretching band (2200–2000 cm^{-1}) can be measured in aqueous solution, since it is outside of these regions. Thus solution equilibria of cyano complexes have been studied extensively by using aqueous infrared spectroscopy (Sec. III-5). Fronaeus and Larsson[581] have extended similar studies to thiocyanato complexes that exhibit the C≡N stretching bands in the same region. They[582] have also studied the solution equilibria of oxalato complexes in the 1500–1200 cm^{-1} region, where the CO stretching bands of the coordinated oxalato group appear. Larsson[583] studied the infrared spectra of metal glycolato complexes in aqueous solution. In this case, the C—OH stretching band near 1060 cm^{-1} was used to elucidate the structure of the complex ions in equilibria.

If D_2O is used instead of H_2O, it is possible to observe infrared spectra in the regions 4000–2900, 2000–1300, and 1100–900 cm^{-1}. The COO stretching bands of NTA, EDTA, and their metal complexes appear between 1750 and 1550 cm^{-1} (Sec. III-18). Nakamoto and co-workers,[584] therefore, studied the solution (D_2O) equilibria of NTA, EDTA, and related ligands in this frequency region. By combining the results of potentiometric studies with the spectra obtained as a function of the pH(pD) of the solution, it was possible to establish the following COO stretching frequencies:

Type A, un-ionized carboxyl ($R_2N\text{-}CH_2COOH$), 1730–1700 cm^{-1}
Type B, α-ammonium carboxylate ($R_2N^+H\text{-}CH_2COO^-$), 1630–1620 cm^{-1}
Type C, α-aminocarboxylate ($R_2N\text{-}CH_2COO^-$), 1595-1575 cm^{-1}

As stated in Sec. III-18, the coordinated (ionized) COO group absorbs at 1650–1620 cm^{-1} for Cr(III) and Co(III), and at 1610–1590 cm^{-1} for Cu(II) and Zn(II). Thus it is possible to distinguish the coordinated COO group from those of Types B and C if a proper metal ion is selected.

Tomita et al.[585] studied the complex formation of NTA with Mg(II) by aqueous infrared spectroscopy. Figure III-29 shows the infrared spectra of equimolar mixtures of NTA and $MgCl_2$ at concentrations about 5–10% by weight. The spectra of the mixture from pD 3.2 to 4.2 exhibit a single band at 1625 cm^{-1}, which is identical to that of the free NTA^{3-} ion in the same pD range.[586] This result indicates that no complex formation occurs in this pD range, and that the 1625 cm^{-1} band is due to the NTA^{3-} ion (Type B). If the pD is raised to 4.2, a new band appears at 1610 cm^{-1}, which is not observed for the free NTA solution over the entire pD range investigated. Figure III-29 shows that this 1610 cm^{-1} band becomes stronger and the 1625 cm^{-1} band becomes weaker as the pD increases. It was concluded that this change is due mainly to the shift of the following equilibrium in the direction of complex formation:

$$\mathrm{H\overset{+}{N}}\begin{cases}\mathrm{CH_2COO^-}\\ \mathrm{CH_2COO^-}\\ \mathrm{CH_2COO^-}\end{cases} + \mathrm{Mg^{2+}} \rightleftharpoons \left[\mathrm{N}\begin{cases}\mathrm{CH_2COO^-}\cdots\\ \mathrm{CH_2COO^-}\cdots \mathrm{Mg}\\ \mathrm{CH_2COO^-}\cdots\end{cases}\right]^- + \mathrm{H^+}$$

$1625\ \mathrm{cm^{-1}}$ (Type B) $\qquad$ $1610\ \mathrm{cm^{-1}}$

By plotting the intensity of these two bands as a function of pD, the stability constant of the complex ion was calculated to be 5.24. This value is in good agreement with that obtained from potentiometric titration (5.41).

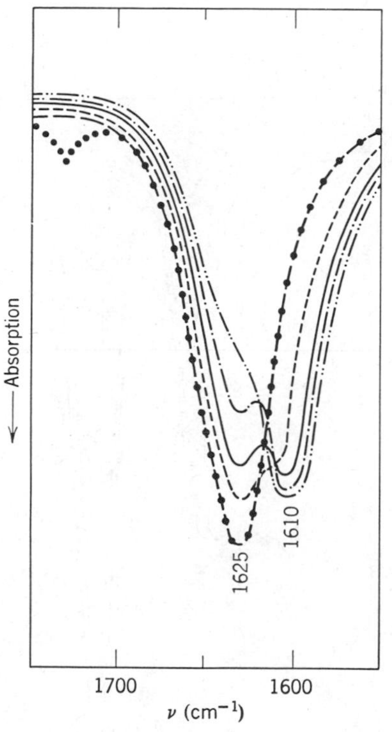

Fig. III-29. Infrared spectra of Mg—NTA complex in D_2O solutions: . . . . . , pD 3.2; ———, pD 4.2; - - - - -, pD 5.5; ———, pD 6.8; —·—·—, pD 10.0; ··—··— pD 11.6. (Ref. 586.)

Kim and Martell[587] have carried out an extensive study on solution equilibria involving the formation of Cu(II) complexes with various polypeptides. As an example, the glycylglycino-Cu(II) system is discussed below. Figure III-30 illustrates the infrared spectra of free glycylglycine in D_2O solution as a function of pD. The observed spectral changes were interpreted in terms of solution equilibria shown below:

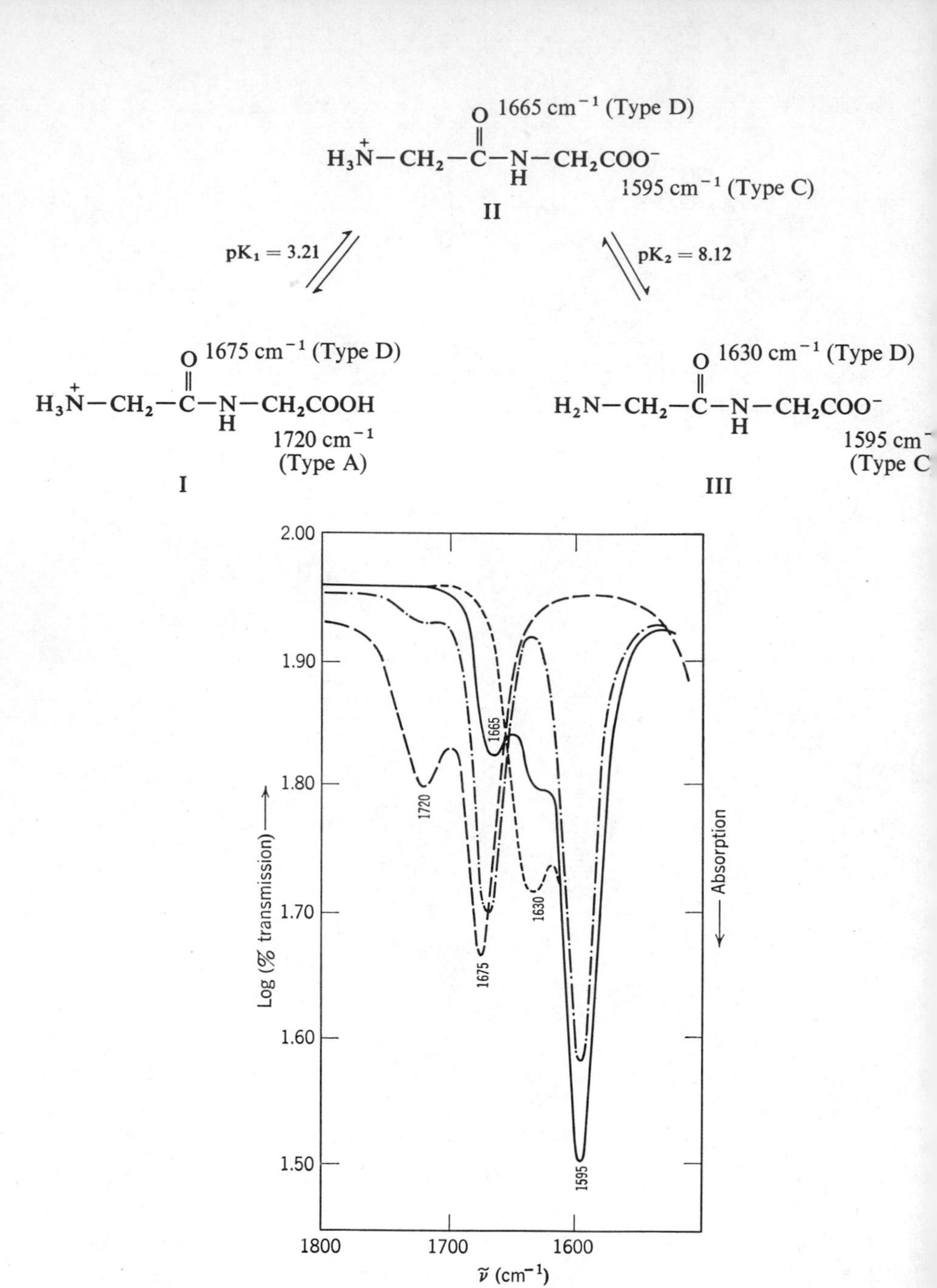

Fig. III-30. Infrared absorption spectra of glycylglycine in D_2O solution at 0.288 *M* concentration and ionic strength 1.0, adjusted with KC1; ———, pD 1.75; -·-·-·-, pD4.31; ———, pD 8.77; --------, pD 10.29. (Ref. 587.)

Band assignments have been made by using the criteria given previously. In addition, type D frequency (1680–1610 cm^{-1}) was introduced to denote the peptide carbonyl group. The exact frequency of this group depends upon the nature of the neighboring groups.

Figure III-31 shows the infrared spectra of glycylglycine mixed with copper chloride at equimolar ratio in D_2O solution.[588] At pD = 3.58, the ligand exhibits three bands at 1720, 1675, and 1595 cm^{-1} (Fig. III-30). This result indicates that I and II are in equilibrium. At the same pD value, however, the mixture exhibits one extra band at 1625 cm^{-1}. This band was attributed to the metal complex (IV), which was formed by the reaction:

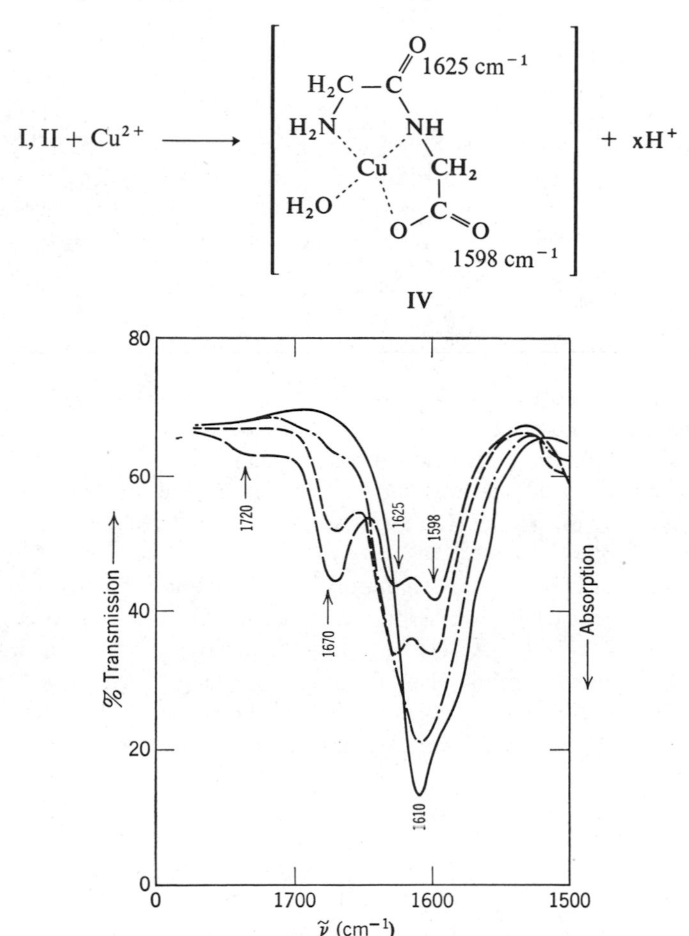

Fig. III 31. Infrared spectra of Cu(II)-glycylglycine complexes in aqueous (D_2O) solutions (1:1). — — —, pD = 3.58; - - - -, pD = 4.24; —·—·—, pD = 5.18; ————, pD = 10.65. Total concentration of ligand and metal is 0.2333*M*, and ionic strength is 1.0 adjusted with KCl).[588]

At pD = 5.18, the solution exhibits one broad band at about 1610 cm^{-1}. This result was interpreted as an indication that the following equilibrium was shifted almost completely to the right-hand side, and that the 1610 cm^{-1} band is an overlap of two bands at 1610 and 1598 cm^{-1}.

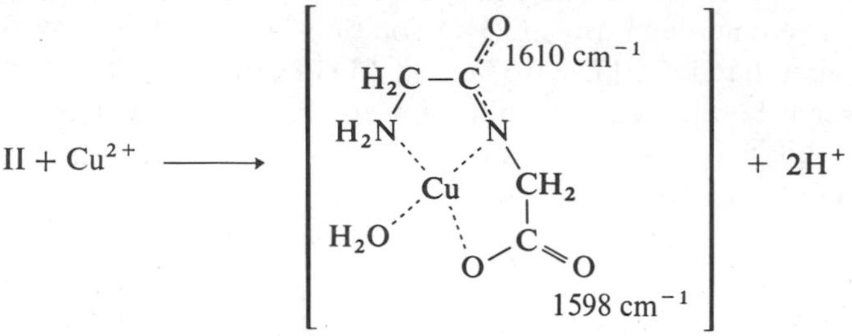

$$II + Cu^{2+} \longrightarrow [\ldots] + 2H^+$$

O
1610 cm^{-1}
H_2C-C
H_2N N
Cu CH_2
H_2O C
O O
1598 cm^{-1}

V

The shift of the peptide carbonyl stretching band from 1625(IV) to 1610 cm^{-1}(V) may indicate the ionization of the peptide NH hydrogen, since such an ionization results in the resonance of the O—C—N system as indicated by the dotted line in structure V. Kim and Martell[589] also studied the triglycine and tetraglycine Cu(II) systems.

III-20. COMPLEXES OF OXALIC ACID AND RELATED COMPOUNDS

The infrared spectra of oxalato ($[ox]^{2-}$) complexes have been studied extensively. Empirical band assignments have been made by using the results of normal coordinate analysis on the free oxalate ion.[590–593] Fujita et al.[594] have carried out a normal coordinate analysis on the metal-chelate ring of oxalato complexes. Table III-48 gives the observed frequencies and band assignments obtained.

X-ray analysis[595] indicates that the four C–O bonds in the free ion are equivalent, the bond distance being 1.27A. The C–O stretching force constant in this ion is 7.20 mdyn/A (UBF).[592] In the Cr(III) complex, however, the two C—O_I bonds coordinated to the metal are lengthened (1.39A) and the two C—O_{II} bonds free from coordination are shortened (1.17A).[596] Accordingly, the C—O_I stretching force constant is reduced to 5.30 mdyn/A, whereas

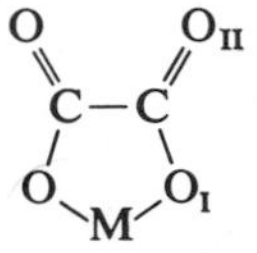

O O_{II}
C—C
O O_I
M

the C—O_{II} stretching force constant is increased to 9.40 mdyn/A (UBF).[594] Also, the C—C distance in the free ion is 1.56A (2.50 mdyn/A) and that in the complex is 1.25A (3.60 mdyn/A).

TABLE III-48. FREQUENCIES AND BAND ASSIGNMENTS IN VARIOUS OXALATO COMPLEXES (cm^{-1})[594]

$K_2[Zn(ox)_2]\cdot 2H_2O$	$K_2[Cu(ox)_2]\cdot 2H_2O$	$K_2[Pd(ox)_2]\cdot 2H_2O$	$K_2[Pt(ox)_2]\cdot 3H_2O$	$K_3[Fe(ox)_3]\cdot 3H_2O$	$K_3[V(ox)_3]\cdot 3H_2O$	$K_3[Cr(ox)_3]\cdot 3H_2O$	$K_3[Co(ox)_3]\cdot 3H_2O$	$K_3[Al(ox)_3]\cdot 3H_2O$	$[Cr(NH_3)_4(ox)]\cdot Cl$	Band Assignment	
1632	(1720) 1672	1698	1709	1712	1708	1708	1707	1722	1704	$\nu_{as}(C{=}O)$	ν_7
–	1645	1675, 1657	1674	1677, 1649	1675, 1642	1684, 1660	1670	1700, 1683	1668	$\nu_{as}(C{=}O)$	ν_1
1433	1411	1394	1388	1390	1390	1387	1398	1405	1393	$\nu_s(C{-}O$ $+\nu(C-C)$	ν_2
1302	1277	1245 (1228)	1236	1270, 1255	1261	1253	1254	1292, 1269	1258	$\nu_s(C{-}O)$ $+\delta(O{-}C{=}O)$	ν_8
890	886	893	900	885	893	893	900	904	914, 890	$\nu_s(C{-}O)$ $+\delta(O{-}C{=}O)$	ν_3
785	795	818	825	797, 785	807, 797	810, 798	822, 803	820, 803	804	$\delta(O{-}C{=}O)$ $+\nu(M{-}O)$	ν_9
622	593	610	–	580	581	595	–	–	–	crystal water?	
519	541	556	575, 559	528	531	543	565	587	545	$\nu(M{-}O)$ $+\nu(C{-}C)$	ν_4
519	481	469	469	498	497	485	472	436	486, 469	ring def. $+\delta(O{-}C{=}O)$	ν_{10}
428, 419	420	417	405	366	368	415	446	485	366	$\nu(M{-}O)$ +ring def.	ν_{11}
377, 364	382, 370	368	370	340	336	358	364	364	347	$\delta(O-C{=}O)$	ν_5
291	339	350	323	–	–	313	332	–	328	π	

These results suggest that, as the M—O bond becomes stronger, the C—O_I becomes weaker and the C—O_{II} and the C—C bonds become stronger. Thus a shift of the M—O stretching band to a higher frequency is expected to accompany a shift of the C—O_I stretching band to a lower frequency, together with a shift of the C—O_{II} stretching band to a higher frequency. According to normal coordinate analysis,[594] both C—O_I and M—O stretching vibrations are coupled with other modes (Table III-48). Nevertheless, linear relationships exist between ν_4 (predominantly M—O stretching) and the average of ν_1 and ν_7 (both are C—O_{II} stretching), and between ν_4 and either ν_2 or ν_8 (both are predominantly C—O_I stretching). These correlations are illustrated in Fig. III-32 for the oxalato complexes of divalent metals.

In oxalato complexes of trivalent metals, strict linear relationships between the frequencies do not exist. It is found, however, that ν_4 and ν_{11} (both are predominantly M—O stretching) increase along the series Fe < V < Cr < Co < Al, which is the same order as that observed in a series of acetylacetonato complexes (Sec. III-21).

The Raman spectra of metal oxalato complexes have also been examined to investigate the solution equilibria and the nature of the M—O bond.[597]

Kuroda et al.[598] have interpreted the infrared spectra of metal oxamido complexes on the basis of the structure (V_h symmetry): Armendarez and

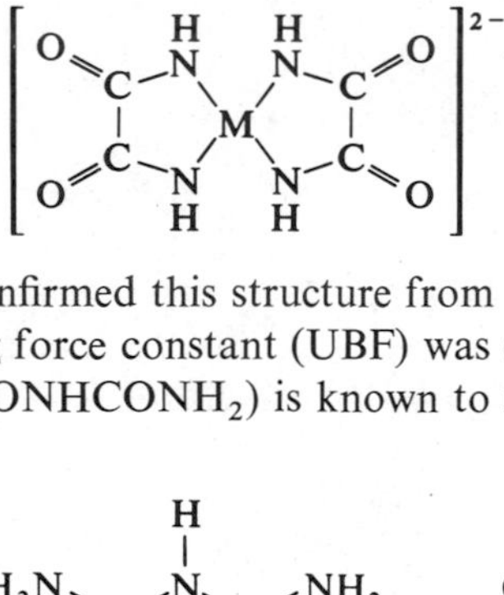

M = Ni or Cu

Nakamoto[599] confirmed this structure from normal coordinate analysis; the Ni—N stretching force constant (UBF) was estimated to be 0.73 mdyn/A.

Biuret ($NH_2CONHCONH_2$) is known to form the following two types of chelate rings:

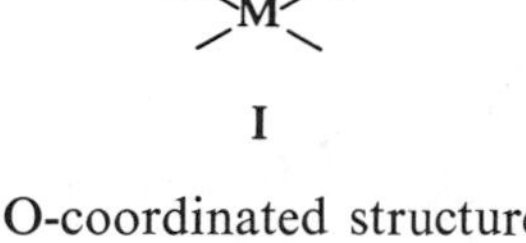

I 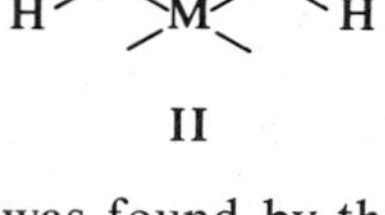 II

The bidentate, O-coordinated structure (I) was found by the x-ray analysis of bis(biuret)copper chloride[600] and bis(biuret)zinc(II) chloride.[601] The bidentate, N-coordinated structure (II) was found by the x-ray analysis of potassium bis(biureto)cuprate(II) tetrahydrate[600] whose violet color is responsible for the well-known " biuret reaction." Aida et al.[602] made the

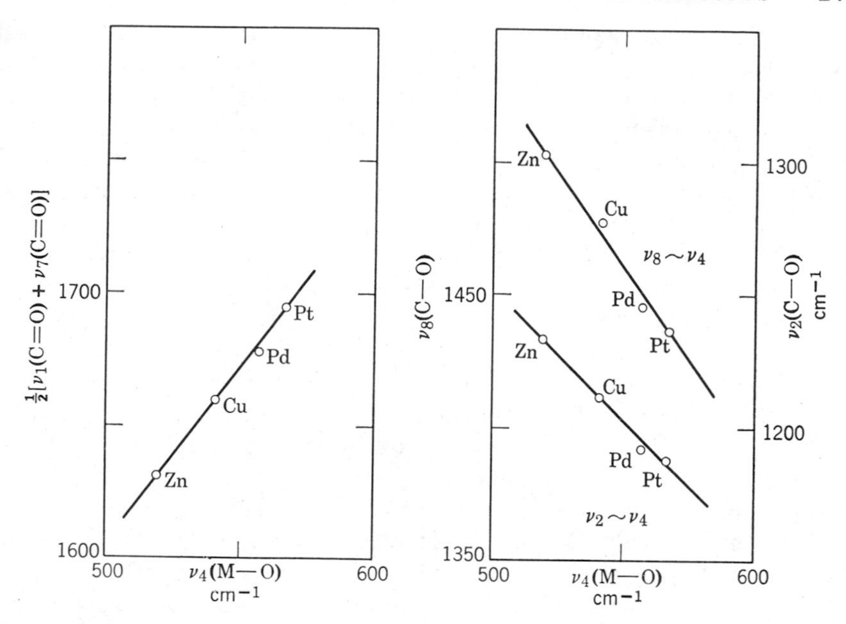

Fig. III-32. M—O stretching frequency vs. C═O and C—O stretching frequencies in oxalato complexes of divalent metals.

empirical band assignments on the infrared spectrum of the N–bonded Cu(II) complex. McLellan and Melson [603] have shown that Cu(II) and Ni(II) form both N-bonded and O-bonded isomers, whereas Mn(II) gives only the N-bonded form. Kedzia et al.[604] have carried out normal coordinate analyses on the Cu(II) complexes of both forms.

III-21. COMPLEXES OF β-DIKETONES

(1) Complexes of Acetylacetone

A number of β-diketones form metal chelate rings of type A:

$$\begin{array}{c} R_{II} \\ | \\ R_I-C \overset{C}{\cdots} C-R_{III} \\ \vdots \qquad \vdots \\ O \diagdown_{M} \diagup O \end{array} \qquad \text{(Type A)}$$

Among them, acetylacetone(acacH) is most common ($R_I = R_{III} = CH_3$ and $R_{II} = H$). According to x-ray analysis on $Fe(acac)_3$,[605] the chelate ring is planar and symmetrical ($\mathbf{C}_{2v}$ symmetry), and the two C$\overset{...}{-}$O bonds are equivalent, as are the two C$\overset{...}{-}$C bonds in the ring.

The infrared spectra of metal acetylacetonate complexes have been studied extensively. In 1956, Mecke and Funck[606] assigned the infrared spectra of acetylacetone (keto and enol forms) and its metal complexes on an empirical basis. In 1960, Nakamoto and Martell[607] carried out an approximate normal coordinate analysis on $Cu(acac)_2$. At the time of this earlier research, no spectral data were available below 400 cm^{-1} and no isotopically substituted acetylacetonato complexes were studied. Since then, a number of investigators[608–610a] have extended their measurements to the far-infrared region. The spectra of metal complexes containing C^{13}- and O^{18}-labeled acetyacetones have been obtained.[611,612] A rigorous normal coordinate analysis considering all the atoms in the molecule has been made for $M(acac)_2$ and $M(acac)_3$-type complexes.[613] Table III-49a lists the observed frequencies, band assignments, and M—O stretching force constants obtained by Mikami et al.[613] Figure III-33a shows the infrared spectra of six acetylacetonato complexes.

The nature of the 1577 and 1529 cm^{-1} bands of $Cu(acac)_2$ has been a subject of controversy. Originally, Nakamoto and Martell[607] assigned the former to a C⋯C stretching and the latter to a C⋯O stretching mode. Mikami et al.[613] also gave similar assignments although they showed that these two modes are coupled slightly with each other. However, the results of the C^{13} and O^{18} experiments [611,612] suggest that the former is a pure C⋯O stretching and the latter is a pure C⋯C stretching. A further theoretical study is needed to solve this discrepancy.

The M—O stretching vibrations of acetylacetonato complexes are most important since they provide direct information about the strength of the M—O bonds. Thus far, they have been assigned from normal coordinate analyses[607,613] without experimental evidence. As stated in Sec. III-9, the metal isotope technique[389] provides a unique method to detect the metal-ligand vibrations. Nakamoto et al.[614a], therefore, applied this technique to acetylacetonato complexes of Fe(III), Cr(III), Pd(II), Cu(II), and Ni(II). Their results indicate that previous band assignments on the M—O stretching vibrations are essentially correct. As an example, Table III-49b gives the observed frequencies, isotopic shifts, and band assignments for $Cr(acac)_3$. Figure III-33b shows an actual tracing of the infrared spectra of $Cr^{50}(acac)_3$ and its Cr^{53} analog. Two bands at 463 and 358 cm^{-1} of the former must be assigned to the Cr—O stretching modes since they give large shifts relative to others. On the other hand, Pinchas et al.[611] assigned the 592 cm^{-1} band to a pure Cr—O stretching since it gives the largest isotopic shift (19 cm^{-1}) by the O^{16}—O^{18} substitution of $Cr(acac)_3$. However, this band is shifted by only 0.7 cm^{-1} by the Cr^{50}—Cr^{53} substitution, and cannot be assigned to a pure Cr—O stretching mode. It may be due to a ligand vibration in which the oxygen atom moves appreciably (e.g., C⋯O out-of-plane bending). It

TABLE III-49a. OBSERVED FREQUENCIES AND BAND ASSIGNMENTS OF ACETYLACETONATO COMPLEXES[613] (CM^{-1})

$Cu(acac)_2$	$Pd(acac)_2$	$Fe(acac)_3$	Predominant Mode
3072	3070	3062	$\nu(CH)$
2987, 2969, 2920	2990, 2965, 2920	2895, 2965, 2920	$\nu(CH_3)$
1577	1569	1570	$\nu(C\cdots C) + \nu(C\cdots O)$
1552	1549		combination
1529	1524	1525	$\nu(C\cdots O) + \nu(C\cdots C)$
1461	(1425)	1445	$\delta(CH) + \nu(C\cdots C)$
1413	1394	1425	$\delta_d(CH_3)$
1353	1358	1385, 1360	$\delta_s(CH_3)$
1274	1272	1274	$\nu(C-CH_3) + \nu(C\cdots C)$
1189	1199	1188	$\delta(CH) + \nu(C-CH_3)$
1019	1022	1022	$\rho_r(CH_3)$
936	937	930	$\nu(C\cdots C) + \nu(C\cdots O)$
780	786, 779	801, 780, 771	$\pi(CH)$
684	700	671, 667[a]	$\nu(C-CH_3)$ + ring def. + $\nu(M-O)$
653	678	656	$\pi\left(CH_3-C\genfrac{}{}{0pt}{}{\diagup C}{\diagdown O}\right)$
612	661	559, 548[a]	ring def. + $\nu(M-O)$
451	463	433	$\nu(M-O) + \nu(C-CH_3)$
431	441	415, 408	ring def.
291	294	298	$\nu(M-O)$
1.45	1.85	1.30	K(M—O), mdyn/A (UBF)

[a] Pure ring deformation.

should be noted that an isotopic substitution of the α-atom of the ligand (atom bonded directly to a metal) shifts the metal-ligand as well as ligand vibrations involving the motion of the α-atom. Thus, the metal-ligand vibrations cannot be uniquely assigned by this method.

Behnke and Nakamoto[615] prepared three deutero analogs of $K[Pt(acac)Cl_2]$, and calculated a set of UBF force constants to fit the observed frequencies of four isotopically substituted species. Table III-50 lists the observed

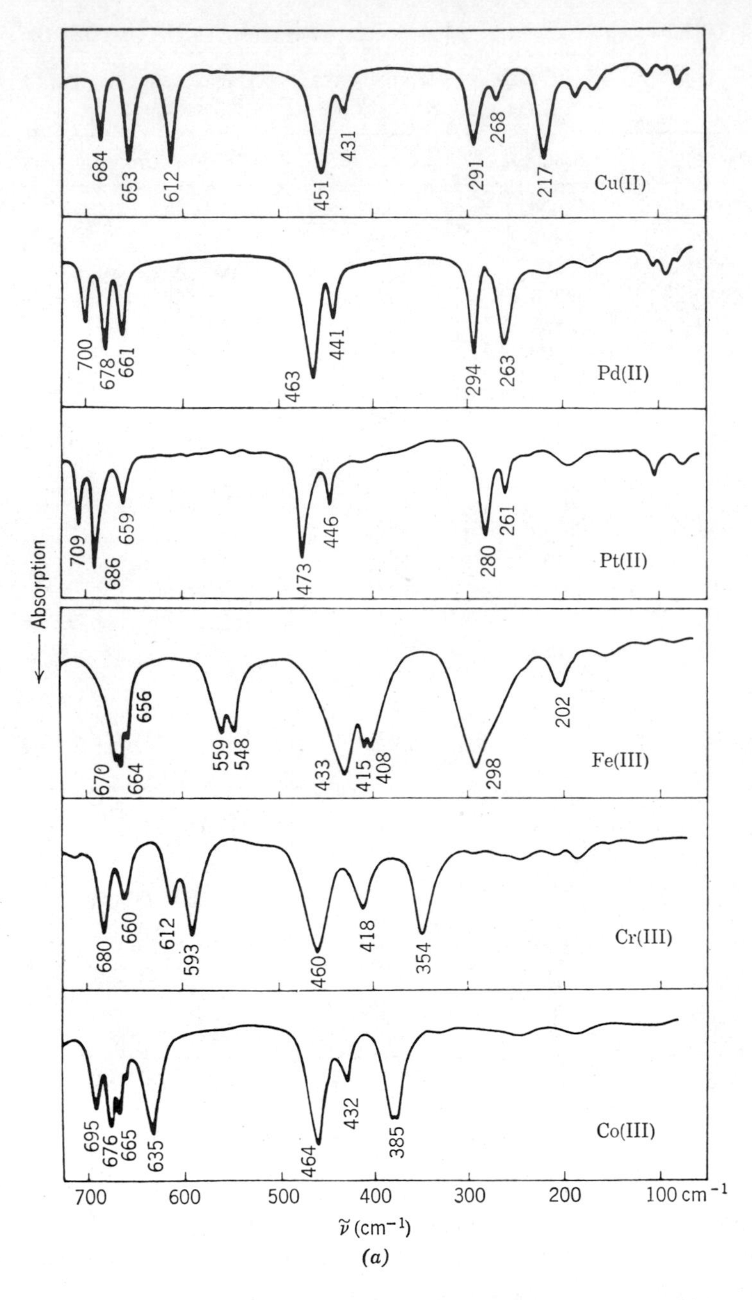

Fig. III-33a. Far-infrared spectra of bis- and tris-acetylacetonato complexes.

TABLE III-49b. OBSERVED FREQUENCIES, ISOTOPIC SHIFTS AND BAND ASSIGNMENTS FOR $Cr(acac)_3$ (cm^{-1})

Pinchas et al.[611]			Nakamoto et al.[614a]		
$\tilde{\nu}$	Shift[a]	Assignment[c]	$\tilde{\nu}$	Shift[b]	Assignment[c]
801 } 766 }		π(CH)	791.5 }	0.0	
			776.0 }	0.0	π(CH)
			772.0 }	0.0	
678	11	ring + ν(CrO)	681.0	0.5	ring + ν(CrO)
661	3	π(ring)	659.5	0.0	π(ring)
592	19	ν(CrO)	596.5	0.7	ring + ν(CrO)
456	5	δ(CR) + ν(CrO)	463.4	3.0	ν(CrO) + δ(CR)
416	8	δ(OCrO)	417.8	0.0	ring
			358.4	3.9	ν(CrO)

[a] $\tilde{\nu}(O^{16}) - \tilde{\nu}(O^{18})$.
[b] $\tilde{\nu}(Cr^{50}) - \tilde{\nu}(Cr^{53})$.
[c] Ring, ring deformation; R, CH_3.

frequencies and band assignments for the nondeuterated species. According to their calculation, the 1563 cm^{-1} band of $K[Pt(acac)Cl_2]$ is the C$\cdots$O stretching and the next band at 1538 cm^{-1} is the C$\cdots$C stretching. Furthermore, another C$\cdots$O stretching band appears strongly at 1380 cm^{-1} with partial overlap on the CH_3 symmetric deformation mode at 1363 cm^{-1}.

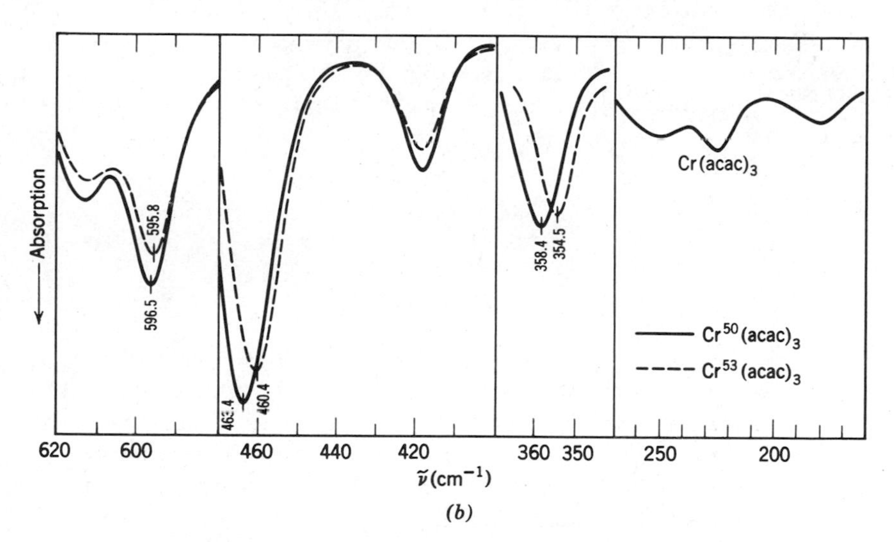

Fig. III-33b. Infrared spectra of $Cr^{50}(acac)_3$ and its Cr^{53} analog.[614a]

TABLE III-50. OBSERVED FREQUENCIES, BAND ASSIGNMENTS, AND FORCE CONSTANTS FOR $K[Pt(acac)Cl_2]$ AND $Na_2[Pt(acac)_2Cl_2] \cdot 2H_2O$

$K[Pt(acac)Cl_2]$ (O-bonded, Type A)	$Na_2[Pt(acac)_2Cl_2] \cdot 2H_2O$ (C-bonded, Type B)	Band Assignment
–	1652, 1626	$\nu(C{=}O)$
1563, 1380	–	$\nu(C\cdots O)$
1538, 1288	–	$\nu(C\cdots C)$
–	1352, 1193	$\nu(C—C)$
1212, 817	1193, 852	$\delta(C—H)$ or $\pi(C—H)$
650, 478	–	$\nu(Pt—O)$
–	567	$\nu(Pt—C)$
$K(C\cdots O) = 6.50$	$K(C{=}O) = 8.84$	
$K(C\cdots C) = 5.23$	$K(C—C) = 2.52$	UBF Force
$K(C—CH_3) = 3.58$	$K(C—CH_3) = 3.85$	Constant (mdyn/A)
$K(Pt—O) = 2.46$	$K(Pt—C) = 2.50$	
$K(C—H) = 4.68$	$K(C—H) = 4.48$	
$\rho = 0.43$[a]		

[a] The stretching-stretching interaction constant (ρ) was used for type A because of the presence of resonance in the chelate ring.

Vibrational spectra of acetylacetonato complexes have been studied by many other investigators. Only the references are listed here: Raman spectra of tris(acac) complexes of Group IIIA elements (616); vibrational analysis of electronic spectrum of $Cr(acac)_3$ (617); infrared spectra of acetylacetonato complexes of lanthanides (618) and Group IVA elements (619).

The acetylacetonate anion can also coordinate to a metal through its γ carbon atom:

```
  H3C
     \
  H   C=O
   \ /
    C          (Type B)
   / \
  M   C=O
     /
  H3C
```

Type B was first found by the x-ray analysis on $K[Pt(acac)_2Cl]$,[620] in which one acetylacetonate ligand is chelated (type A) and the other is C-bonded to the Pt atom (type B). [See Fig. III-34]. The CO and CC distances are markedly different between two types; those of type B are close to the pure C=O and C—C distances, respectively, whereas those of type A are between the pure single and pure double bond distances.

Lewis and his co-workers[621] have carried out an extensive study on the

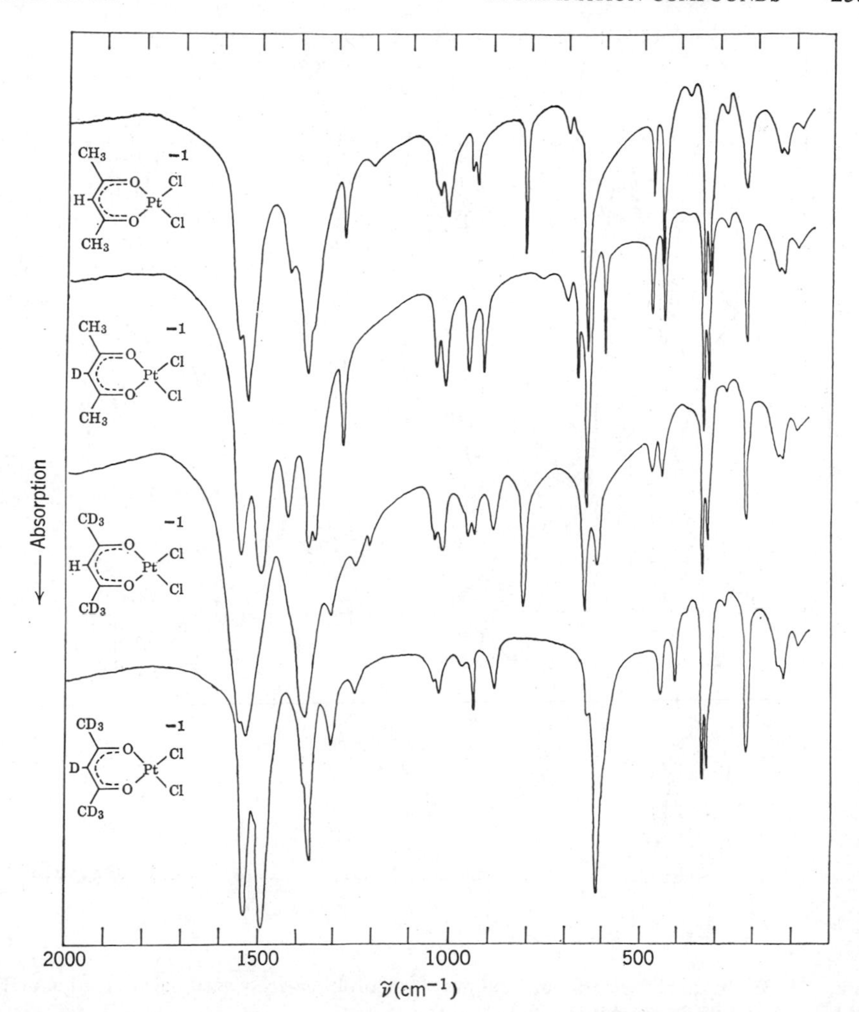

Fig. III-34. Infrared spectra of potassium dichloro(acetylacetonato) platinate(II) and its three deuterated analogs.

infrared and NMR spectra of the C-bonded acetylacetonate complexes of Pt(II). Behnke and Nakamoto[622] carried out a normal coordinate analysis on $Na_2[Pt(acac)_2Cl_2]$, which contains two C-bonded acetylacetonato ions in the trans position. Figure III-35 shows the infrared spectra of $Na_2[Pt(acac)_2Cl_2]$ and its deutero analogs, and Table III-50 gives the observed frequencies and band assignments obtained for the nondeuterated species. These results indicate that (1) the two C—O stretchings appear at higher frequencies than

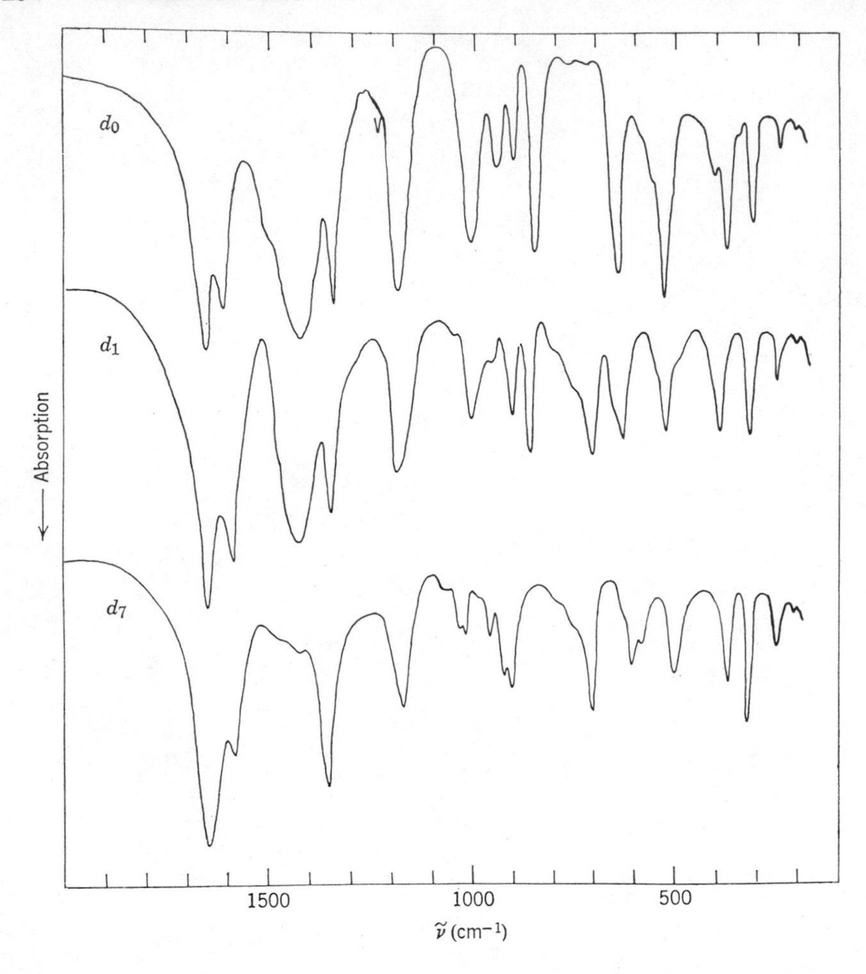

Fig. III-35. Infrared spectra of sodium dichlorobis(γ-acetylacetonato) platinum(II) dihydrate and its deuterated analogs.

those of type A, (2) the two C—C stretchings are at lower frequencies than those of type A, and (3) the Pt—C stretching is at 567 cm^{-1}. The Pt—C stretching force constant obtained (2.50 mdyn/A) is almost the same as the Pt—O stretching force constant (2.46 mdyn/A) of type A complexes. They[623] have also shown that the spectrum of K[Pt(acac)$_2$Cl] can be explained as a superposition of type A and B spectra. It was suggested from the spectra that the structure of K[Pt(acac)$_3$] originally prepared by Lewis et al.[621] may contain one type A ring, one type B ligand, and one type B′ ligand in which two C—O bonds are transoid:

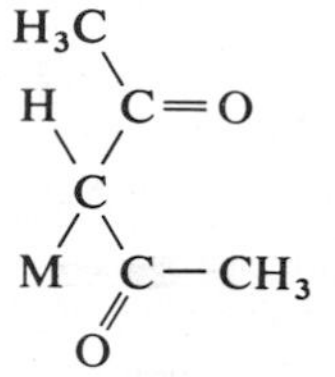

```
    H3C
       \
  H    C=O
   \  /
    C              (Type B')
   / \
  M   C—CH3
     //
    O
```

If a solution of K[Pt(acac)$_2$Cl] is acidified, the C-bonded ligand is converted into type C shown below:

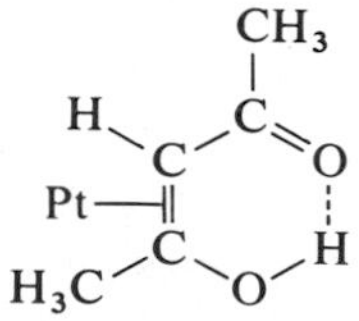

```
              CH3
               |
     H        C
      \     /   \\
        C         O
Pt—||             :        (Type C)
        C         H
      /   \     /
  H3C        O
```

This structure was first suggested by Allen et al.[624] based on NMR evidence. Behnke and Nakamoto[625] have shown that the infrared spectrum of this compound ([Pt(acac)(acacH)Cl]) can be interpreted as a superposition of the type A spectrum with that of the enol form of acetylacetone. The bands characteristic of type C are: C=O stretching (1627 cm^{-1}), C=C stretching (1550 cm^{-1}), O—H ··· O stretching (2905 cm^{-1}) and O—H ··· O bending (1450 cm^{-1}).

The keto form of acetylacetone also coordinates to a metal to give a chelate ring of type D:

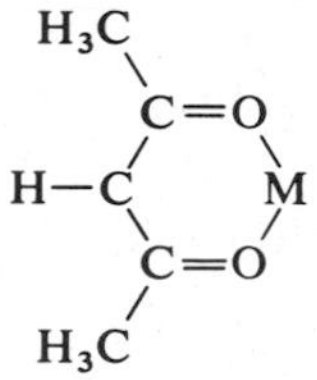

```
    H3C
       \
        C=O
       /    \
  H—C        M      (Type D)
       \    /
        C=O
       /
    H3C
```

Type D was found by van Leeuwen[626] in [Ni(acacH)$_3$](ClO$_4$)$_2$ and its derivatives, and by Nakamura and Kawaguchi[627] in Co(acacH)Br$_2$, They were prepared in acidic or neutral media, and exhibit strong C=O stretching bands near 1700 cm^{-1}.

Other types of coordination are also possible. Unidentate coordination such as the following has been suggested for *bis*(dipivaloylmethanido) mercury.[628]

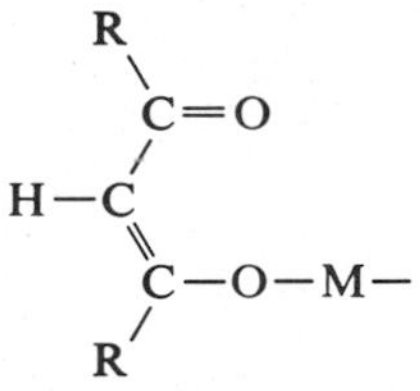

```
      R
       \
        C=O
       /
  H—C
       \\
        C—O—M—
       /
      R
```

In [Ni(acac)$_2$]$_3$ and [Co(acac)$_2$]$_4$, the oxygen atoms of the acetylacetonato anion serve as a bridge between two metal atoms.[629] However, no detailed infrared studies have been made on these polymeric complexes.

Some metal acetylacetonates form addition compounds with water, pyridine, and other bases. If there is bonding between the metal and the basic atom of the donor molecule, a marked shift of the M—O (acetylacetone) stretching band is anticipated. In VO $(acac)_2$, the V—O stretching band at 480 cm^{-1} is in fact shifted to 463 cm^{-1} by forming an addition compound with pyridine.[630] Infrared studies on other addition compounds are found in References 630–633.

(2) Complexes of Other β-Diketones

Belford et al.[634] have studied the effect of substitution of CF_3, OC_2H_5, and $N(C_2H_5)_2$ for the CH_3 groups of $Cu(acac)_2$. Dryden and Winston[635] have discussed the relation between the CO stretching frequency and the electronic effect of the substituent. Nakamoto and co-workers[636] have found that substitution of CF_3 for CH_3 causes marked shifts of the C$\overset{\cdots}{-}$O and C$\overset{\cdots}{-}$C stretching bands to higher frequencies and of the M—O stretching band to a lower frequency. This result indicates that the strong positive inductive effect of the CF_3 group strengthens the former two bonds and weakens the M—O bond. They have also suggested from infrared spectra that the M—O bond is slightly strengthened by the phenyl substitution. This result is, however, contradictory to that of an ESR study by Kuska and Rogers.[637] Thus a more detailed infrared study is desirable. Infrared studies on metal complexes of other β-diketones are: hexafluoroacetylacetonato complexes[638], and complexes of polymethacroylacetone and its low molecular weight analogs.[639]

III-22. COMPLEXES OF SULFUR-CONTAINING LIGANDS

Allkins and Hendra[640] have studied the infrared spectra of metal complexes of the type, *trans*-MX_2Y_2, where M is Pd(II) or Pt(II); X is Cl^-, Br^-, or I^-; and Y is $(CH_3)_2S$, $(CH_3)_2Se$, or $(CH_3)_2Te$. The bands at 350–300, 240–170, and 230–165 cm^{-1} were assigned to the M—S, M—Se, and M—Te stretching modes, respectively.

Chatt and co-workers[641] investigated the infrared spectra of dithiocarbamato complexes, the structure of which may be represented by a hybrid of the structures:

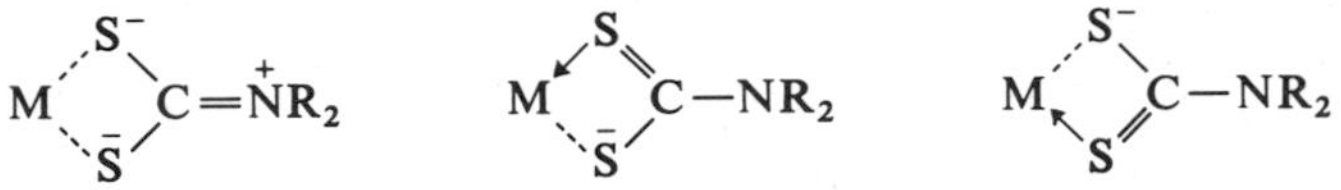

$$\mathrm{M}\begin{matrix}\cdot\cdot S^-\\ \cdot\cdot \bar{S}\end{matrix}\!\!>\!\mathrm{C{=}\overset{+}{N}R_2} \qquad \mathrm{M}\begin{matrix}\leftarrow S\\ \cdot\cdot \bar{S}\end{matrix}\!\!>\!\mathrm{C{-}NR_2} \qquad \mathrm{M}\begin{matrix}\cdot\cdot S^-\\ \leftarrow S\end{matrix}\!\!>\!\mathrm{C{-}NR_2}$$

I II III

They noted that the C$\overset{\cdots}{-}$N stretching band at 1540–1480 cm^{-1} shows very little dependency on the nature of the metal or an organic group, R. From normal coordinate analysis, Nakamoto et al.[642] have assigned the Pt—S stretching bands of the Pt(II) complex at 375 and 228 cm^{-1}. Cotton and McCleverty[643] have measured the infrared spectra of metal carbonyls containing dialkyldithiocarbamato groups. They noted that the dithiocarbamato group has no marked effect on metal-CO bonding. Moore and Larson[644] have assigned the Mo—S stretching bands of dithiocarbamato complexes at 515–460 cm^{-1}.

The electronic structure of the xanthato ($[S_2COR]^-$) is represented by a hybrid of structures similar to those shown above for carbamato complexes. However, their CO stretching bands (1100–1000 cm^{-1}) appear in the region of the C—O single bond stretching frequency. Therefore, the contribution of the resonance structure involving the C=O bond (similar to I) may not be significant in xanthato complexes. This is supported by the observation that the CS stretching frequencies of xanthato complexes (1260–1170 cm^{-1}) are higher than those of dithiocarbamato complexes (1000–600 cm^{-1}).[645,646] The infrared spectra of methyl- and ethyl-xanthato complexes have been obtained by Watt and McCormick.[647] They assigned the M—S stretching bands at 360–310 cm^{-1} for Pt(II), Pd(II), Ni(II), Cr(III), and Co(III).

From a normal coordinate analysis of the chelate ring of the dithiooxalato Pt(II) complex, Fujita and Nakamoto[648] have assigned the Pt—S stretching bands at 440–420 and 320 cm^{-1}. Chaston and Livingston[649] assigned the M—O and M—S stretching bands of bis(monothioacetylacetonato) complexes of Co(II), Ni(II), and Cu(II) at 500–485 and 395–335 cm^{-1}, respectively.

Shindo and Brown[650] studied the infrared spectra of metal complexes of L-cysteine(HS—CH_2—CH(NH_2)—COOH). Since L-cysteine has three coordination sites (S, N, and O), it is interesting to determine which donor atoms are involved in coordination. For example, the Zn(II) complex shows no S—H bands and its carboxylate frequency indicates the presence of the free COO^- groups Thus, they proposed the structure:

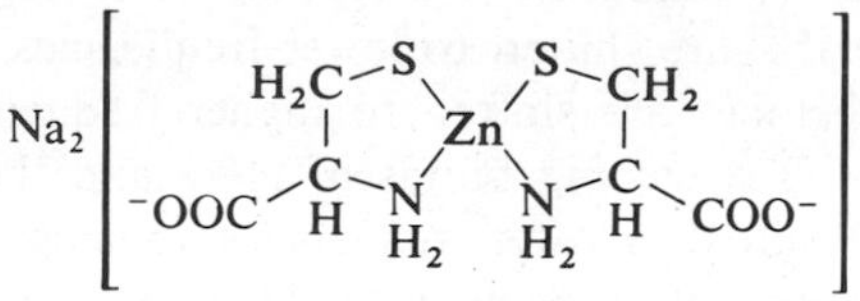

```
        ┌                                    ┐
        │      H2C─S         S─CH2           │
  Na2   │       │    \     /    │            │
        │       │      Zn       │            │
        │       C    /     \    C            │
        │ -OOC  H  N         N  H  COO-      │
        │          H2        H2              │
        └                                    ┘
```

McAuliffe et al.[651] have studied the infrared spectra of metal complexes of methionine (CH_3—S—CH_2—CH_2—CH(NH_2)—COOH). They have found that most of the metals they studied (except Ag(I)) coordinate through the NH_2 and COO^- groups, and that the CH_3S groups of these complexes are available for further coordination to other metals. McAuliffe[652] suggested that complexes of the type, M(methionine)Cl_2 (M = Pd(II) and Pt(II)), take the polymeric structure:

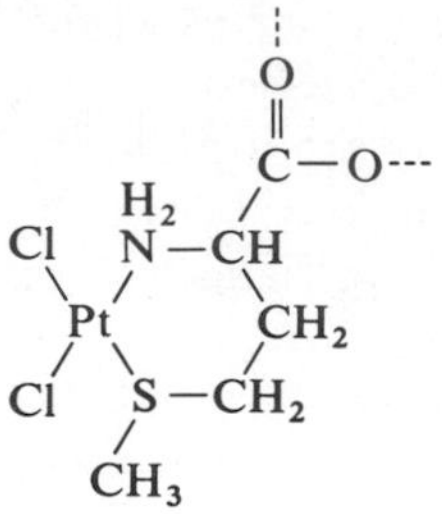

```
                  :
                  O
                  ‖
                  C—O···
          H2     /
     Cl   N—CH
       \ /     \
       Pt       CH2
       / \     /
     Cl   S—CH2
         /
       CH3
```

where the noncoordinated COOH group of the monomer is hydrogen-bonded to the neighboring molecule. Infrared spectra of metal complexes of dithizone ($C_6H_5N{=}N$—C(S)—NH—NHC_6H_5) have been measured by Dyfverman.[653]

III-23. MISCELLANEOUS INTERACTIONS

(1) Molecular Complexes

A large number of molecular complexes are formed as a result of the interaction between an electron acceptor or Lewis acid (A) and an electron donor or Lewis base (B). If A and B are relatively simple molecules, it is possible to carry out a normal coordinate analysis on the entire complex and to estimate the strength of the A—B bond in terms of the A—B stretching force constant. Table III-51 lists the observed frequencies and the force constants for the A—B stretching vibrations. It is to be noted that AlH_3 and BH_3 exist only as molecular complexes.

A variety of organic molecules (B) form molecular complexes with metal halides, hydrogen halides, and halogens (A). Some bands of organic molecules are shifted to higher, and other bands are shifted to lower, frequencies as a result of complex formation. For example, the C=O stretching bands of *ketones*[663] and *esters*[664] are shifted to lower frequencies, whereas the C=O bending bands of ketones are shifted to higher frequencies.[665] In $(CH_3)_2O{\cdot}HCl$, the two C—O stretching bands of *ether* and the H—Cl stretching band of the acid are shifted to lower frequencies upon complex formation.[666] In molecular complexes of *pyridine* with $TiCl_4$ and $ZrCl_4$,[667] some ring stretching frequencies decrease by 10–20 cm^{-1}, whereas some ring deformation frequencies increase by 20 cm^{-1}. In pyridine-I_2 complex, the I—I stretching band at 213 cm^{-1} of free iodine is shifted to 184 cm^{-1},[668] and the 990 cm^{-1} band of free pyridine is shifted to a higher frequency by complex formation.[669] The N—I stretching bands of pyridine ·ICl and pyridine ·IBr were assigned at 170 and 160 cm^{-1}, respectively.[670] For molecular complexes of pyridine with other metal halides, see Refs. 671–673.

TABLE III-51. FREQUENCIES AND FORCE CONSTANTS OF COORDINATE BOND STRETCHING VIBRATIONS

Compound (B—A)	Observed Frequency (cm^{-1})	Force Constant (mdyn/A)	References
$2(CH_3)_3N—AlH_3$[a]	456(IR)	–	654–656
$H_3N—AlCl_3$	750(R)	2.2	657
$(CH_3)_3N—BH_3$	676(R)	2.79	658
$H_3N—BF_3$	735(IR)	–	659
$CH_3CN—BF_3$	630(IR)	2.027	660, 661
$F_3P—BH_3$	607(R)	–	662

[a] The $N—AlH_3—N$ skeleton is assumed to be trigonal-bipyramidal.

The C≡N stretching bands of *nitriles* are shifted to higher frequencies by forming molecular complexes with metal halides.[674–677] The infrared spectrum of $CH_3CN{\cdot}(HBr)_2$ indicates the presence of the acethalogenimidium ion.[678]:

$$\left[CH_3-C\begin{matrix}\nearrow NH_2 \\ \searrow Br\end{matrix} \right]^+ Br^-$$

The P=O stretching bands of $POCl_3$ and $POBr_3$ are shifted to lower frequencies by forming molecular complexes with metal halides.[679–680] For the far-infrared studies on molecular complexes involving metal halides, see Refs. 681–683.

In "loose" molecular complexes such as benzene—$SbCl_3$ and benzene—$AgClO_4$, no marked band shifts of benzene are observed. Instead, the appearance of new lines and a redistribution of intensities are observed in the Raman spectrum.[684,685] It was shown that the symmetry of the complex, $SbCl_3 \cdot \frac{1}{2}(C_6H_6)$ is $\mathbf{C}_{3v}$ in solution and $\mathbf{C}_{2v}$ in the solid state.[686]

(2) Interactions in the Alkali Halide Pellet

The pressed pellet technique using alkali halide powder is widely employed to obtain the infrared spectra of solids. Several investigators have noted that the spectrum obtained by using this technique must be interpreted with caution. For example, Ketelaar et al.[687] have noted that the ν_3 frequency of the $[HF_2]^-$ ion is markedly sensitive to the kind of alkali halide used: 1599, 1570, 1527, and 1478 cm^{-1} in NaBr, KCl, KBr, and KI, respectively. On the other hand, the spectra of $NaHF_2$, KHF_2, and NH_4HF_2 in KCl pellets are the same. These results imply the formation of a mixed crystal in which the halide ion is replaced by the $[HF_2]^-$ ion. Thus the surrounding alkali halide lattice induces a dipole moment in the $[HF_2]^-$ ion that results in a marked change in the ν_3 frequency. Jones and Chamberlain[688] have found that over a period of time the bands of $K[Au(CN)_2]$ at 2141 and 427 cm^{-1} are gradually replaced by new bands at 2154 and 448 cm^{-1} in KBr and KI pellets. This result also indicates the replacement of a Br—K—Br chain by the $[Au(CN)_2]^-$ ion. The $[Au(CN)_2]^-$ ion in the KBr pellet exhibits the same spectrum as that in the KI pellet; it differs in this respect from the $[HF_2]^-$ ion. Strasheim and Buijs[689] have found that the ν_3 frequency of the $[NO_3]^-$ ion increases linearly with the lattice energy of the alkali halide used.

Meloche and Kalbus[690] have noted that the major factor involved in an exchange of anions between KBr and inorganic salts is the moisture adsorbed on the surface of the sample and of the KBr. According to Buchanan and Bowen,[691] carefully prepared water-free alkali halide pellets of LiOH exhibit the infrared spectrum of $LiOH \cdot H_2O$ as soon as they absorb water vapor from the air. An alkali halide pellet, immediately after pressing, is in a stressed state. Vrátný[692] has noted that the rate of recovery from this stressed state is governed by the humidity of the environment of the pellet as well as by the residual water in the pressed pellet. Farmer[693] has studied the infrared spectra of *phenols* and *organic salts* in alkali halide pellets and has found that they are adsorbed on the surface of the alkali halide crystals through their hydroxyl groups. It is interesting to note that the carboxylic acids are adsorbed principally as monomers. Durie and Szewczyk[694] have made the unusual observation that nonhydroxylic compounds such as anthracene exhibit an abormal OH stretching band (3400 cm^{-1}) in KCl pellets, which is definitely not due to adsorbed moisture. They suggested that the interaction of water and KCl during grinding causes the formation of the $[OH]^-$ ions.

(3) Chemisorbed Molecules

Infrared studies of chemisorbed molecules afford valuable information on: (1) the structure of chemisorbed molecules and the surface structure of solid catalyst; (2) the interacting force between chemisorbed molecules and solid catalyst; (3) the mechanism of catalytic reaction. Some of these investigations will be described briefly since an excellent review is available.[695a]

Eischens and co-workers[695] have found that the band of free CO at 2143 cm^{-1} is shifted to 2128, 2070, 2053, and 2033 cm^{-1} when CO is adsorbed on the surface of metallic Cu, Pt, Pd, and Ni, respectively. In Ni and Pd, additional bands are observed at 1908 and 1916 cm^{-1}, respectively, possibly because of the existence of a bridged molecule such as

$$\begin{matrix} M \searrow \\ \quad C{=}O \\ M \nearrow \end{matrix}$$

If the surface of the catalyst is homogeneous, the intensity of the band due to CO is expected to increase in proportion to the exposed area. If this does not occur, the surface may not be homogeneous. Using this principle, Eischens and colleagues have found that Pd is not homogeneous, whereas Pt is homogeneous. For more work on chemisorbed CO, see Refs. 696–702.

Blyholder[703] has studied the infrared spectra of $C^{16}O$ and $C^{18}O$ adsorbed on Fe, and assigned the very weak, broad band near 500 cm^{-1} to the Fe—C stretch. Blyholder and Allen[704] studied the infrared spectra of NO adsorbed on Ni and Fe. The bands observed at 1840, 650, and 625 cm^{-1} for NO on Ni were assigned to NO stretch, Ni—N stretch, and Ni—N—O bend, respectively. Ward and Habgood[705] studied the infrared spectra of CO_2 adsorbed on Zeolite X.

NH_3 molecules adsorbed on cracking catalysts exhibit both NH_3 and $[NH_4]^+$ absorption. Since the latter bands are much weaker than the former, Mapes and Eischens[706] concluded that the surface structure of the catalyst is of the Lewis acid type and not of the Brönsted acid type.

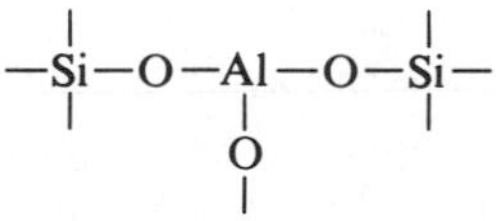

```
                                              |
                                             —Si—
                                              |
                                             H⁺O
    |          |                    |         |        |
  —Si—O—Al—O—Si—                  —Si—O—Al⁻—O—Si—
    |    |     |                    |         |        |
         O                                    O
         |                                    |
                                             —Si—
                                              |
```

Lewis acid 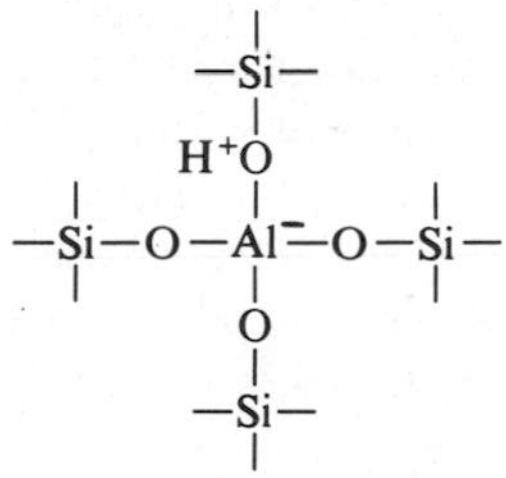 Brönsted acid

Folman and Yates[707] studied the infrared spectra of Vycor porous silica glass during the process of NH_3 adsorption and concluded that two types of adsorption sites are available on the glass surface; free hydroxyl groups and silicon or oxygen atoms on the surface. Ammonia adsorbed on the former site shows the NH stretching bands at 3400 and 3320 cm^{-1}.[708] If NH_3 is adsorbed on chlorinated Vycor glass, it reacts with the ⋛Si—Cl group to give the aminated surface (⋛Si—NH_2), which absorbs at 3520 and 3445 cm^{-1}.[709] In addition, Folman[710] observed three bands at 3150, 3050, and 2805 cm^{-1}, which are due to NH_4Cl on the surface formed during the process of amination.

Since chemisorbed molecules have lower symmetry than those in the free state, the appearance of new bands and the splitting of degenerate vibrations are anticipated. Sheppard and Yates[711] have found that the symmetry of CH_4, C_2H_4, and H_2 adsorbed on porous silica glass may be $\mathbf{C}_{3v}$, $\mathbf{C}_{2v}$, and $\mathbf{C}_{2v}$, respectively. The infrared spectra of chemisorbed olefins on silica-supported Ni have been studied by Pliskin and Eischens.[712] From the observation of the CH_2 and CH_3 deformation vibrations, they have found that ethylene molecules adsorbed on the catalyst exist predominantly as H_2C^*—C^*H_2 (C^*; C atom bonded to a metal) and are converted to H_2C^*—CH_3 by half-hydrogenation, whereas acetylene molecules are transformed readily into H_2C^*—CH_3 by self-hydrogenation. For investigations of other chemisorbed molecules, see Refs. 713–721.

III-24. COMPLEXES OF ALKENES AND ALKYNES*

In 1953, Chatt and Duncanson[724] elucidated the structure of Pt olefin complexes using infrared spectroscopy. Asymmetrical olefins such as propylene exhibit the C=C stretching bands near 1650 cm^{-1}. This band is shifted to 1504 cm^{-1} in $K[PtCl_3(C_3H_6)]\cdot H_2O$. Although symmetrical olefins such as ethylene do not exhibit the C=C stretching band in the infrared spectrum, this vibration may become infrared active if the field around the ethylene molecule becomes asymmetric through coordination. It was found, however, that the C=C stretching band cannot be observed even in complexes such as Zeise's salt ($K[PtCl_3(C_2H_4)]\cdot H_2O$). This result suggests that the ethylene molecule is symmetrically coordinated to the Pt atom as shown in Fig. III-36 (structure I). Later, this arrangement of the olefin in the complex was confirmed by x-ray analysis.[725]

By comparing the infrared spectrum of Zeise's salt with that of ethylene sulfide, Powell and Sheppard[726] attempted to assign all the observed bands of Zeise's salt on the basis of $\mathbf{C}_{2v}$ symmetry, that is, with the four hydrogen atoms deviating equally from the C_2H_4 plane. They concluded that the fundamental frequencies of ethylene change only slightly upon coordination to the metal, and suggested that bonding scheme A which involves C—C and Pt—C single bonds[727] can be ruled out.

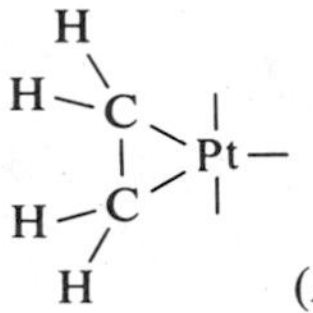

```
   H
    \
H—C    |
   |  >Pt—
H—C    |
   /
  H          (A)
```

Proton magnetic resonance studies[728] also indicate that the chemical shift of the ethylene hydrogen of the complex is much closer to that of free ethylene than to that of ethylene oxide.

According to Chatt et al.,[729] two types of bonding are involved in the Pt-ethylene bond: (1) the σ-type bond is formed by the overlap of the filled bonding $2p\pi$ molecular orbital of the olefin with the vacant dsp^2 bonding

*For comprehensive reviews on infrared spectra of organometallic compounds, see Refs. 722 and 723.

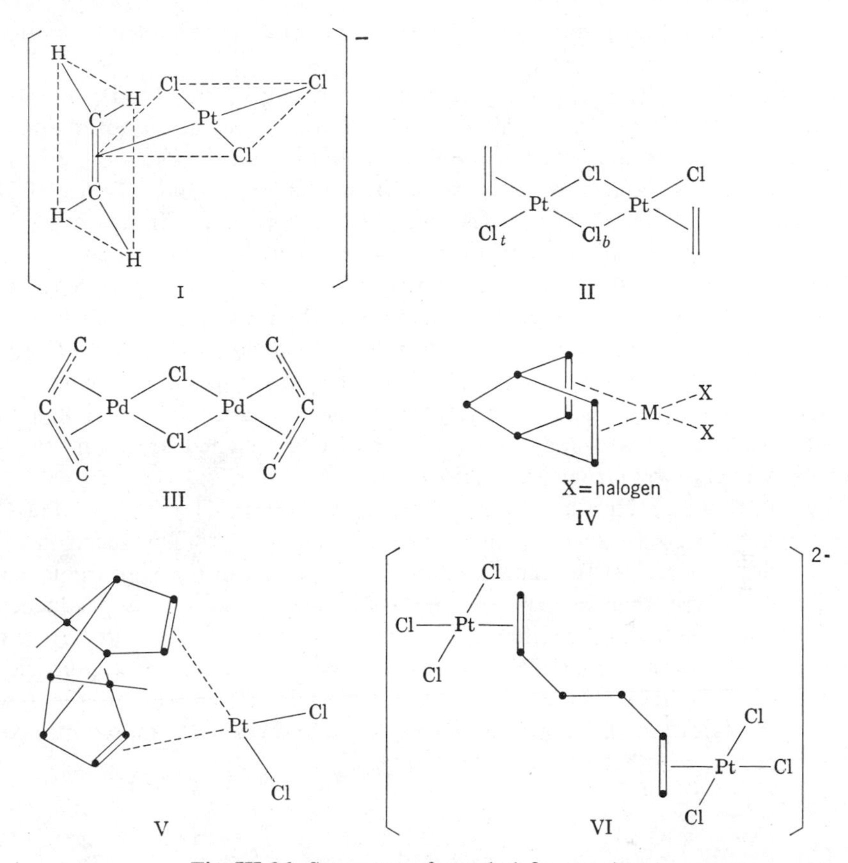

Fig. III-36. Structures of metal olefin complexes.

orbital of the metal, and (2) the π-type bond is formed by the overlap of the $2p\pi^*$ antibonding orbital of the olefin with a filled *dp* hybrid orbital of the metal. In scheme B below, the σ-type bonding is predominant. On the other hand, scheme C emphasizes the π-type bonding.

(B) (C)

The relative importance of these two types of bonding is not clear at present. It will depend upon the nature of the metal and the olefin. The π bonding becomes more important as the oxidation state of the metal becomes lower, since the metal must back-donate more electrons to the olefin through π-bonds.

Assuming that the π-bonding is not significant in the Pt(II) complex (scheme B),[625] Grogan and Nakamoto[730] have carried out an approximate normal coordinate analysis on Zeise's salt and its deutero analog. Figure III-37 illustrates the infrared spectra of these two compounds. Table III-52 compares the observed frequencies and band assignments for free ethylene and Zeise's salt. In their calculations, the band at 407 cm^{-1} of the non-deuterated compound was assigned to the Pt-ethylene stretching mode (force constant, 2.23 mdyn/A). On the other hand, Paradilla-Sorzano and Fackler[731] carried out a normal coordinate analysis based on bonding scheme C and assigned the bands at 491 and 403 cm^{-1} to the symmetric and antisymmetric Pt—C stretching modes, respectively. The former band was attributed to crystal water by Powell and Sheppard[726] and was assigned to a non-fundamental vibration by Grogan and Nakamoto.[732]

The infrared spectra of Zeise's dimer ($[Pt(C_2H_4)Cl_2]_2$) and its Pd(II) analog have been assigned by Grogan and Nakamoto.[732] The spectrum of Zeise's dimer is almost the same as that of Zeise's salt in the high frequency region. However, their spectra are markedly different in the low frequency region, since the former exhibits the bands characteristic of the Pt_2Cl_2 ring (see structure II of Fig. III-36). Table III-53 gives the observed frequencies and band assignments of Zeise's dimer and its Pd(II) analog in the low frequency region. Infrared spectra of Zeise's salt and related compounds have been reported by other investigators.[733,734]

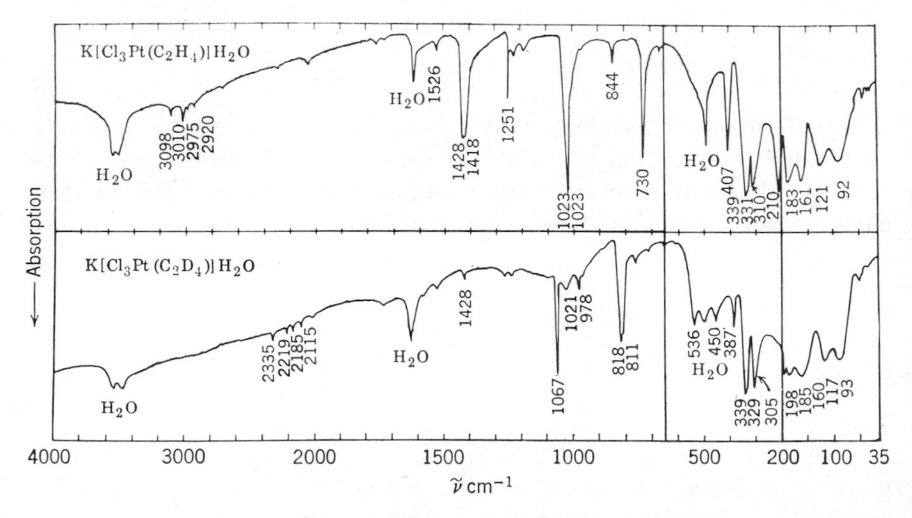

Fig. III-37. Infrared spectra of Zeise's salt and Zeise's salt-d_4 from 4000 to 33 cm^{-1}.[729]

TABLE III-52. OBSERVED FREQUENCIES OF FREE ETHYLENE, ZEISE'S SALT AND THEIR DEUTERATED ANALOGS (CM^{-1})[729]

	C_2H_4	Zeise's Salt	C_2D_4	Zeise's Salt-d_4	Assignment[a]
Coordinated ethylene	3019	2920	2251	2115	ν(C—H)
	1623	1526	1515	1428	ν(C=C), $\delta_s(CH_2)$[b]
	1342	1418	981	978	$\delta_s(CH_2)$, ν(C=C)[b]
	3108	2975	2304	2219	ν(C—H)
	1236	1251	1009	1021	$\rho_r(CH_2)$
	3106	3098	2345	2335	ν(C—H)
	810	844	586	536	$\rho_r(CH_2)$
	2990	3010	2200	2185	ν(C—H)
	1444	1428	1078	1067	$\delta_{as}(CH_2)$
	1007	730	726	450	$\rho_t(CH_2)$
	943	1023	780	811	$\rho_w(CH_2)$
	949	1023	721	818	$\rho_w(CH_2)$
Square-planar skeleton		331		329	ν_s(Pt—Cl)
		407		387	$\nu(Pt—C_2H_4)$
		310		305	$\nu(Pt—Cl_t)$[c]
		183		185	δ(Cl—Pt—Cl)
		339		339	ν_{as}(Pt—Cl)
		210		198	$\delta(ClPtC_2H_4) + \delta(ClPtCl)$
		161		160	$\delta(ClPtCl_t) + (ClPtC_2H_4)$
		121		117	$\pi(C_2H_4PtCl_t)$
		92		92	π(ClPtCl)

[a] Band assignments are for nondeuterated Zeise's salt.
[b] This coupling does not exist for Zeise's salt-d_4.
[c] Cl_t denotes the Cl atom *trans* to C_2H_4.

The infrared spectra of metal allyl complexes such as $[Pd(C_3H_5)Cl]_2$ and $[Pd(C_3H_5)Br]_2$ have been assigned empirically by Fritz[735] (see structure III of Fig. III-36). Norbornadiene (NBD, C_7H_8) is known to form metal complexes through its two C=C bonds (structure IV of Fig. III-36). Infrared spectra of these NBD complexes have been obtained by Alexander et al. in the high frequency region.[736] Similar to NBD, dicyclopentadiene ($C_{10}H_{12}$) coordinates to a metal as a bidentate ligand (structure V). The infrared spectrum of the Pt(II) complex has been reported.[737,738] Hendra and Powell[739] report the infrared spectra of 1,5-hexadiene complexes, such as $Pt(C_6H_{10})Cl_2$ and $K_2[(PtCl_3)_2(C_6H_{10})]$. The spectrum of the latter compound is simple and similar to that of the free ligand (*trans* conformation). Thus, they suggested structure VI for this compound. The spectrum of the former

TABLE III-53. FAR-INFRARED SPECTRA OF ZEISE'S DIMER AND ITS DEUTERO AND PALLADIUM ANALOGS (CM^{-1})[732]

$[Pt(C_2H_4)Cl_2]_2$	$[Pt(C_2D_4)Cl_2]_2$	$[Pd(C_2H_4)Cl_2]_2$	Assignment[a]
408	392	427	$\nu(Pt—C_2H_4)$
364	364	{357 333	$\nu(Pt—Cl_t)$
321	322	306	$\nu(Pt—Cl_b) + \nu(Pt—Cl_t)$ + ring def.
293	294	271	$\nu(Pt—Cl_b)$ + $\delta(C_2H_4—Pt—Cl_t)$
197	–[b]	198	$\delta(C_2H_4—Pt—Cl_t)$
138	–[b]	131	ring def.

[a] Band assignments are given for $[Pt(C_2H_4)Cl_2]_2$.
[b] This region has not been examined.

compound is more complex than that of the free ligand. The ligand is probably chelated to the Pt atom in this case. Hendra and Powell[740] also studied the infrared spectra of 1,5-cyclooctadiene (C_8H_{12}) complexes such as $Pt(C_8H_{12})Cl_2$ and $Pd(C_8H_{12})Cl_2$. They show spectra that are complex and similar to that of the free ligand in the solid state (tub conformation). These complexes probably contain the tub conformation of the olefin whose two C=C bonds are chelated to the metal.

Slade and Jonassen[741] obtained the infrared spectra of butadiene (C_4H_6) complexes of Pt(II) and Pd(II). Hendra and Powell,[742] from infrared studies, suggested structures I and II of Fig. III-38 for $K_2[(PtCl_3)_2(C_4H_6)]$ and $[PdCl_2(C_4H_6)]_2$, respectively. In structure I, butadiene is cisoid and lies between two Pt atoms, each C atom being bound symmetrically to the metal atom. In structure II of the Pd(II) complex, butadiene is twisted, that is, intermediate between the cisoid and transoid forms. On the other hand, Grogan and Nakamoto[743] proposed structure III for the Pt(II) complex. They noted that the far-infrared spectrum of the Pt(II) complex is very similar to that of Zeise's salt and that the spectrum of the Pt(II) complex in the high frequency region is similar to that of free butadiene (transoid configuration) and markedly different from that of $Fe(CO)_3(C_4H_6)$ where butadiene is definitely planar-cisoid.[744] These observations can be accounted for by structure III in which two units of Zeise's anion are bonded through the central C—C bond to give $\mathbf{C}_{2h}$ symmetry for the whole anion.

Doyle et al.[745] concluded that complexes of the type $R_2Pt(C_8H_8)PtR_2$ (C_8H_8 = cyclooctatetraene, R = CH_3 or phenyl) take the structure IV of Fig. III-38, since the bands at 1635 and 1609 cm^{-1} (C=C stretching) of the free ligand are absent in the complexes. According to x-ray analysis,[746] $Fe(C_8H_8)(CO)_3$ takes structure V in the crystalline state. Bailey et al.[747]

I

II

III

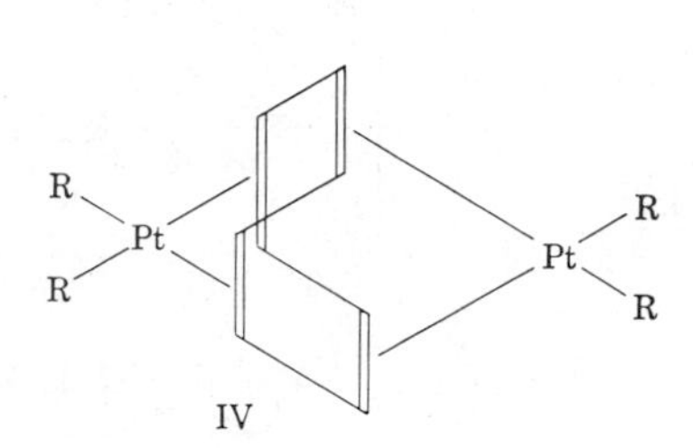

IV

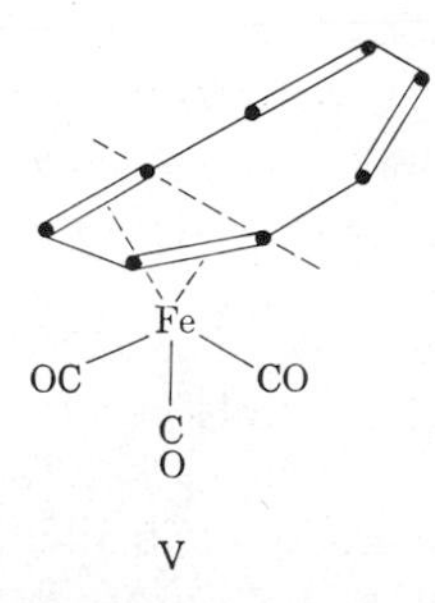

V

Fig. III-38. Metal complexes with conjugated olefins.

assigned the uncoordinated C=C stretching at 1562 cm^{-1} and the coordinated C=C stretching bands at 1490 and 1460 cm^{-1}. Infrared spectra of other cyclic oligoolefin complexes such as $Ru(C_7H_8)(C_8H_{12})$ and $Ru(C_7H_8)$(NBD) have been reported.[748]

The C≡C stretching bands of disubstituted alkynes, RC≡CR′ (R and R′ are alkyl groups), appear between 2260 and 2190 cm^{-1}. Chatt and coworkers[749] have noted that these frequencies are lowered to about 1700 cm^{-1} in complexes of the type Pt(RC≡CR′) $(PPh_3)_2$, and to about 2000 cm^{-1} in complexes of the types Na[Pt(RC≡CR′) Cl_3] and $[Pt(RC{\equiv}CR')Cl_2]_2$. The Ag complexes of alkynes also exhibit the C≡C stretching bands at 2070–1930

cm^{-1}.[750] In di(hydroxyalkyl)acetylene complexes of Pt(II), the O—H stretching bands are shifted by 140–110 cm^{-1} to lower frequencies upon coordination to the metal. Chatt et al.[751] interpreted this shift as an indication of intramolecular OH ··· Pt-type hydrogen bonds such as shown in structure A. On the other hand, Allen and Theophanides[752] attributed it to the Pt ··· 0 interaction shown in structure B.

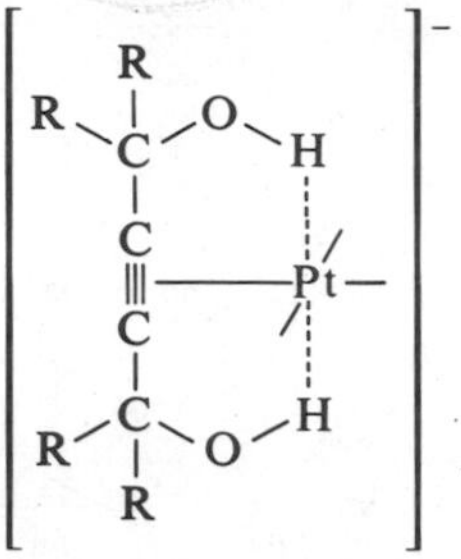

(A) 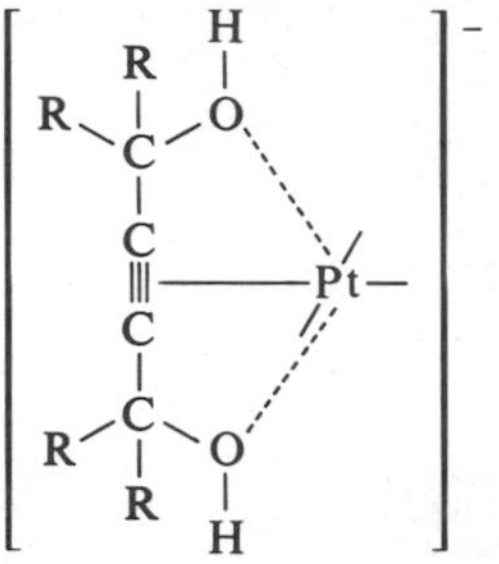 (B)

III-25. METAL SANDWICH COMPOUNDS

(1) Cyclopentadienyl Complexes

Since Kealy and Pauson[753] discovered ferrocene in 1951, a great number of metal complexes containing the cyclopentadienlyl group have been prepared; their infrared spectra have been reviewed by Fritz.[754] According to Fritz, they can be classified into four groups: (a) ionically bonded complexes, (b) centrally σ-bonded complexes, (c) centrally π-bonded complexes, and (d) diene-type (σ-bonded) complexes.

Some alkali and alkaline earth metals (K, Rb, Ca, Sr, etc.) form type (a) complexes in which the metal is ionically bonded to the ring. In this case, the C_5H_5 ring exhibits a spectrum which is essentially the same as that of the $C_5H_5^-$ ion ($\mathbf{D}_{5h}$ symmetry). Other alkali and alkaline earth metals Li, Na, Mg, etc.) and metals such as T1, Sn, and Pbform type (b) complexes in which the metal is bonded to the center of the ring through a σ-bond. In this case, the spectrum of the ring is interpreted on the basis of $\mathbf{C}_{5v}$ symmetry. In addition, type (b) complexes exhibit the metal-ring stretching bands in the far-infrared region.

A large number of transition metals form type (c) complexes in which the metal is bonded to the center of the ring through a π-bond. The spectra of the C_5H_5 rings in these compounds are similar to those of type (b) complexes. However, the CH out-of-plane bending frequencies of type (c) complexes (880–760 cm^{-1}) are higher than those of type (b) complexes (800–700 cm^{-1}). In addition, type (c) complexes exhibit metal-ring stretching, as well as other skeletal modes in the far-infrared region. Figure III-39 illustrates the infrared

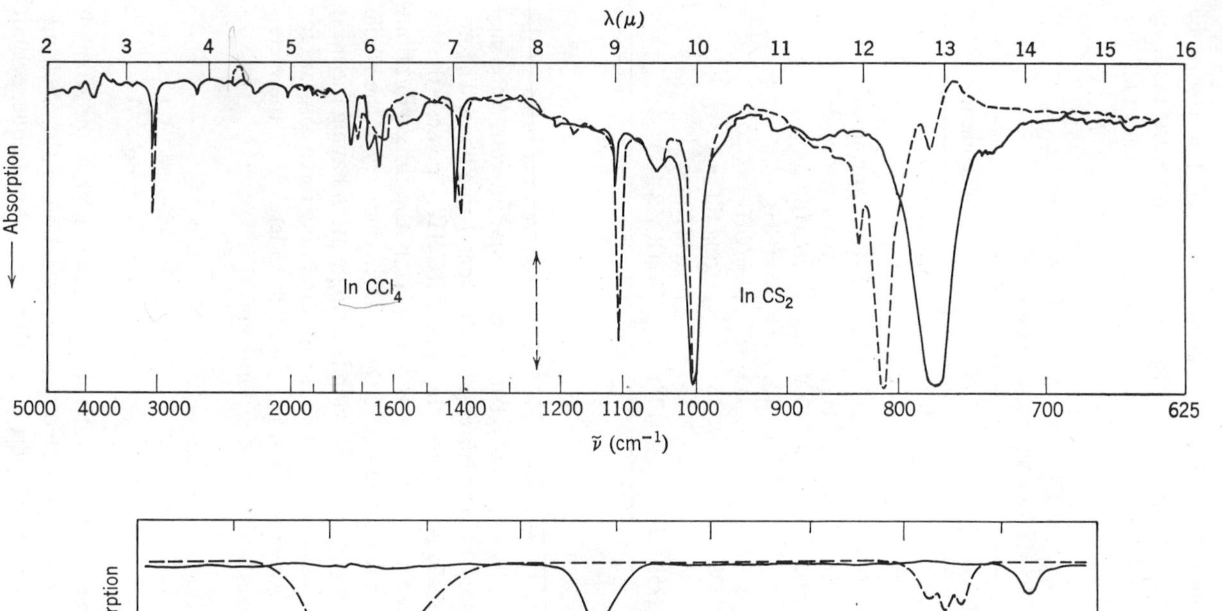

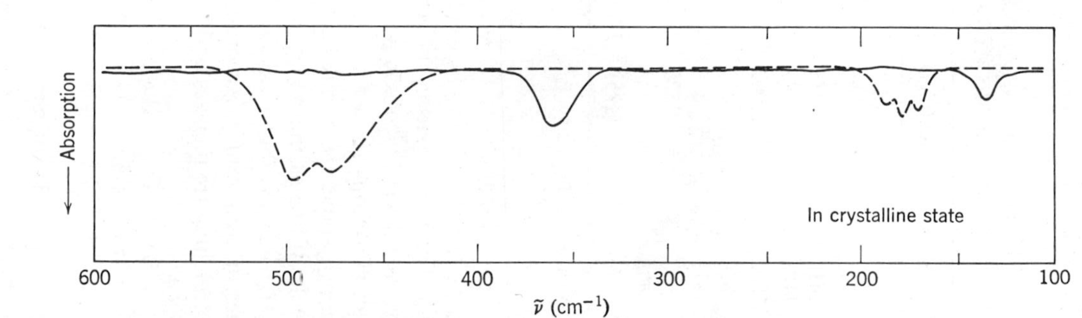

Fig. III-39. Infrared spectra of $Ni(C_5H_5)_2$ (solid line) and $Fe(C_5H_5)_2$ (dotted line).[755,766]

spectra of $Fe(C_5H_5)_2$ and $Ni(C_5H_5)_2$ in the high-[755] and low frequency regions. Empirical band assignments on these compounds were first made by Lippincott and Nelson.[756] Table III-54 lists the observed frequencies and band

TABLE III-54. INFRARED FREQUENCIES AND BAND ASSIGNMENTS OF $M(C_5H_5)_2$-TYPE COMPLEXES[756]

$Fe(C_5H_5)_2$	$Ru(C_5H_5)_2$	$Ni(C_5H_5)_2$	Assignment
3085	3100	3075	$\nu(CH)$
3075	3100	3075	$\nu(CH)$
1411	1413	1430	$\nu(CC)$ or ring def.
1108	1103	1110	$\nu(CC)$ or ring def.
1002	1002	1000	$\delta(CH)$
834	835	800	$\pi(CH)$
811	806	773	$\pi(CH)$
492	528	355	ring tilt (ν_5)
478	446	355	$\nu(MR)$ (ν_3)
170	(185)	(125)	$\delta(RMR)(\nu_6)$

R denotes the C_5H_5 ring.

assignments obtained by these investigators. Of the vibrations listed in the table, the last three bands (ν_5, ν_3, and ν_6) are of particular interest, since they provide direct information about the strength of the metal-ring bond. Theoretically, metal sandwich compounds of the type $M(C_5H_5)_2$ possess the six skeletal vibrations shown in Fig. III-40. However, only the three vibrations mentioned above are infrared active under $\mathbf{D}_{5d}$ or $\mathbf{D}_{5h}$ symmetry of the whole molecule. If one assumes that the C_5H_5 ring is a single atom having the mass of C_5H_5, it is possible to calculate the metal-ring stretching force constant for a series of compounds. According to Fritz and Schneider,[754,757] the metal-ring stretching frequencies and force constants are

Os		Fe		Ru		Cr		Co		V		Ni		Zn
353		478		379		408		355		379		355		345 (cm^{-1})
2.8	$>$	2.7	$>$	2.4	$\gg$	1.6	$>$	1.5	$\sim$	1.5	$\sim$	1.5	$\sim$	1.5 (mdyn/A)

The above order of metals may be interpreted as the order of the strength of the metal-ring bond. Fritz also noted that the deviation of the electronic structure of the metal from a filled inert gas shell causes a low frequency shift of the metal-ring stretching band. For example, the 478 cm^{-1} band of $Fe(C_5H_5)_2$ is shifted to 423 cm^{-1} in the $[Fe(C_5H_5)_2]^+$ ion. Apparently, the metal-ring bond of the latter is weakened by the deviation from inert gas configuration.

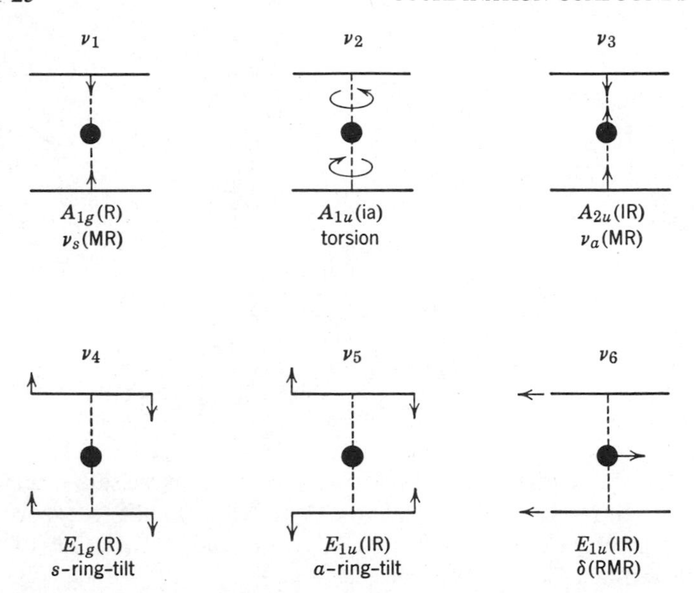

Fig. III-40. Skeletal vibrations of dicyclopentadienyl metal complexes ($\mathbf{D}_{5d}$ symmetry).[754]

Several metals such as Hg, Cu, and Si form type (d) complexes in which the metal is bonded to one of the ring carbon atoms through a σ-bond.

This structure was first suggested by Wilkinson and Piper[758] on the basis of both spectral and chemical evidence. The spectrum of this group is similar to that of C_5H_6 and is entirely different from those of other groups. Figure III-41 illustrates the infrared spectrum of $Hg(C_5H_5)_2$. Recently, Maslowsky and Nakamoto[759] have assigned the infrared and Raman spectra of this compound. According to Fritz,[754] complexes of the type $M(C_5H_5)_4$ where M is Hf, Nb, Ta, and Mo, exhibit the bands characteristic of both types (b) and (d). They may contain these two types of bonding. The infrared spectra of complexes having the general formula $M(C_5H_5)_n$ have been reported by many other investigators.[760–765]

(2) Cyclopentadienyl Complexes Containing Other Groups

Cotton et al.[766] have studied the infrared spectra of $Ni(C_5H_5)NO$, $Co(C_5H_5)(CO)_2$, $Mn(C_5H_5)(CO)_3$, and $V(C_5H_5)(CO)_4$. The number of infrared active CO stretching bands of these compounds was in good agreement with those predicted from the selection rule derived from a consideration

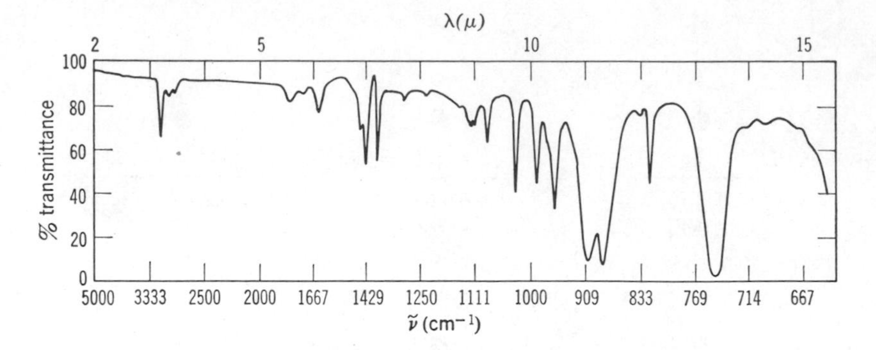

Fig. III-41. Infrared spectrum of $Hg(C_5H_5)_2$, 35 mg/ml in CS_2 (2 ~ 6μ) and (7.1 ~ 15.5μ), in $CHCl_3$ (6 ~ 6.6 μ) and in CCl_4 (6.6 ~ 7.1 μ).[758]

of only the local symmetry of the $M(CO)_n$ portion of these molecules. For example, in $Mn(C_5H_5)(CO)_3$, only two CO stretching bands are expected from the local symmetry ($\mathbf{C}_{3v}$) of the $Mn(CO)_3$ group, and, in fact, only two bands were observed at 1930 and 1923 cm^{-1}.

$Fe(C_5H_5)_2(CO)_2$ was first prepared by Hallam and Pauson[767] who suggested that both rings are π-bonded to the metal. Piper and Wilkinson,[768] on the other hand, proposed that one ring is π-bonded and the other σ-bonded to the metal. They[769] have demonstrated their structure to be correct by proton magnetic resonance and infrared studies. Later, x-ray analysis on this compound[770] confirmed their structure. Davidson and co-workers[771] reported the infrared spectra of compounds of the type $[M^n(C_5H_5)(CO)_3]^{n-1}$, where M is V^-, Cr^0, Mn^+, or Fe^{2+}, and noted that the CO stretching frequency increases in this order of metals.

Some cyclopentadienyl complexes are polymerized through the CO bridges. The infrared spectrum of $Fe_2(C_5H_5)_2(CO)_4$ was first reported by Piper et al,[766] and its structure was shown to be

O ‖ C / Fe ... Fe \ C ‖ O ; OC, CO ($\mathbf{C}_{2h}$)

from x-ray analysis.[772] This compound exhibits one terminal and one bridging CO stretching bands at 1942 and 1770 cm^{-1}, respectively. In nonpolar solvents such as octane, it exhibits three CO stretching bands at 2005, 1961, and 1794 cm^{-1}. This observation, together with the large value of the dipole moment (3.1D in benzene) suggests the presence of a tautomer in which the two rings are in the *cis* position.[773] An equilibrium study of the *trans* and *cis* isomers in solution has been made as a function of temperature.[774]

According to x-ray analysis,[775] the structure of $[Mo(C_5H_5)(CO)_3]_2$ is

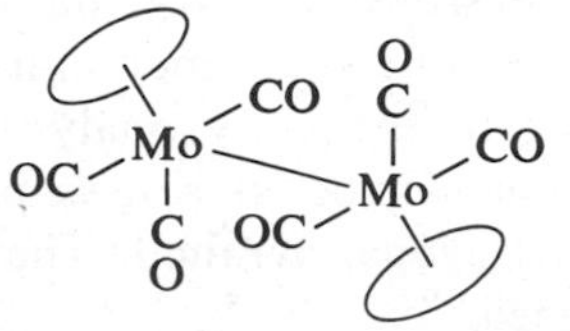

(C_{2h})

In agreement with this structure, its infrared spectrum exhibits two terminal CO stretching bands at 1942 and 1862 cm^{-1}.[776] The infrared spectra of analogous W[776] and Cr[777] compounds and of $[Fe(C_5H_5)(CO)]_4$[778] are also available.

Piper and Wilkinson[779] prepared $Mn_2(C_5H_5)_3(NO)_3$ and observed two NO stretching bands at 1732 and 1510 cm^{-1}. Attributing the former to the terminal and the latter to the bridging NO group, they proposed the following structure:

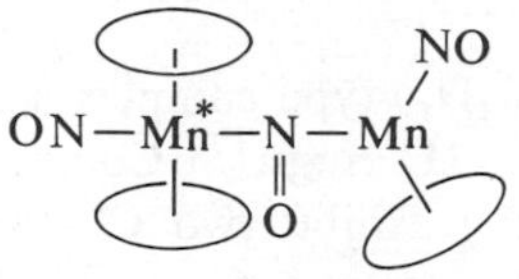

(One of the rings on Mn* is σ-bonded)

Cyclopentadienyl hydrido complexes, $ReH(C_5H_5)_2$ and $WH_2(C_5H_5)_2$ exhibit sharp metal-hydrogen stretching bands between 2100 and 1800 cm^{-1}.[780–782,777] Cyclopentadienyl complexes containing metal-halogen bonds are expected to show the metal-halogen stretching bands in the regions mentioned in Sec. III-12. The infrared spectra of $Ti(C_5H_5)_2X_2$ (X = F, Cl, Br, and I)[754] and $Mo(C_5H_5)(CO)_2X_3$ (X = Cl, Br, and I)[783] have been reported. The infrared spectra are also reported for $Pt(C_5H_5)(CH_3)_3$[784] and $Ni(C_5H_5)(CO)R_f$ ($R_f = CF_3$ and C_2F_5).[785]

A number of substituted cyclopentadienyl groups also form metal complexes with transition metals. References on infrared spectra of some of these complexes are found in Refs. 763, 786–788. Fischer and Zahn[789] and Green et al.[763] have noted that, in complexes of the type $M(C_5H_5)(C_5H_6)$ (M = Rh, Ir, and Co), the endo (α) hydrogen of the C_5H_6 ring exhibits its C—H stretching band at 2770–2740 cm^{-1}, whereas the exo (β) hydrogen shows it at 2950–2940 cm^{-1}. The infrared spectrum of $Fe(C_5H_6)(CO)_2(PPh_3)$ is available.[771]

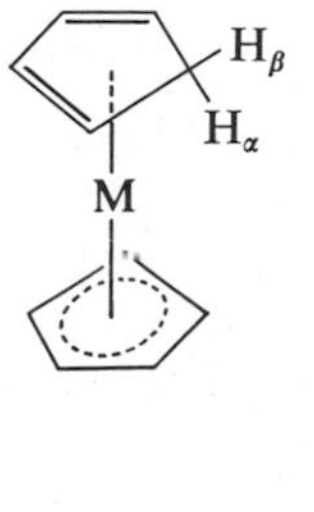

(3) Other Metal Sandwich Compounds

The infrared spectrum of $Ag(C_4H_4)NO_3$ has been reported by Fritz et al.[790] However, the structure of this compound has not yet been determined unambiguously. According to the x-ray analysis of $Ni(C_4(CH_3)_4)Cl_2$,[791] the C_4 skeleton of the four-membered ring is planar and the Ni atom is situated at the apex of a tetragonal pyramid. The infrared spectrum of this compound has been assigned.[792]

Fischer and Hafner[793] first prepared dibenzene chromium, $Cr(C_6H_6)_2$, and suggested a ferrocene-like structure on the basis of infrared and magnetic data. This was confirmed by x-ray analysis ($\mathbf{D}_{6h}$ symmetry).[794] Fritz et al.[795] and Snyder[796] have made complete band assignments for $Cr(C_6H_6)_2$. Table III-55 gives the observed frequencies and band assignments obtained by Fritz et al. According to Snyder, the metal-ring stretching force constant is 2.39 mdyn/A, which is smaller than that of ferrocene (2.7 mdyn/A). Fisher and Seus[797] have measured the infrared spectrum of the dibiphenyl chromium cation, $[Cr(C_{12}H_{10})_2]^+$.

The infrared spectra of $M(C_6H_6)(C_5H_5)$-type complexes show the bands characteristic of both the C_6H_6 and C_5H_5 rings.[777] Complexes of the type $M(C_6H_6)(CO)_3$ (M = Cr, Mo, and W) exhibit two CO stretching bands, since the local symmetry of the $M(CO)_3$ group is $\mathbf{C}_{3v}$.[798] This has also been confirmed by the spectra of a series of compounds of the type (substituted benzene)$Cr(CO)_3$.[799] For the infrared spectra of other sandwich compounds containing the CO groups, see Refs. 763, 800, and 801.

TABLE III-55. INFRARED FREQUENCIES AND BAND ASSIGNMENTS OF $Cr(C_6H_6)_2$ AND $Cr(C_6D_6)_2$.[795]

$Cr(C_6H_6)_2$	$Cr(C_6D_6)_2$	Assignment
3037	2252	ν(CH)
1426	1378	ν(CC) or ring def.
999	–	δ(CH)
971	928	ν(CC) or ring def.
833	–	π(CH)
794	–	π(CH)
490	479	ring tilt (ν_5)[b]
459	423	ν(MR)(ν_3)[a,b]
(140)	–	δ(RMR)(ν_6)[a,b]

[a] R denotes the C_6H_6 ring.
[b] See Fig. III-40.

References

1. I. Nakagawa and S. Mizushima, *Bull. Chem. Soc. Jap.*, **28**, 589 (1955).
2. S. Mizushima, I. Nakagawa, and J. V. Quagliano, *J. Chem. Phys.*, **23**, 1367 (1955).
3. J. Fujita, K. Nakamoto, and M. Kobayashi, *J. Amer. Chem. Soc.*, **78**, 3295 (1956).
4. A. Nakahara, Y. Saito, and H. Kuroya, *Bull. Chem. Soc. Jap.*, **25**, 331 (1952).
5. G. F. Svatos, C. Curran, and J. V. Quagliano, *J. Amer. Chem. Soc.*, **77**, 6159 (1955).
6. G. F. Svatos, D. M. Sweeny, S. Mizushima, C. Curran, and J. V. Quagliano, *J. Amer. Chem. Soc.*, **79**, 3313 (1957).
7. J. K. Wilmshurst, *Can. J. Chem.*, **38**, 467 (1960).
8. D. B. Powell and N. Sheppard, *J. Chem. Soc.*, **1956** 3108.
9. K. Nakamoto, Y. Morimoto, and J. Fujita, *Proc.* 7th *I.C.C.C.*, Stockholm, 1962, p. 15.
10. K. Nakamoto, J. Fujita, and H. Murata, *J. Amer. Chem. Soc.*, **80**, 4817 (1958).
11. G. M. Barrow, R. H. Krueger, and F. Basolo, *J. Inorg. Nucl. Chem.*, **2**, 340 (1956).
12. E. P. Bertin, I. Nakagawa, S. Mizushima, T. J. Lane, and J. V. Quagliano, *J. Amer. Chem. Soc.*, **80**, 525 (1958).
13. F. A. Cotton, *Acta Chem. Scand.*, **10**, 1520 (1956).
14. H. Block, *Trans. Faraday Soc.*, **55**, 867 (1959).
15. T. Shimanouchi and I. Nakagawa, *Spectrochim. Acta*, **18**, 89 (1962).
16. I. Nakagawa and T. Shimanouchi, *Spectrochim. Acta*, **22**, 759 (1966).
17. J. M. Terrasse, H. Poulet, and J. P. Mathieu, *Spectrochim. Acta*, **20**, 305 (1964).
18. T. E. Haas and J. R. Hall, *Spectrochim. Acta*, **22**, 988 (1966).
19. W. P. Griffith, *J. Chem. Soc.*, **A,1966**, 899.
20. I. Nakagawa, *Kagaku No Ryoiki*, **72**, 87 (1965).
21. L. Sacconi, A. Sabatini, and P. Gans, *Inorg. Chem.*, **3**, 1772 (1964).
22. A. D. Allen and C. V. Senoff, *Can. J. Chem.*, **45**, 1337 (1967).
23. N. Tanaka, M. Kamada, J. Fujita, and E. Kyuno, *Bull. Chem. Soc. Jap.*, **37**, 222 (1964).
24. T. Shimanouchi and I. Nakagawa, *Inorg. Chem.*, **3**, 1805 (1964).
25. G. W. Watt and D. S. Klett, *Inorg. Chem.*, **3**, 782 (1964).
26. H. Poulet, P. Delorme, and J. P. Mathieu, *Spectrochim. Acta*, **20**, 1855 (1964).
27. P. J. Hendra, *Spectrochim. Acta*, **23A**, 1275 (1967).
28. G. W. Leonard, E. R. Lippincott, R. D. Nelson, and D. E. Sellers, *J. Amer. Chem. Soc.*, **77**, 2029 (1955).
29. I. Nakagawa and T. Shimanouchi, *Spectrochim. Acta*, **22**, 1707 (1966).
30. A. F. Schreiner and J. A. McLean, *J. Inorg. Nucl. Chem.*, **27**, 253 (1965).
31. C. H. Perry, D. P. Athans, E. F. Young, J. R. Durig, and B. R. Mitchell, *Spectrochim. Acta*, **23A**, 1137 (1967).
32. R. Layton, D. W. Sink, and J. R. Durig, *J. Inorg. Nucl. Chem.*, **28**, 1965 (1966).
33. J. R. Durig, R. Layton, D. W. Sink, and B. R. Mitchell, *Spectrochim. Acta*, **21**, 1367 (1965).
34. D. B. Powell, *J. Chem. Soc.*, **1956**, 4495.
35. J. R. Durig and B. R. Mitchell, *Appl. Spectroscopy*, **21**, 221 (1967).
36. P. J. Hendra and N. Sadasivan, *Spectrochim. Acta*, **21**, 1271 (1965).
37. R. C. Leech, D. B. Powell, and N. Sheppard, *Spectrochim. Acta*, **21**, 559 (1965).
38. K. Nakamoto, P. J. McCarthy, J. Fujita, R. A. Condrate, and G. T. Behnke, *Inorg. Chem.*, **4**, 36 (1965).
39. S. Mizushima, I. Nakagawa, and D. M. Sweeny, *J. Chem. Phys.*, **25**, 1006 (1956); I. Nakagawa, R. B. Penland, S. Mizushima, T. J. Lane, and J. V. Quagliano, *Spectrochim. Acta*, **9**, 199 (1957).
40. K. Brodersen and H. J. Becher, *Chem. Ber.*, **89**, 1487 (1956).

41. A. Novak, J. Portier, and P. Bouclier, *Compt. rend.*, **261**, 455 (1965).
42. G. W. Watt, B. B. Hutchinson, and D. S. Klett, *J. Amer. Chem. Soc.*, **89**, 2007 (1967).
43. Y. Y. Kharitonov, I. K. Dymina, and T. Leonova, *Izv. Akad. Nauk SSSR, Ser. Khim.*, 2057 (1966).
44. M. Goldstein and E. F. Mooney, *J. Inorg. Nucl. Chem.*, **27**, 1601 (1965).
45. J. Chatt, L. A. Duncanson, and L. M. Venanzi, *J. Chem. Soc.*, 4456, 4461, **1955**; 2712, **1956**; *J. Inorg. Nucl. Chem.*, **8**, 67 (1958).
46. L. A. Duncanson and L. M. Venanzi, *J. Chem. Soc.*, **1960**, 3841.
47. Y. Y. Kharitonov, M. A. Sarukhanov, I. B. Baranovskii, and K. U. Ikramov, *Opt. Spektrosk.* **19**, 460 (1965).
48. L. Sacconi and A. Sabatini, *J. Inorg. Nucl. Chem.*, **25**, 1389 (1963).
49. K. Brodersen, *Z. Anorg. Allg. Chem.*, **290**, 24 (1957).
50. I. Nakagawa, T. Shimanouchi, and K. Yamasaki, *Inorg. Chem.*, **3**, 772 (1964).
51. C. J. H. Schutte, *Z. Anorg. Allg. Chem.*, **334**, 304 (1965).
52. M. N. Hughes and W. R. McWhinnie, *Spectrochim. Acta*, **22**, 987 (1966).
53. R. A. Krause, A. E. Wickenden, and C. R. Ruggles, *Inorg. Chem.*, **5**, 936 (1966).
54. H. Elliott, B. J. Hathaway, and R. C. Slade, *Inorg. Chem.*, **5**, 669 (1966).
55. M. LePostollec, J. P. Mathieu, and H. Poulet, *J. Chim. Phys.*, **60**, 1319 (1963).
56. A. V. Babaeva and Y. Y. Kharitonov, *Zh. Neorg. Khim.*, **5**, 1196 (1960).
57. M. LePostolle, *J. Chim. Phys.*, **62**, 67 (1965).
58. C. J. H. Schutte, *Z. Naturforsch.*, **A17**, 98 (1962).
59. I. R. Beattie and H. J. V. Tyrrell, *J. Chem. Soc.*, **1956**, 2849.
60. J. P. Faust and J. V. Quagliano, *J. Amer. Chem. Soc.*, **76**, 5346 (1954).
61. P. E. Merritt and S. E. Wiberley, *J. Phys. Chem.*, **59**, 55 (1955).
62. J. Chatt, L. A. Duncanson, B. M. Gatehouse, J. Lewis, R. S. Nyholm, M. L. Tobe, P. F. Todd, and L. M. Venanzi, *J. Chem., Soc.*, **1959**, 4073.
63. M. L. Morris and D. H. Busch, *J. Amer. Chem. Soc.*, **82**, 1521 (1960).
64. C. O'Connor, *J. Chem. Soc.*, **1964**, 509.
65. I. Nakagawa and T. Shimanouchi, *Spectrochim. Acta*, **23A**, 2099 (1967).
66. G. Blyholder and A. Kittila, *J. Phys. Chem.*, **67**, 2147 (1963).
67. M. J. Cleare and W. P. Griffith, *J. Chem. Soc.*, **A**, **1967**, 1144.
68. D. M. L. Goodgame and M. A. Hitchman, *Inorg. Chem.*, **3**, 1389 (1964).
69. L. El-Sayed and R. O. Ragsdale, *Inorg. Chem.*, **6**, 1644 (1967).
70. R. B. Penland, T. J. Lane, and J. V. Quagliano, *J. Amer. Chem. Soc.*, **78**, 887 (1956).
71. I. R. Beattie and D. P. N. Satchell, *Trans. Faraday Soc.*, **52**, 1590 (1956).
72. F. Basolo and G. S. Hammaker, *J. Amer. Chem. Soc.*, **82**, 1001 (1960); *Inorg. Chem.*, **1**, 1 (1962).
73. D. M. L. Goodgame and M. A. Hitchman, *Inorg. Chem.*, **4**, 721 (1965).
74. D. M. L. Goodgame and M. A. Hitchman, *Inorg. Chem.*, **6**., 813 (1967).
75. P. J. Lucchesi and W. A. Glasson, *J. Amer. Chem. Soc.*, **78**, 1347 (1956).
76. C. Vassas-Dubuisson, *Compt. rend.*, **233**, 374 (1951); *J. Chim. Phys.*, **50**, 98 (1953).
77. A. Weil-Marchand, *Compt. rend.*, **242**, 93 (1956); **236**, 2147 (1953).
78. L. C. Mathieu and J. P. Mathieu, *Acta Cryst.*, **5**, 571 (1952).
79. J. Chapelle and A. Galy, *Compt. rend.*, **236**, 1653 (1953); **233**, 1181 (1951).
80. A. Weil-Marchand, *Compt. rend.* **242**, 1791 (1956); **241**, 1456 (1955); **240**, 509 (1955).
81. R. Lafont, *Ann. Phys.*, **4**, 905 (1959); *J. Chim. Phys.*, **50**, C91 (1953); *Compt. rend.*, **244**, 1481 (1957).
82. E. Schwarzmann and O. Glemser, *Z. Anorg. Allg. Chem.*, **312**, 45 (1961).
83. J. van der Elsken and D. W. Robinson, *Spectrochim. Acta*, **17**, 1249 (1961).
84. J. Fujita, K. Nakamoto, and M. Kobayashi, *J. Amer. Chem. Soc.*, **78**, 3963 (1956).

85. I. Gamo, *Bull. Chem. Soc. Jap.*, **34**, 760, 765, 1430, 1433 (1961).
86. G. Sartori, C. Furlani, and A. Damiani, *J. Inorg. Nucl. Chem.*, **8**, 119 (1958).
87. I. Nakagawa and T. Shimanouchi, *Spectrochim. Acta*, **20**, 429 (1964).
88. R. E. Hester and R. A. Plane, *Inorg. Chem.*, **3**, 768 (1964).
89. H. Boutin, G. J. Safford, and H. R. Danner, *J. Chem. Phys.*, **40**, 2670 (1964); H. J. Prask and H. Boutin, *ibid.*, **45**, 699, 3284 (1966).
90. J. R. Durig, W. A. McAllister, and E. E. Mercer, *J. Inorg. Nucl. Chem.*, **29**, 1441 (1967).
91. G. Blyholder and S. Vergez, *J. Phys. Chem.*, **67**, 2149 (1963).
92. D. Scargill, *J. Chem. Soc.*, **1961**, 4440.
93. T. Dupuis, C. Duval, and J. Lecomte, *Compt. rend.*, **257**, 3080 (1963).
94. V. Lorenzelli, T. Dupuis, and J. Lecomte, *Compt. rend.*, **259**, 1057 (1964).
95. K. Nakamoto, J. Fujita, and Y. Morimoto. To be published.
96. J. R. Ferraro and W. R. Walker, *Inorg. Chem.*, **4**, 1382 (1965).
97. W. R. McWhinnie, *J. Inorg. Nucl. Chem.*, **27**, 1063 (1965).
98. J. R. Ferraro, R. Driver, W. R. Walker, and W. Wozniak, *Inorg. Chem.*, **6**, 1586 (1967).
99. G. Blyholder and N. Ford, *J. Phys. Chem.*, **68**, 1496 (1964).
100. V. A. Maroni and T. G. Spiro, *J. Amer. Chem. Soc.*, **89**, 45 (1967).
101. D. J. Hewkin and W. P. Griffith, *J. Chem. Soc.*, **A,1966**, 472.
102. K. Nakamoto, J. Fujita, S. Tanaka, and M. Kobayashi, *J. Amer. Chem. Soc.*, **79**, 4904 (1957).
103. B. M. Gatehouse, S. E. Livingston, and R. S. Nyholm, *J. Chem. Soc.*, **1958**, 3137.
104. J. Fujita, A. E. Martell, and K. Nakamoto, *J. Chem. Phys.*, **36**, 339 (1962).
105. R. E. Hester and W. E. L. Grossman, *Inorg. Chem.*, **5**, 1308 (1966).
106. H. Elliott and B. J. Hathaway, *Spectrochim. Acta*, **21**, 1047 (1965).
107. B. M. Gatehouse, S. E. Livingston, and R. S. Nyholm, *J. Chem. Soc.*, **1957**, 4222; *J. Inorg. Nucl. Chem.*, **8**, 75 (1958).
108. N. F. Curtis and Y. M. Curtis, *Inorg. Chem.*, **4**, 804 (1965).
109. A. B. P. Lever, *Inorg. Chem.*, **4**, 1042 (1965).
110. R. V. Biagetti and H. M. Haendler, *Inorg. Chem.*, **5**, 383 (1966).
111. C. C. Addison and W. B. Simpson, *J. Chem. Soc.*, **1965**, 598.
112. J. I. Bullock, *J. Inorg. Nucl. Chem.*, **29**, 2257 (1967).
113. J. R. Ferraro, A. Walker, and C. Cristallini, *J. Inorg. Nucl. Chem.* Letters, **1**, 25 (1965).
114. G. Topping, *Spectrochim. Acta*, **21**, 1743 (1965).
115. J. R. Ferraro and A. Walker, *J. Chem. Phys.*, **42**, 1273 (1965).
116. A. Walker and J. R. Ferraro, *J. Chem. Phys.*, **43**, 2689 (1965).
117. J. R. Ferraro and A. Walker, *J. Chem. Phys.*, **45**, 550 (1966).
118. G. J. Janz and D. W. James, *J. Chem. Phys.*, **35**, 739 (1961); G. J. Janz, T. R. Kozlowski, and S. C. Wait, *ibid.*, **39**, 1809 (1963).
119. C. G. Barraclough and M. L. Tobe, *J. Chem. Soc.*, **1961**, 1993.
120. J. E. Finholt, R. W. Anderson, J. A. Fyfe, and K. G. Caulton, *Inorg. Chem.*, **4**, 43 (1965).
121. W. R. McWhinnie, *J. Inorg. Nucl. Chem.*, **26**, 21 (1964).
122. R. Eskenazi, J. Raskovan, and R. Levitus, *J. Inorg. Nucl. Chem.*, **28**, 521 (1966).
123. J. R. Ferraro and A. Walker, *J. Chem. Phys.*, **42**, 1278 (1965).
124. N. Tanaka, H. Sugi, and J. Fujita, *Bull. Chem. Soc. Jap.*, **37**, 640 (1964).
125. B. J. Hathaway and A. E. Underhill, *J. Chem. Soc.*, **1961**, 3091.
126. B. J. Hathaway, D. G. Holah, and M. Hudson, *J. Chem. Soc.*, **1963**, 4586.
127. M. E. Farago, J. M. James, and V. C. G. Trew, *J. Chem. Soc.*, *A*,**1967**, 820.
128. A. E. Wickenden and R. A. Krause, *Inorg. Chem.*, **4**, 404 (1965).

129. L. E. Moore, R. B. Gayhart, and W. E. Bull, *J. Inorg. Nucl. Chem.*, **26**, 896 (1964).
130. H. Siebert, *Z. Anorg. Allg. Chem.*, **296**, 280 (1958); **298**, 51 (1959).
131. D. E. Peters and R. J. M. Fraser, *J. Amer. Chem. Soc.*, **87**, 2758 (1965).
132. J. A. Costamagna and R. Levitus, *J. Inorg. Nucl. Chem.*, **28**, 1116 (1966).
133. F. A. Cotton and R. Francis, *J. Amer. Chem. Soc.*, **82**, 2986 (1960).
134. G. Newman and D. B. Powell, *Spectrochim. Acta*, **19**, 213 (1963).
135. M. E. Baldwin, *J. Chem. Soc.*, **1961**, 3123.
136. V. Caglioti, G. Sartori, and M. Scrocco, *Atti Accad. Naz. Lincei, Rend., Cl. Sci. Fis. Mat. Nat.*, **22**, 266 (1957).
137. G. B. Bonino and G. Fabbri, *Atti. Accad. Naz. Lincei, Rend., Cl. Sci. Fis. Mat. Nat.*, **20**, 414 (1956); **19**, 386 (1955); G. Fabbri, *ibid*, **22**, 488 (1957); G. B. Bonino and O. Salvetti, *Ric. Sci.*, **26**, 3627 (1956).
138. O. Salvetti, *Ric. Sci.*, **29**, 1228 (1959).
139. E. F. Herington and W. Kynaston, *J. Chem. Soc.*, **1955**, 3555.
140. G. Bor, *J. Inorg. Nucl. Chem.*, **17**, 174 (1961).
141. W. P. Griffith and G. Wilkinson, *J. Chem. Soc.*, **1959**, 2757; *J. Inorg. Nucl. Chem.*, **7**, 297 (1958).
142. M. F. A. El-Sayed and R. K. Sheline, *J. Inorg. Nucl. Chem.*, **6**, 187 (1958).
143. M. F. A. El-Sayed and R. K. Sheline, *J. Amer. Chem. Soc.*, **80**, 2047 (1958).
144. M. F. A. El-Sayed and R. K. Sheline, *J. Amer. Chem. Soc.*, **78**, 702 (1956).
145. L. H. Jones, *J. Chem. Phys.*, **29**, 463 (1958).
146. L. H. Jones, *Spectrochim. Acta*, **17**, 188 (1961).
147. H. Poulet and J. P. Mathieu, *Spectrochim. Acta*, **15**, 932 (1959).
148. E. G. Brame, F. A. Johnson, E. M. Larsen, and V. W. Meloche, *J. Inorg. Nucl. Chem.*, **6**, 99 (1958).
149. G. B. Bonino and G. Fabbri, *Atti Accad. Naz. Lincei, Rend., Cl. Sci. Fis., Mat. Nat.*, **20**, 566 (1956).
150. L. H. Jones, *J. Chem. Phys.*, **26**, 1578 (1957); **25**, 379 (1956).
151. D. M. Sweeny, I. Nakagawa, S. Mizushima, and J. V. Quagliano, *J. Amer. Chem. Soc.*, **78**, 889 (1956).
152. R. L. McCullough, L. H. Jones, and G. A. Crosby, *Spectrochim. Acta*, **16**, 929 (1960).
153. L. H. Jones, *J. Chem. Phys.*, **27**, 468 (1957); **21**, 1891 (1953); **22**, 1135 (1954).
154. H. Poulet and J. P. Mathieu, *Compt. rend.*, **248**, 2079 (1959).
155. L. H. Jones and R. A. Penneman, *J. Chem. Phys.*, **22**, 965 (1954).
156. R. A. Penneman and L. H. Jones, *J. Chem. Phys.*, **24**, 293 (1956).
157. R. A. Penneman and L. H. Jones, *J. Inorg. Nucl. Chem.*, **20**, 19 (1961).
158. R. L. McCullough, L. H. Jones, and R. A. Penneman, *J. Inorg. Nucl. Chem.*, **13**, 286 (1960).
159. G. W. Chantry and R. A. Plane, *J. Chem. Phys.*, **33**, 736 (1960); **34**, 1268 (1961); **35**, 1027 (1961).
160. V. Caglioti, G. Sartori, and C. Furlani, *J. Inorg. Nucl. Chem.*, **13**, 22 (1960); **8**, 87 (1958).
161. L. H. Jones, *J. Mol. Spectrosc.*, **8**, 105 (1962); *J. Chem. Phys.*, **36**, 1209 (1962).
162. L. H. Jones, *J. Chem. Phys.*, **41**, 856 (1964).
163. D. Bloor, *J. Chem. Phys.*, **41**, 2573 (1964).
164. I. Nakagawa and T. Shimanouchi, *Spectrochim. Acta*, **18**, 101 (1962).
165. L. H. Jones, *Inorg. Chem.*, **2**, 777 (1963).
166. L. H. Jones, *J. Mol. Spectrosc.*, **8**, 105 (1962); **9**, 130 (1963).
167. L. H. Jones and J. M. Smith, *Inorg. Chem.*, **4**, 1677 (1965).
168. L. Tosi and J. Danon, *Inorg. Chem.*, **3**, 150 (1964).

169. J. L. Hoard and H. H. Nordsieck, *J. Amer. Chem. Soc.*, **61**, 2853 (1939).
170. H. Stammreich and O. Sala, *Z. Elektrochem.*, **64**, 741 (1960); **65**, 149 (1961).
171. S. F. A. Kettle and R. V. Parish, *Spectrochim. Acta*, **21**, 1087 (1965).
172. R. V. Parish, *Spectrochim. Acta*, **22**, 1191 (1966).
173. C. W. F. T. Pistorius, *Z. Phys. Chem.*, **23**, 197 (1960).
174. L. H. Jones, *J. Chem. Phys.*, **27**, 665 (1957).
175. L. H. Jones and J. M. Smith, *J. Chem. Phys.*, **41**, 2507 (1964).
176. L. H. Jones, *Spectrochim. Acta*, **19**, 1675 (1963).
177. V. Lorenzelli and P. Delorme, *Spectrochim. Acta*, **19**, 2033 (1963).
178. L. H. Jones, *Inorg. Chem.*, **3**, 1581 (1964); **4**, 1472 (1965); J. M. Smith, L. H. Jones, I. K. Kressin, and R. A. Penneman, *ibid.*, **4**, 369 (1965).
179. J. C. Coleman, H. Peterson, and R. A. Penneman, *Inorg. Chem.*, **4**, 135 (1965).
180. R. Nast, H. Ruppert-Mesche and M. Helbig-Neubauer, *Z. Anorg. Allg. Chem.*, **312**, 314 (1961).
181. A. Haim and W. K. Wilmarth, *J. Amer. Chem. Soc.*, **83**, 509 (1961).
182. D. A. Dows, A. Haim, and W. K. Wilmarth, *J. Inorg. Nucl. Chem.*, **21**, 33 (1961).
183. D. F. Shriver, *J. Amer. Chem. Soc.*, **84**, 4610 (1962); **85**, 1405 (1963); D. F. Shriver and J. Posner, *ibid.*, **88**, 1672 (1966).
184. D. F. Shriver, S. A. Shriver, and S. E. Anderson, *Inorg. Chem.*, **4**, 725 (1965).
185. R. A. Walton, *Spectrochim. Acta*, **21**, 1795 (1965); *Can. J. Chem.*, **44**, 1480 (1966).
186. J. C. Evans and G. Y-S. Lo, *Spectrochim. Acta.* **21**, 1033 (1965).
187. M. Kubota and D. L. Johnston, *J. Inorg. Nucl. Chem.*, **29**, 769 (1967).
188. J. Reedijk and W. L. Groeneveld, *Rec. Trav. Chim., Pays-Bas*, **86**, 1127 (1967).
189. F. A. Cotton and F. Zingales, *J. Amer. Chem. Soc.*, **83**, 351 (1961).
190. A. Sacco and F. A. Cotton, *J. Amer. Chem. Soc.*, **84**, 2043 (1962).
191. S. Ahrland, J. Chatt, and N. R. Davies, *Quart. Rev.*, **12**, 265 (1958).
192. P. C. H. Mitchell and R. J. P. Williams, *J. Chem. Soc.*, **1960**, 1912.
193. A. Turco and C. Pecile, *Nature*, **191**, 66 (1961).
194. J. Lewis, R. S. Nyholm, and P. W. Smith, *J. Chem. Soc.*, **1961**, 4590.
195. A. Sabatini and I. Bertini, *Inorg. Chem.*, **4**, 959 (1965).
196. C. Pecile, *Inorg. Chem.*, **5**, 210 (1966).
197. R. J. H. Clark and C. S. Williams, *Spectrochim. Acta*, **22**, 1081 (1966).
198. S. M. Nelson and T. M. Shepherd, *J. Inorg. Nucl. Chem.*, **27**, 2123 (1965).
199. M. A. Bennett, R. J. H. Clark, and A. D. J. Goodwin, *Inorg. Chem.*, **6**, 1625 (1967).
200. M. A. Jennings and A. Wojcicki, *Inorg. Chem.*, **6**, 1854 (1967).
201. M. M. Chamberlain and J. C. Bailar, *J. Amer. Chem. Soc.*, **81**, 6412 (1959).
202. M. E. Baldwin, *J. Chem. Soc.*, **1960**, 4369; **1961**, 471.
203. F. Basolo, J. L. Burmeister, and A. J. Poe, *J. Amer. Chem. Soc.*, **85**, 1700 (1963).
204. J. L. Burmeister and F. Basolo, *Inorg. Chem.*, **3**, 1587 (1964).
205. A. Sabatini and I. Bertini, *Inorg. Chem.*, **4**, 1665 (1965).
206. K. N. Raymond and F. Basolo, *Inorg. Chem.*, **5**, 1632 (1966).
207. P. C. Jain and E. C. Lingafelter, *J. Amer. Chem. Soc.*, **89**, 724 (1967).
208. I. Bertini and A. Sabatini, *Inorg. Chem.*, **5**., 1025 (1966).
209. J. Chatt and L. A. Duncanson, *Nature*, **178**, 997 (1956).
210. J. Chatt, L. A. Duncanson, F. A. Hart, and P. G. Owston, *Nature*, **181**, 43 (1958).
211. P. G. Owston and J. M. Rowe, *Acta Cryst.*, **13**, 253 (1960).
212. J. Chatt and F. A. Hart, *J. Chem. Soc.*, **1961**, 1416.
213. C. Pecile, A. Turco, and G. Pizzolotto, *Ric. Sci.*, **31**, 2A, 247 (1961).
214. F. A. Cotton, D. M. L. Goodgame, M. Goodgame, and T. E. Haas, *Inorg. Chem.*, **1**, 565 (1962).

215. J. L. Burmeister and L. E. Williams, *Inorg. Chem.*, **5**, 1113 (1966).
216. M. E. Farago and J. M. James, *Inorg. Chem.*, **4**, 1706 (1965).
217. A. Turco, C. Pecile, and M. Nicolini, *J. Chem. Soc.*, **1962**, 3008.
218. J. L. Burmeister and Y. Al-Janabi, *Inorg. Chem.*, **4**, 962 (1965).
219. D. Forster and D. M. L. Goodgame, *Inorg. Chem.* **4**, 1712 (1965).
220. J. L. Burmeister and H. *J.* Gysling, *Inorg. Chim. Acta.*, **1**, 100 (1967).
221. J. L. Burmeister and H. J. Gysling, *Chem. Comm.* 543 (1967).
222. D. Forster and D. M. L. Goodgame *J. Chem. Soc.*, **1965**, 1286.
223. D. Forster and W. D. Horrocks, *Inorg. Chem.* **6**, 339 (1967).
224. D. Forster and D. M. L. Goodgame, *J. Chem. Soc.*, **1965**, 262.
225. W. Beck, W. P. Fehlhammer, P. Pöllmann, E. Schuierer, and K. Feldl, *Chem. Ber.* **100**, 2335 (1967); *Angew. Chem.*, **77**, 458 (1965).
226. D. Forster and W. D. Horrocks, *Inorg. Chem.*, **5**, 1510 (1966).
227. G. W. Bethke and M. K. Wilson, *J. Chem. Phys.*, **26**, 1118 (1957).
228. R. C. Taylor, *J. Chem. Phys.*, **26**, 1131; **27**, 979 (1957).
229. S. Sundaram and F. F. Cleveland, *J. Chem. Phys.*, **32**, 166 (1960).
230. R. D. Cowan, *J. Chem. Phys.*, **18**, 1101 (1950); **17**, 218 (1949).
231. L. H. Jones, *J. Chem. Phys.*, **28**, 1215 (1958); **23**, 2448 (1955); *J. Mol. Spectrosc.*, **5**, 133 (1960); *Spectrochim. Acta*, **19**, 1899 (1963).
232. B. L. Crawford and P. C. Cross, *J. Chem. Phys.*, **6**, 525 (1938); B. L. Crawford and W. Horwitz, *ibid.*, **16**, 147 (1948).
233. C. W. F. T. Pistorius, *Spectrochim. Acta*, **15**, 717 (1959).
234. M. Bigorgne, *J. Inorg. Nucl. Chem.*, **8**, 113 (1958); Compt. rend., **246**, 1685 (1958).
235. H. Murata and K. Kawai, *J. Chem. Phys.*, **26**, 1355 (1957).
236. R. S. Nyholm and L. N. Short, *J. Chem. Soc.*, **1953**, 2670.
237. H. Stammreich, K. Kawai, O. Sala, and P. Krumholz, *J. Chem. Phys.*, **35**, 2168, 2175 (1961).
238. H. Stammreich, K. Kawai, Y. Tavares, P. Krumholz, J. Behmoiras, and S. Bril, *J. Chem. Phys.*, **32**, 1482 (1960).
239. R. A. Friedel, I. Wender, S. L. Shufler, and H. W. Sternberg, *J. Amer. Chem. Soc.*, **77**, 3951 (1955).
240. W. F. Edgell, J. Huff, J. Thomas, H. Lehman, C. Angell, and G. Asato, *J. Amer. Chem. Soc.*, **82**, 1254 (1960); W. F. Edgell, W. E. Wilson, and R. Summitt, *Spectrochim. Acta*, **19**, 863 (1963).
241. W. G. Fateley and E. R. Lippincott, *Spectrochim. Acta*, **10**, 8 (1957); F. T. King and E. R. Lippincott, *J. Amer. Chem. Soc.*, **78**, 4192 (1956).
242. C. W. F. T. Pistorius and P. C. Haarhoff, *J. Chem. Phys.*, **31**, 1439 (1959).
243. H. Stammreich, O. Sala, and Y. Tavares, *J. Chem. Phys.*, **30**, 856 (1959).
244. H. Murata and K. Kawai, *J. Chem. Phys.*, **28**, 516 (1958).
245. F. A. Cotton, A. Danti, J. S. Waugh, and R. W. Fessenden, *J. Chem. Phys.*, **29**, 1427 (1958).
246. R. K. Sheline and K. S. Pitzer, *J. Amer. Chem. Soc.*, **72**, 1107 (1950).
247. L. H. Jones and R. S. McDowell, *Spectrochim. Acta*, **20**, 248 (1964).
248. M. F. O'Dwyer, *J. Mol. Spectrosc.*, **2**, 144 (1958).
249. L. H. Jones, *Spectrochim. Acta*, **19**, 329 (1963); *J. Chem. Phys.*, **36**, 2375 (1962).
250. R. L. Amster, R. B. Hannan, and M. C. Tobin, *Spectrochim. Acta*, **19**, 489 (1963).
251. N. J. Hawkins, H. C. Mattraw, W. W. Sabol, and D. R. Carpenter, *J. Chem. Phys.*, **23**, 2422 (1955).
252. S. L. Shufler, H. W. Sternberg, and R. A. Friedel, *J. Amer. Chem. Soc.*, **78**, 2687 (1956).
253. A. Danti and F. A. Cotton, *J. Chem. Phys.*, **28**, 736 (1958).
254. H. Murata and K. Kawai, *J. Chem. Phys.*, **27**, 605 (1957); *Bull. Chem. Soc. Jap.*, **33**, 1008 (1960).

255. A. W. Hanson, *Acta Cryst.*, **15**, 930 (1962).
256. L. H. Jones, *J. Mol. Spectrosc.*, **8**, 105 (1962); **9**, 130 (1962); *J. Chem. Phys.*, **36**, 2375 (1962).
257. R. J. H. Clark and B. Crociani, *Inorg. Chim. Acta*, **1**, 12 (1967).
258. G. Bor, *Spectrochim. Acta*, **18**, 817 (1962).
259. W. F. Edgell, A. T. Watts, J. Lyford, and W. M. Risen, *J. Amer. Chem. Soc.*, **88**, 1815 (1966).
260. H. W. Sternberg, I. Wender, R. A. Friedel, and M. Orchin, *J. Amer. Chem. Soc.*, **75**, 2717 (1953).
261. W. F. Edgell, C. Magee, and G. Gallup, *J. Amer. Chem. Soc.*, **78**, 4185, 4188 (1956).
262. F. A. Cotton, *J. Amer. Chem. Soc.*, **80**, 4425 (1958).
263. T. C. Farrar, F. E. Brinckman, T. D. Coyle, A. Davison, and J. W. Faller, *Inorg. Chem.*, **6**, 161 (1967).
264. F. A. Cotton and G. Wilkinson, *Chem. Ind.* (*London*), 1305 (1956).
265. W. F. Edgell and R. Summitt, *J. Amer. Chem. Soc.*, **83**, 1772 (1961).
266. F. Zingales, F. Canziani, and A. Chiesa, *Inorg. Chem.*, **2**, 1303 (1963).
267. E. O. Bishop, J. L. Down, P. R. Emtage, R. E. Richards, and G. Wilkinson, *J. Chem. Soc.*, **1959**, 2484.
268. W. E. Wilson, *Z. Naturforsch.*, **B13**, 349 (1958).
269. F. A. Cotton, J. L. Down, and G. Wilkinson, *J. Chem. Soc.*, **1959**, 833.
270. S. J. LaPlaca, W. C. Hamilton, and J. A. Ibers, *Inorg. Chem.*, **3**, 1491 (1964); *J. Amer. Chem. Soc.*, **86**, 2288 (1964).
271. D. K. Huggins and H. D. Kaesz, *J. Amer. Chem. Soc.*, **86**, 2734 (1964).
272. P. S. Braterman, R. W. Harrill, and H. D. Kaesz, *J. Amer. Chem. Soc.*, **89**, 2851 (1967).
273. T. C. Farrar, W. Ryan, A. Davison, and J. W. Faller, *J. Amer. Chem. Soc.*, **88**, 184 (1966).
274. W. Beck, W. Hieber, and G. Braun, *Z. Anorg. Allg. Chem.*, **308**, 23 (1961).
275. D. K. Huggins, W. Fellmann, J. M. Smith, and H. D. Kaesz, *J. Amer. Chem. Soc.*, **86**, 4841 (1964).
276. J. M. Smith, W. Fellmann, and L. H. Jones, *Inorg. Chem.*, **4**, 1361 (1965).
277. R. G. Hayter, *J. Amer. Chem.*, *Soc.*, **88**, 4376 (1966).
278. S. S. Bath and L. Vaska, *J. Amer. Chem. Soc.*, **85**, 3500 (1963).
279. J. Chatt, N. P. Johnson, and B. L. Shaw, *J. Chem. Soc.*, **1964**, 1625.
280. L. Vaska, *J. Amer. Chem.*, *Soc.*, **88**, 4100 (1966).
281. J. W. Cable, R. S. Nyholm, and R. K. Sheline, *J. Amer. Chem. Soc.*, **76**, 3373 (1954).
282. H. W. Sternberg, I. Wender, R. A. Friedel, and M. Orchin, *J. Amer. Chem. Soc.*, **75**, 3148 (1953).
283. F. A. Cotton and R. R. Monchamp, *J. Chem. Soc.*, **1960**, 1882.
284. G. Bor and L. Markó, *Spectrochim. Acta*, **15**, 747 (1959); **16**, 1105 (1960).
285. G. G. Summer, H. P. Klug, and L. E. Alexander, *Acta Cryst.*, **17**, 732 (1964).
286. K. Noack, *Spectrochim. Acta*, **19**, 1925 (1963).
287. G. Bor, *Spectrochim. Acta*, **19**, 1209 (1963).
288. G. Bor, *Spectrochim. Acta*, **19**, 2065 (1963).
289. P. Corradini, *J. Chem. Phys.*, **31**, 1676 (1959).
290. E. O. Brimm, M. A. Lynch, and W. J. Sesny, *J. Amer. Chem. Soc.*, **76**, 3831 (1954).
291. F. A. Cotton, A. O. Liehr, and G. Wilkinson, *J. Inorg. Nucl. Chem.*, **2**, 141 (1956).
292. L. F. Dahl, E. Ishishi, and R. E. Rundle, *J. Chem. Phys.*, **26**, 1750 (1957); L. F. Dahl and R. E. Rundle, *Acta Cryst.*, **16**, 419 (1963).
293. H. M. Powell and R. V. G. Ewens, *J. Chem. Soc.*, **1939**, 286.
294. R. K. Sheline, *J. Amer. Chem. Soc.*, **73**, 1615 (1951).
295. F. A. Cotton and G. Wilkinson, *J. Amer. Chem. Soc.*, **79**, 752 (1957).

296. O. S. Mills, *Chem. Ind. (London)*, 73 (1957).
297. L. F. Dahl and R. E. Rundle, *J. Chem. Phys.*, **26**, 1751 (1957).
298. L. F. Dahl and R. E. Rundle, *J. Chem. Phys.*, **27**, 323 (1957).
299. D. A. Brown, *J. Inorg. Nucl. Chem.*, **5**, 289 (1958).
300. R. H. Herber, W. R. Kingston, and G. K. Wertheim, *Inorg. Chem.*, **2**, 153 (1963).
301. G. R. Dobson and R. K. Sheline, *Inorg. Chem.*, **2**, 1313 (1963).
302. L. F. Dahl and J. F. Blount, *Inorg. Chem.*, **4**, 1373 (1965).
303. N. E. Erickson and A. W. Fairhall, *Inorg. Chem.*, **4**, 1320 (1965).
304. E. R. Corey and L. F. Dahl, *Inorg. Chem.*, **1**, 521 (1962).
305. D. K. Huggins, N. Flitcroft, and H. D. Kaesz, *Inorg. Chem.*, **4**, 166 (1965).
306. H. Stammreich, K. Kawai, O. Sala, and P. Krumholz, *J. Chem. Phys.*, **35**, 2175 (1961).
307. N. Flitcroft, D. K. Huggins, and H. D. Kaesz, *Inorg. Chem.*, **3**, 1123 (1964).
308. L. E. Orgel, *Inorg. Chem.*, **1**, 25 (1962).
309. F. A. Cotton and C. S. Kraihanzel, *J. Amer. Chem. Soc.*, **84**, 4432 (1962); *Inorg. Chem.*, **2**, 533 (1963); **3**, 702 (1964); **4**, 1328 (1965); **6**, 1357 (1967).
310. L. H. Jones, *Inorg. Chem.*, **6**, 1269 (1967).
311. R. J. Irving and E. A. Magnusson, *J. Chem. Sec.*, **1956**, 1860; **1958**, 2283.
312. M. A. El-Sayed and H. D. Kaesz, *Inorg. Chem.*, **2**, 158 (1963).
313. C. W. Garland and J. R. Wilt, *J. Chem. Phys.*, **36**, 1094 (1962).
314. L. F. Dahl, C. Martell, and D. L. Wampler, *J. Amer. Chem. Soc.*, **83**, 1761 (1961).
315. M. A. El-Sayed and H. D. Kaesz, *J. Mol. Spectrosc.*, **9**, 310 (1962).
316. M. A. Bennett and R. J. H. Clark, *J. Chem. Soc.*, *Suppl.*, 556 (1964).
317. H. D. Kaesz, R. Bau, D. Hendrickson, and J. M. Smith, *J. Amer. Chem. Soc.*, **89**, 2844 (1967).
318. A. R. Manning and J. R. Miller, *J. Chem. Soc.*, **A,1966**, 1521.
319. E. W. Abel and I. S. Butler, *Trans. Faraday Soc.*, **63**, 45 (1967).
320. W. Hieber and W. Beck, *Z. Anorg. Allg. Chem.*, **305**, 265, 274 (1960).
321. E. W. Abel, M. A. Bennett, and G. Wilkinson, *J. Chem. Soc.*, **1959**, 2323.
322. H. L. Nigam, R. S. Nyholm, and M. H. B. Stiddard, *J. Chem. Soc.*, **1960**, 1803; H. L. Nigam, R. S. Nyholm, and D. V. R. Rao, *ibid.*, **1959**, 1397.
323. F. A. Cotton and R. V. Parish, *J. Chem. Soc.*, **1960**, 1440; F. A. Cotton and F. Zingales, *J. Amer. Chem. Soc.*, **83**, 351 (1961).
324. L. S. Meriwether and M. L. Fiene, *J. Amer. Chem. Soc.*, **81**, 4200 (1959).
325. R. J. Angelici and M. D. Malone, *Inorg. Chem.*, **6**, 1731 (1967).
326. I. W. Stolz, G. R. Dobson, and R. K. Sheline, *Inorg. Chem.*, **2**, 322 (1963).
327. M. F. Farona, J. G. Grasselli, and B. L. Ross, *Spectrochim. Acta*, **23A**, 1875 (1967).
328. R. C. Taylor and W. D. Horrocks, *Inorg. Chem.*, **3**, 584 (1964).
329. W. F. Edgell and M. P. Dunkle, *Inorg. Chem.*, **4**, 1629 (1965).
330. R. S. McDowell, W. D. Horrocks, and J. T. Yates, *J. Chem. Phys.*, **34**, 530 (1961).
331. W. D. Horrocks and R. C. Taylor, *Inorg. Chem.*, **2**, 723 (1963).
332. G. Bor, *Inorg. Chim. Acta*, **1**, 81 (1967).
333. A. P. Hagen and A. G. MacDiarmid, *Inorg. Chem.*, **6**, 1941 (1967).
334. R. J. H. Clark, *Inorg. Chem.*, **3**, 1395 (1964).
335. E. W. Abel and G. Wilkinson, *J. Chem. Soc.*, **1959**, 1501
336. G. Winkhaus and G. Wilkinson, *J. Chem. Soc.*, **1961**, 602
337. J. Lewis, R. J. Irving, and G. Wilkinson, *J. Inorg. Nucl. Chem.*, **7**, 32 (1958).
338. W. P. Griffith, J. Lewis, and G. Wilkinson, *J. Chem. Soc.*, **1959**, 872, 1632, 1775; **1961**, 775.
339. P. Gans, *Chem. Comm.*, **144** (1965).
340. W. P. Griffith, J. Lewis, and G. Wilkinson, *J. Inorg. Nucl. Chem.*, **7**, 38 (1958).

341. D. Dale and D. C. Hodgkin, *J. Chem. Soc.*, **1965**, 1364.
342. D. Hall and A. A. Taggart, *J. Chem. Soc.*, **1965**, 1359.
343. E. P. Bertin, S. Mizushima, T. J. Lane, and J. V. Quagliano, *J. Amer. Chem. Soc.*, **81**, 3821 (1959).
344. E. E. Mercer, W. A. McAllister, and J. R. Durig, *Inorg. Chem.*, **6**, 1816 (1967).
345. B. F. Hoskins, F. D. Whillans, D. H. Dale, and D. C. Hodgkins, *Chem. Comm.*, 69 (1969).
346. P. Gans, A. Sabatini, and L. Sacconi, *Inorg. Chem.*, **5**, 1877 (1966).
347. J. R. Durig, W. A. McAllister, J. N. Willis, and E. E. Mercer, *Spectrochim. Acta*, **22**, 1091 (1966).
348. E. E. Mercer, W. A. McAllister, and J. R. Durig, *Inorg. Chem.*, **5**, 1881 (1966).
349. E. Miki, T. Ishimori, H. Yamatera, and H. Okuno, *J. Chem. Soc. Japan*, **87**, 703 (1966).
350. A. Jahn, *Z. Anorg. Allg. Chem.*, **301**, 301 (1959); W. Hieber and A. Jahn, *Z. Naturforsch.*, **B13**, 195 (1958).
351. W. Beck and K. Lottes, *Chem. Ber.* **98**, 2657 (1965).
352. W. Beck and K. Lottes, *Z. Anorg. Allg. Chem.*, **335**, 258 (1965).
353. T. C. Waddington and F. Klanberg, *Z. Anorg. Allg. Chem.*, **304**, 185 (1960).
354. J. Chatt, *Proc. Chem. Soc.*, 318 (1962).
355. J. Chatt, L. A. Duncanson, and B. L. Shaw, *Chem. Ind.* (*London*), 859 (1958).
356. J. C. Bailar and H. Itatani, *Inorg. Chem.*, **4**, 1618 (1965).
357. T. Kruck, *Angew. Chem. Internat. Ed.*, **6**, 53 (1967).
358. J. Chatt and R. G. Hayter, *J. Chem. Soc.*, **1961**, 5507.
359. L. Vaska and J. W. Diluzio, *J. Amer. Chem. Soc.*, **84**, 4989 (1962).
360. J. Chatt and R. G. Hayter, *J. Chem. Soc.*, **1961**, 2605.
361. A. D. Allen and C. V. Senoff, *Chem. Comm.*, 621 (1965).
362. A. D. Allen, F. Bottomley, R. O. Harris, V. P. Reinsalu, and C. V. Senoff, *J. Amer. Chem. Soc.*, **89**, 5595 (1967).
363. F. Bottomley and S. C. Nyberg, *Chem. Comm.*, 897 (1966).
364. J. P. Collman and J. W. Kang, *J. Amer. Chem. Soc.*, **88**, 3459 (1966).
365. A. Yamamoto, S. Kitazume, L. S. Pu, and S. Ikeda, *Chem. Comm.*, 79 (1967); A. Misono, Y. Uchida, and T. Saito, *Bull. Chem. Soc. Jap.*, **40**, 700 (1967).
366. A. Sacco and M. Rossi, *Chem. Comm.*, 316 (1967).
367. J. H. Enemark, B. R. Davis, J. A. McGinnety, and J. A. Ibers, *Chem. Comm.*, 96 (1968).
368. L. Vaska, *Science*, **140**, 809 (1963).
369. S. J. LaPlaca and J. A. Ibers, *J. Amer. Chem. Soc.*, **87**, 2581 (1965).
370. P. B. Chock and J. Halpern, *J. Amer. Chem. Soc.*, **88**, 3511 (1966).
371. L. Vaska and D. L. Catone, *J. Am. Chem. Soc.*, **88**, 5324 (1966).
372. K. Hirota and M. Yamamoto, *Chem. Comm.*, 533 (1968).
373. L. Vaska and S. S. Bath, *J. Amer. Chem. Soc.*, **88**, 1333 (1966).
374. L. Vaska, *J. Amer. Chem. Soc.*, **88**, 5325 (1966).
375. L. Vaska, *Chem. Comm.*, 614 (1966).
376. Misono, Y. Uchida, T. Saito, and K. M. Song, *Chem. Comm.*, 419 (1967).
377. K. Shobatake, C. Postmus, J. R. Ferraro, and K. Nakamoto, *Appl. Spectroscopy*, **23**, 12 (1969).
378. V. Albano, P. L. Bellon, and V. Seatturin, *Chem. Comm.*, 507 (1966).
379. R. J. Goodfellow, P. L. Goggin, and L. M. Venanzi, *J. Chem. Soc.*, **A,1967**, 1897.
380. W. F. Edgell and M. P. Dunkle, *Inorg. Chem.*, **4**, 1629 (1965).
381. J. Bradbury, K. P. Forest, R. H. Nuttall, and D. W. A. Sharp, *Spectrochim. Acta*, **23A**, 2701 (1967).
382. G. B. Deacon and J. H. Green, *Chem. Comm.*, 629 (1966).

383. A. A. Chalmers, J. Lewis, and R. Whyman, *J. Chem. Soc.*, **A,1967**, 1817.
384. J. H. S. Green, *Spectroschim. Acta*, **24A**, 137 (1968).
385. P. L. Goggin and R. J. Goodfellow, *J. Chem. Soc.*, **A,1966**, 1462.
386. D. M. Adams and P. J. Chandler, *Chem. Comm.*, 69 (1966).
387. G. D. Coates and C. Parkin, *J. Chem. Soc.*, **1963**, 421.
388. M. A. Bennett, R. J. H. Clark, and A. D. J. Goodwin, *Inorg. Chem.*, **6**, 1625 (1967).
389. K. Nakamoto, K. Shobatake and B. Hutchinson, *Chem. Comm.*, 1451 (1969).
390. K. Shobatake and K. Nakamoto, *J. Amer. Chem. Soc.*, to be published.
391. T. Kruck, *Angew. Chem. Internat. Ed.*, **6**, 53 (1967).
392. L. A. Woodward and J. R. Hall, *Nature*, **181**, 831 (1958); *Spectrochim. Acta*, **16**, 654 (1960).
393. A. Loutellier and M. Bigorgne, *Bull. Soc. Chim. Fr.*, **1965**, 3186.
394. T. Kruck and A. Prasch, *Z. Anorg. Allg. Chem.*, **356**, 118 (1968).
395. F. A. Cotton, R. D. Barnes, and E. Bannister, *J. Chem. Soc.*, **1960**, 2199.
396. D. M. L. Goodgame and F. A. Cotton, *J. Chem. Soc.*, **1961**, 2298, 3735.
397. G. A. Rodley, D. M. L. Goodgame, and F. A. Cotton, *J. Chem. Soc.*, **1965**, 1499.
398. R. B. Penland, S. Mizushima, C. Curran, and J. V. Quagliano, *J. Amer. Chem. Soc.*, **79**, 1575 (1957).
399. A. Yamaguchi, T. Miyazawa, T. Shimanouchi, and S. Mizushima, *Spectrochim. Acta*, **10**, 170 (1957).
400. A. Yamaguchi, R. B. Penland, S. Mizushima, T. J. Lane, C. Curran, and J. V. Quagliano, *J. Amer. Chem. Soc.*, **80**, 527 (1958).
401. R. A. Bailey and T. R. Peterson, *Can. J. Chem.*, **45**, 1135 (1967).
402. K. Swaminathan and H. M. N. H. Irving, *J. Inorg. Nucl. Chem.*, **26**, 1291 (1964).
403. C. D. Flint and M. Goodgame, *J. Chem. Soc.*, **A,1966**, 744.
404. P. J. Hendra and Z. Jović, *J. Chem. Soc.*, **A,1967**, 735.
405. D. M. Adams and J. B. Cornell, *J. Chem. Soc.*, **A,1967**, 884.
406. R. Rivest, *Can. J. Chem.*, **40**, 2234 (1962).
407. T. J. Lane, A. Yamaguchi, J. V. Quagliano, J. A. Ryan, and S. Mizushima, *J. Am. Chem. Soc.*, **81**, 3824 (1959).
408. M. Schafer and C. Curran, *Inorg. Chem.*, **5**, 265 (1966).
409. R. K. Gosavi and C. N. R. Rao, *J. Inorg. Nucl. Chem.*, **29**, 1937 (1967).
410. F. A. Cotton, R. Francis, and W. D. Horrocks, *J. Phys. Chem.*, **64**, 1534 (1960).
411. R. S. Drago and D. W. Meek, *J. Phys. Chem.*, **65**, 1446 (1961); D. W. Meek, D. K. Straub, and R. S. Drago, *J. Amer. Chem. Soc.*, **82**, 6013 (1960).
412. P. W. N. M. van Leeuwen, *Rec. Trav. Chim. Pays-Bas*, **86**, 201 (1967); P. W. N. M. van Leeuwan and W. L. Groeneveld, *ibid.*, **86**, 721 (1967).
413. B. F. G. Johnson and R. A. Walton, *Spectrochim. Acta*, **22**, 1853 (1966).
414. T. Tanaka, *Inorg. Chim. Acta*, **1**, 217 (1967).
415. R. Francis and F. A. Cotton, *J. Chem. Soc.*, **1961**, 2078.
416. N. S. Gill, R. H. Nuttall, D. E. Scaife, and D. W. A. Sharp, *J. Inorg. Nucl. Chem.*, **18**, 79 (1961).
417. R. J. H. Clark and C. S. Williams, *Inorg. Chem.*, **4**, 350 (1965).
418. C. Postmus, J. R. Ferraro, A. Quattrochi, K. Shobatake, and K. Nakamoto, *Inorg. Chem.*, **8**, 1851 (1969).
419. J. R. Allan, D. H. Brown, R. H. Nuttall, and D. W. A. Sharp, *J. Inorg. Nucl. Chem.*, **27**, 1305 (1965).
420. N. S. Gill and H. J. Kingdon, *Aust. J. Chem.*, **19**, 2197 (1966).
421. S. Buffagni, L. M. Vallarino, and J. V. Quagliano, *Inorg. Chem.*, **3**, 480 (1964).
422. W. R. McWhinnie, *J. Inorg. Nucl. Chem.*, **27**, 2573 (1965).

423. F. Farha and R. T. Iwamoto, *Inorg. Chem.*, **4**, 844 (1965).
424. D. G. Brewer and P. T. T. Wong, *Can. J. Chem.*, **44**, 1407 (1966).
425. D. Cook, *Can. J. Chem.*, **43**, 741 (1965).
426. M. Goldstein, E. F. Mooney, A. Anderson, and H. A. Gebbie, *Spectrochim. Acta*, **21**, 105 (1965).
427. S. Kida, J. V. Quagliano, J. A. Walmsley, and S. Y. Tyree, *Spectrochim. Acta*, **19**, 189 (1963).
428. Y. Kakiuchi, S. Kida, and J. V. Quagliano, *Spectrochim. Acta*, **19**, 201 (1963).
429. R. Whyman and W. E. Hatfield, *Inorg. Chem.*, **6**, 1859 (1967).
430. R. J. H. Clark, "Review of Metal-Halogen Vibrational Frequencies," in *Halogen Chemistry*, Academic Press, 1967.
431. R. H. Nuttall, *Talanta Review*, **15**, 157 (1968).
432. R. J. H. Clark, *Spectrochim. Acta*, **21**, 955 (1965).
433. I. Wharf and D. F. Shriver, *Inorg. Chem.*, **8**, 914 (1969).
434. A. D. Allen and T. Theophanides, *Can. J. Chem.*, **42**, 1551 (1964).
435. F. H. Herbelin, J. D. Herbelin, J. P. Mathieu, and H. Poulet, *Spectrochim. Acta*, **22** 1515 (1966).
436. D. M. Adams, J. Chatt, J. Gerratt, and A. D. Westland, *J. Chem. Soc.*, **1964**, 734.
437. C. Postmus, K. Nakamoto, and J. R. Ferraro, *Inorg. Chem.*, **6**, 2194 (1967).
438. J. R. Durig, W. A. McAllister, and E. E. Mercer, *J. Inorg. Nucl. Chem.*, **29**, 1441 (1967).
439. D. M. Adams and P. J. Chandler, *J. Chem. Soc.*, **A,1967**, 1009.
440. I. R. Beattie, M. Webster, and G. W. Chantry, *J. Chem. Soc.*, *Suppl.*, **1964**, 6172.
441. I. R. Beattie and L. Rule, *J. Chem. Soc.*, **1964**, 3267.
442. G. E. Coates and D. Ridley, J. Chem. *Soc.*, **1964**, 166.
443. R. J. H. Clark and W. Errington, *J. Chem. Soc.*, **A,1967**, 258.
444. A. Sabatini and I. Bertini, *Inorg. Chem.*, **5**, 204 (1966).
445. C. Djordjevic, *Spectrochim. Acta*, **21**, 301 (1965).
446. D. M. Adams, P. J. Chandler, and R. G. Churchill, *J. Chem. Soc.*, **A**, **1967**, 1274.
447. D. M. Adams, M. Goldstein, and E. F. Mooney,, *Trans. Faraday Soc.* **59**, 2228 (1963).
448. P. M. Boorman and B. P. Staughan, *J. Chem. Soc.*, **A,1966**, 1514.
449. R. J. H. Clark, D. L. Kepert, R. S. Nyholm, and G. A. Rodley, *Spectrochim. Acta*, **22**, 1697 (1966).
450. D. M. Adams and P. J. Lock, *J. Chem. Soc.*, **A,1967**, 620.
451. M. J. Campbell, M. Goldstein, and R. Grezeskowiak, *Chem. Comm.*, 778 (1967).
452. F. A. Cotton, R. M. Wing, and R. A. Zimmerman, *Inorg. Chem.*, **6**, 11 (1967).
453. J. N. van Niekerk and F. R. L. Schoening, *Acta Cryst.*, **6**, 227 (1953).
454. W. H. Watson and J. Waser, *Acta Cryst.*, **11**, 689 (1958).
455. D. F. Shriver and M. P. Johnson, *Inorg. Chem.*, **6**, 1265 (1967).
456. F. A. Cotton and R. M. Wing, *Inorg. Chem.*, **4**, 1329 (1965).
457. N. A. D. Carey and H. C. Clark, *Chem. Comm.*, 292 (1967).
458. N. A. D. Carey and H. C. Clark, *Inorg. Chem.*, **7**, 94 (1968).
459. H. M. Gager, J. Lewis, and M. L. Ware, *Chem. Comm.*, 616 (1966).
460. D. M. Adams, J. B. Cornell, J. L. Dawes, and R. D. W. Kemmitt, *Inorg. Nucl. Chem. Letters*, **3**, 437 (1967).
461. H. Kriegsmann and K. Licht, *Z. Elektrochem.*, **62**, 1163 (1958).
462. C. G. Barraclough, D. C. Bradley, J. Lewis, and I. M. Thomas, *J. Chem. Soc.*, **1961**, 2601.
463. R. W. Adams, R. L. Martin, and G. Winter, *Aust. J. Chem.*, **20**, 773 (1967).
464. C. T. Lynch, K. S. Mazdiyasni, J. S. Smith, and W. J. Crawford, *Anal. Chem.*, **36**, 2332 (1964).

465. D. L. Guertin, S. E. Wiberley, W. H. Bauer, and J. Goldenson, *J. Phys. Chem.*, **60**, 1018 (1956).
466. P. W. N. M. van Leeuwen, *Rec. Trav. Chim. Pays-Bas*, **86**, 247 (1967).
467. A. Miyake, *Bull. Chem. Soc. Jap.*, **32**, 1381 (1959).
468. C. Duval, H. Gerding, and J. Lecomte, *Rec. Trav. Chim., Pays-Bas*, **69**, 391 (1950).
469. M. J. Schmelz, I. Nakagawa, S. Mizushima, and J. V. Quagliano, *J. Amer. Chem. Soc.*, **81**, 287 (1959).
470. K. Itoh and H. J. Bernstein, *Can. J. Chem.*, **34**, 170 (1956).
471. L. H. Jones and E. McLaren, *J. Chem. Phys.*, **22**, 1796 (1954).
472. J. K. Wilmshurst, *J. Chem. Phys.*, **23**, 2463 (1955).
473. K. Nakamura, *J. Chem. Soc. Japan*, **79**, 1411, 1420 (1958).
474. K. B. Harvey, B. A. Morrow, and H. F. Shurvell, *Can. J. Chem.*, **41**, 1181 (1963).
475. C. J. H. Schutte and K. Buijs, *Spectrochim. Acta*, **20**, 187 (1964).
476. J. D. Donaldson, J. F. Knifton, and S. D. Ross, *Spectrochim. Acta*, **20**, 847 (1964).
477. J. D. Donaldson, J. F. Knifton, and S. D. Ross, *Spectrochim. Acta*, **21**, 275 (1965).
478. F. Vrátný, C. N. R. Rao, and M. Dilling, *Anal. Chem.*, **33**, 1455 (1961).
479. R. Theimer and O. Theimer, *Monatsh. Chem.*, **81**, 313 (1950).
480. R. E. Kagarise, *J. Phys. Chem.*, **59**, 271 (1955).
481. B. Ellis and H. Pyszora, *Nature*, **181**, 181 (1958).
482. W. H. Zachariasen, *J. Amer. Chem. Soc.*, **62**, 1011 (1940).
483. V. Amirthalingam, and V. M. Padmanabhan, *Acta Cryst.*, **11**, 896 (1958).
484. J. H. Talbot, *Acta Cryst.*, **6**, 720 (1953).
485. W. H. Zachariasen and H. A. Pettinger, *Acta Cryst.*, **12**, 526 (1959).
486. J. N. van Niekerk and F. R. L. Schoening, *Acta Cryst.*, **6**, 501 (1953).
487. W. H. Bragg and G. T. Morgan, *Proc. Roy. Soc.*, **A104**, 437 (1923).
488. H. Koyama and Y. Saito, *Bull. Chem. Soc. Jap.*, **27**, 113 (1954).
489. K. Nakamoto, Y. Morimoto, and A. E. Martell, *J. Amer. Chem. Soc.*, **83**, 4528 (1961).
490. R. D. Gillard, D. M. Harris, and G. Wilkinson, *J. Chem. Soc.*, **1964**, 2838.
491. M. F. Lappert, *J. Chem. Soc.*, **1962**, 542.
492. S. Mizushima, I. Ichishima, I. Nakagawa, and J. V. Quagliano, *J. Phys. Chem.*, **59**. 293 (1955).
493. D. M. Sweeny, S. Mizushima, and J. V. Quagliano, *J. Amer. Chem. Soc.*, **77**, 6521 (1955).
494. D. B. Powell and N. Sheppard, *J. Chem. Soc.*, **1959**, 791 and **1961**, 1112; *Spectrochim Acta*, **16**, 241 (1960) and **17**, 68 (1961).
495. K. Krishnan and R. A. Plane, *Inorg. Chem.*, **5**, 852 (1966).
496. A. D. Allen and C. V. Senoff, *Can. J. Chem.*, **43**, 888 (1965).
497. G. W. Watt and D. S. Klett, *Inorg. Chem.*, **5**, 1278 (1966).
498. N. Tanaka, N. Sato, and J. Fujita, *Spectrochim. Acta*, **22**, 577 (1966).
499. M. E. Baldwin, *J. Chem. Soc.*, **1960**, 4369.
500. J. A. McLean, A. F. Schreiner, and A. F. Laethem, *J. Inorg. Nucl. Chem.*, **26**, 1245 (1964).
501. J. M. Rigg and E. Sherwin, *J. Inorg. Nucl. Chem.*, **27**, 653 (1965).
502. M. N. Hughes and W. R. McWhinnie, *J. Inorg. Nucl. Chem.*, **28**, 1659 (1966).
503. S. Kida, *Bull. Chem. Soc. Jap.*, **39**, 2415 (1966).
504. D. B. Powell and N. Sheppard, *J. Chem. Soc.*, **1959**, 3089.
505. G. Newman and D. B. Powell, *J. Chem. Soc.*, **1961**, 477; **1962**, 3447.
506. K. Brodersen and T. Kahlert, *Z. Anorg. Allg. Chem.*, **348**, 273 (1966).
507. K. Brodersen, *Z. Anorg. Allg. Chem.*, **298**, 142 (1959).
508. Y. Kinoshita, I. Matsubara, and Y. Saito, *Bull. Chem. Soc. Jap.*, **32**, 741 (1959).

509. T. Fujiyama, K. Tokumaru, and T. Shimanouchi, *Spectrochim. Acta*, **20**, 415 (1964).
510. I. Matsubara, *Bull. Chem. Soc. Jap.*, **34**, 1710 (1961).
511. M. Kubota, D. L. Johnston, and I. Matsubara, *Inorg. Chem.*, **5**, 386 (1966).
512. Y. Kinoshita, I. Matsubara, and Y. Saito, *Bull. Chem. Soc. Jap.*, **32**, 1216 (1959).
513. I. Matsubara, *Bull. Chem. Soc. Jap.*, **34**, 1719 (1961); *J. Chem. Phys.*, **35**, 373 (1961).
514. J. K. Brown, N. Sheppard, and D. M. Simpson, *Phil. Trans. Roy. Soc.*, **A247**, 35 (1954).
515. M. Kubota and D. L. Johnston, *J. Amer. Chem. Soc.*, **88**, 2451 (1966).
516. I. Matsubara, *Bull. Chem. Soc. Jap.*, **35**, 27 (1962).
517. G. W. Watt and D. S. Klett, *Spectrochim. Acta*, **20**, 1053 (1964).
518. D. A. Buckingham and D. Jones, *Inorg. Chem.*, **4**, 1387 (1965).
519. A. A. Schilt and R. C. Taylor, *J. Inorg. Nucl. Chem.*, **9**, 211 (1959).
520. A. A. Schilt, *J. Amer. Chem. Soc.*, **81**, 2966 (1959); *Inorg. Chem.*, **3**, 1323 (1964).
521. R. J. H. Clark and C. S. Williams, *Spectrochim. Acta*, **21**, 1861 (1965).
522. S. P. Sinha, *Spectrochim. Acta*, **20**, 879 (1964).
523. R. G. Inskeep, *J. Inorg. Nucl. Chem.*, **24**, 763 (1962).
524. J. R. Durig, B. R. Mitchell, D. W. Sink, J. N. Willis, and A. S. Wilson, *Spectrochim. Acta*, **23A**, 1121 (1967).
525. J. R. Ferraro, L. J. Basile, and D. L. Kovacic, *Inorg. Chem.*, **5**, 391 (1966).
526. C. Postmus, J. R. Ferraro, and W. Wozniak, *Inorg. Chem.*, **6**, 2030 (1967).
527. R. J. H. Clark and C. S. Williams, *Spectrochim. Acta*, **23A**, 1055 (1967).
528. B. Hutchinson, J. Takemoto and K. Nakamoto, *J. Amer. Chem. Soc.*, to be published.
529. R. C. Stoufer and D. H. Busch, *J. Amer. Chem. Soc.*, **82**, 3491 (1960).
530. W. J. Stratton and D. H. Busch, *J. Amer. Chem. Soc.*, **82**, 4834 (1960).
531. R. A. Krause, N. B. Colthup, and D. H. Busch, *J. Phys. Chem.*, **65**, 2216 (1961); P. E. Figgins and D. H. Busch, *J. Phys. Chem.*, **65**, 2236 (1961).
532. H. Shindo, J. L. Walter, and R. J. Hooper, *J. Inorg. Nucl. Chem.*, **27**, 871 (1965).
533. W. R. McWhinnie, *J. Inorg. Nucl. Chem.*, **27**, 1619 (1965).
534. S. F. Mason, *J. Chem. Soc.*, **1958**, 976.
535. D. W. Thomas and A. E. Martell, *J. Amer. Chem. Soc.*, **81**, 5111 (1959); **78**, 1338 (1956).
536. L. J. Boucher and J. J. Katz, *J. Amer. Chem. Soc.*, **89**, 1340 (1967).
537. H. R. Wetherell, M. J. Hendrickson, and A. R. McIntyre, *J. Amer. Chem. Soc.*, **81**. 4517 (1959).
538. I. M. Keen and B. W. Malerbi, *J. Inorg. Nucl. Chem.*, **27**, 1311 (1965).
539. B. I. Knudsen, *Acta, Chem. Scand.*, **20**, 1344 (1966).
540. H. F. Shurvell and L. Pinzuti, *Can. J. Chem.*, **44**, 125 (1966).
541. Y. Murakami and K. Sakata, *Inorg. Chim. Acta*, **2**, 273 (1968).
542. R. E. Rundle and M. Parasol, *J. Chem. Phys.*, **20**, 1487 (1952).
543. K. Nakamoto, M. Margoshes, and R. E. Rundle, *J. Amer. Chem. Soc.*, **77**, 6480 (1955).
544. L. E. Godycki and R. E. Rundle, *Acta Cryst.*, **6**, 487 (1953).
545. J. Fujita, A. Nakahara, and R. Tsuchida, *J. Chem. Phys*· **23** 1541 (1955).
546. A. Nakahara, *Bull. Chem. Soc. Jap.*, **28**, 473 (1955); A. Nakahara, J. Fujita and R. Tsuchida, *ibid.*, **29**, 296 (1956).
547. R. D. Gillard and G. Wilkinson, *J. Chem. Soc.*, **1963**, 6041.
548. E. Frasson, C. Panattoni, and R. Zannetti, *Acta Cryst.*, **12**, 1027 (1959).
549. R. Blinc and D. Hadži, *J. Chem. Soc.*, **1958**, 4536; *Spectrochim. Acta*, **16**, 853 (1960).
550. M. Tsuboi, K. Onishi, I. Nakagawa, T. Shimanouchi, and S. Mizushima, *Spectrochim. Acta*, **12**, 253 (1958).
551. K. Fukushima, T. Onishi, T. Shimanouchi, and S. Mizushima, *Spectrochim. Acta*, **14**, 236 (1959).

552. M. Tsuboi and T. Takenishi, *Bull. Chem. Soc. Jap.*, **32**, 1044 (1959).
553. E. Steger, A. Turcu, and V. Macovei, *Spectrochim. Acta*, **19**, 293 (1963).
554. A. J. Stosick, *J. Amer. Chem. Soc.*, **67**, 365 (1945).
555. B. M. Low, F. L. Hirshfeld, and F. M. Richards, *J. Amer. Chem. Soc.*, **81**, 4412 (1959).
556. A. Rosenberg *Acta Chem. Scand.*, **10**, 840 (1956); **11**, 1390 (1957).
557. D. N. Sen, S. Mizushima, C. Curran, and J. V. Quagliano, *J. Amer. Chem. Soc.*, **77**, 211 (1955); S. Mizushima and J. V. Quagliano, *ibid.*, **75**, 4870 (1953); D. M. Sweeny, C. Curran and J. V. Quagliano, *ibid.*, **77**, 5508 (1955); A. J. Saraceno, I. Nakagawa, S. Mizushima, C. Curran, and J. V. Quagliano, *ibid.*, **80**, 5018 (1958); D. Segnini, C. Curran, and J. V. Quagliano, *Spectrochim. Acta*, **16**, 540 (1960); V. Moreno K. Dittmer and J. V. Quagliano, *ibid.*, **16**, 1368 (1960).
558. K. Tomita, *Bull. Chem. Soc. Jap.*, **34**, 280, 286 (1961).
559. T. J. Lane, J. A. Durkin, and R. J. Hooper, *Spectrochim. Acta*, **20**, 1013 (1964).
560. R. A. Condrate and K. Nakamoto, *J. Chem. Phys.*, **42**, 2590 (1965).
561. F. Basolo, C. J. Ballhausen, and J. Bjerrum, *Acta Chem. Scand.*, **9**, 810 (1955).
562. J. A. Kieft and K. Nakamoto. To be published.
563. G. W. Watt and J. F. Knifton, *Inorg. Chem.*, **6**, 1010 (1967).
564. I. Nakagawa, R. J. Hooper, J. L. Walter, and T. J. Lane, *Spectrochim. Acta*, **21**, 1 (1965).
565. R. J. Hooper, T. J. Lane, and J. L. Walter, *Inorg. Chem.*, **3**, 1568 (1964).
566. J. F. Jackovitz and J. L. Walter, *Spectrochim. Acta*, **22**, 1393 (1966).
567. J. F. Jackovitz, J. A. Durkin, and J. L. Walter, *Spectrochim. Acta*, **23A**, 67 (1967).
568. K. Krishnan and R. A. Plane, *Inorg. Chem.*, **6**, 55 (1967).
569. M. P. Springer and C. Curran, *Inorg. Chem.*, **2**, 1270 (1963).
570. J. A. Kieft and K. Nakamoto, *J. Inorg. Nucl. Chem.*, **29**, 2561 (1967).
571. K. Nakamoto, P. J. McCarthy, and B. Miniatus, *Spectrochim. Acta*, **21**, 379 (1965).
572. J. D. S. Goulden, *Spectrochim. Acta*, **16**, 715 (1960).
573. M. Girard and J. Lecomte, *Compt. rend.*, **241**, 292 (1955); **240**, 415 (1955).
574. S. Kirschner and R. Kiesling, *J. Amer. Chem. Soc.*, **82**, 4174 (1960).
575. D. H. Busch and J. C. Bailar, *J. Amer. Chem., Soc.*, **75**, 4574 (1953); **78**, 716 (1956); M. L. Morris and D. H. Busch, *ibid.*, **78**, 5178 (1956); K. Swaminathan and D. H. Busch, *J. Inorg. Nucl. Chem.*, **20**, 159 (1961); R. E. Sievers and J. C. Bailar, *Inorg. Chem.*, **1**, 174 (1962).
576. D. Chapman, *J. Chem. Soc.*, **1955**, 1766
577. T. Moeller, F. A. Moss, and R. H. Marshall, *J. Amer. Chem. Soc.*, **77**, 3182 (1955).
578. R. D. Gillard and G. Wilkinson, *J. Chem. Soc.*, **1963**, 4271
579. D. Sawyer and P. Paulsen, *J. Amer. Chem. Soc.*, **80**, 1597 (1958); **81**, 816 (1959); **82**, 4191 (1960).
580. Y. Tomita and K. Ueno, *Bull. Chem. Soc. Jap.*, **36**, 1069 (1963).
581. S. Fronaeus and R. Larsson, *Acta Chem. Scand.*, **16**, 1433, 1447 (1962).
582. S. Fronaeus and R. Larsson, *Acta Chem. Scand.*, **14**, 1364 (1960).
583. R. Larsson, *Acta Chem. Scand.*, **19**, 783 (1965).
584. K. Nakamoto, Y. Morimoto, and A. E. Martell, *J. Amer. Chem. Soc.*, **84**, 2081 (1962); **85**, 309 (1963).
585. Y. Tomita, T. Ando, and K. Ueno, *J. Phys. Chem.*, **69**, 404 (1965).
586. Y. Tomita and K. Ueno, *Bull. Chem. Soc. Jap.*, **36**, 1069 (1963).
587. M. K. Kim and A. E. Martell, *J. Amer. Chem. Soc.*, **85**, 3080 (1963).
588. M. K. Kim and A. E. Martell, *Biochem.*, **3**, 1169 (1964).
589. M. K. Kim and A. E. Martell, *J. Amer. Chem. Soc.*, **88**, 914 (1966).
590. H. Murata and K. Kawai, *J. Chem. Phys.*, **25**, 589, 796 (1956).

591. J. Fujita, K. Nakamoto, and M. Kobayashi, *J. Phys. Chem.*, **61**, 1014 (1957).
592. M. J. Schmelz, T. Miyazawa, S. Mizushima, T. J. Lane, and J. V. Quagliano, *Spectrochim. Acta*, **9**, 51 (1957).
593. K. Kawai and H. Murata, *J. Chem. Soc. Japan*, **81**, 997 (1960).
594. J. Fujita, A. E. Martell, and K. Nakamoto, *J. Chem. Phys.*, **36**, 324, 331 (1962).
595. G. A. Jeffrey and G. S. Parry, *J. Chem. Soc.*, **1952**, 4864.
596. J. van Niekerk and F. R. L. Schoening, *Acta Cryst.*, **4**, 35, 381 (1951).
597. R. E. Hester and R. A. Plane, *Inorg. Chem.*, **3**, 513 (1964); E. C. Gruen and R. A. Plane, *ibid.*, **6**, 1123 (1967).
598. Y. Kuroda, M. Kato, and K. Sone, *Bull. Chem. Soc. Jap.*, **34**, 877 (1961).
599. P. X. Armendarez and K. Nakamoto, *Inorg. Chem.*, **5**, 796 (1966).
600. H. C. Freeman, J. E. W. L. Smith, and J. C. Taylor, *Nature*, **184**, 707 (1959); *Acta Cryst.*, **14**, 407 (1961).
601. M. Nardelli, G. Fava, and G. Giraldi, *Acta Cryst.*, **16**, 343 (1963).
602. K. Aida, Y. Musya, and S. Kinumaki, *Inorg. Chem.*, **2**, 1268 (1963).
603. A. W. McLellan and G. A. Melson, *J. Chem. Soc.*, **A,1967**, 137.
604. B. B. Kedzia, P. X. Armendarez, and K. Nakamoto, *J. Inorg. Nucl. Chem.*, **30**, 849 (1968).
605. R. B. Roof, *Acta Cryst.*, **9**, 781 (1956).
606. R. Mecke and E. Funck, *Z. Elektrochem.* ,**60**, 1124 (1956).
607. K. Nakamoto and A. E. Martell, *J. Chem. Phys.*, **32**, 588 (1960).
608. K. E. Lawson, *Spectrochim. Acta*, **17**, 248 (1961).
609. C. Djordjevic, *Spectrochim. Acta*, **17**, 448 (1961).
610. J. P. Dismukes, L. H. Jones, and J. C. Bailar, *J. Phys. Chem.*, **65**, 792 (1961).
610a. R. D. Gillard, H. G. Silver and J. L. Wood, *Spectrochim. Acta*, **20**, 63 (1964).
611. S. Pinchas, B. L. Silver, and I. Laulicht, *J. Chem. Phys.*, **46**, 1506 (1967).
612. H. Musso and H. Junge, *Tetrahedron Lett.*, **33**, 4003, 4009 (1966).
613. M. Mikami, I. Nakagawa, and T. Shimanouchi, *Spectrochim. Acta*, **23A**, 1037 (1967).
614. K. Nakamoto, P. J. McCarthy, A. Ruby, and A. E. Martell, *J. Amer. Chem. Soc.*, **83**, 1066, 1272 (1961).
614a. K. Nakamoto, C. Udovich and J. Takemoto, *J. Amer. Chem. Soc.*, to be published.
615. G. T. Behnke and K. Nakamoto, *Inorg. Chem.*, **6**, 433 (1967).
616. R. E. Hester and R. A. Plane, *Inorg. Chem.*, **3**, 513 (1964).
617. P. X. Armendarez and L. S. Forster, *J. Chem. Phys.*, **40**, 273 (1964).
618. S. Misumi and N. Iwasaki, *Bull. Chem. Soc. Jap.*, **40**, 550 (1967).
619. Y. Kawasaki, T. Tanaka, and R. Okawara, *Spectrochim. Acta*, **22**, 1571 (1966).
620. B. N. Figgis, J. Lewis, R. F. Long, R. Mason, R. S. Nyholm, P. J. Pauling, and G. B. Robertson, *Nature*, **195**, 1278 (1962).
621. J. Lewis, R. F. Long, and C. Oldham, *J. Chem. Soc.*, **1965**, 6740; D. Gibson, J. Lewis, and C. Oldham, *ibid.*, **A,1966**, 1453; J. Lewis and C. Oldham, *ibid.*, **A,1966**, 1456; D. Gibson, J. Lewis, and C. Oldham, *ibid.*, **A,1967**, 72.
622. G. T. Behnke and K. Nakamoto, *Inorg. Chem.*, **6**, 440 (1967).
623. G. T. Behnke and K. Nakamoto, *Inorg. Chem.*, **7**, 330 (1968).
624. G. Allen, J. Lewis, R. F. Long, and C. Oldham, *Nature*, **202**, 589 (1964).
625. G. T. Behnke and K. Nakamoto, *Inorg. Chem.*, **7**, 2030 (1968).
626. P. W. N. M. van Leeuwen, *Rec. Trav. Chim. Pays-Bas*, **87**, 396 (1968).
627. Y. Nakamura and S. Kawaguchi, *Chem. Comm.*, 716 (1968).
628. G. S. Hammond, D. C. Nonhebel, and C. S. Wu, *Inorg. Chem.*, **2**, 73 (1963).
629. F. A. Cotton and R. C. Elder, *J. Amer. Chem. Soc.*, **86**, 2294 (1964); *Inorg. Chem.*, **4**. 1145 (1965).

630. K. Nakamoto, Y. Morimoto, and A. E. Martell, *J. Amer. Chem. Soc.*, **83**, 4533 (1961).
631. L. Sacconi, G. Caroti, and P. Paoletti, *J. Chem. Soc.*, **1958**, 4257 *J. Inorg. Nucl. Chem.*, **8**, 93 (1958).
632. A. E. Comyns, B. M. Gatehouse, and E. Wait, *J. Chem. Soc.*, **1958**, 4655.
633. G. Satori and G. Costa, *Z. Elektrochem.*, **63**, 105 (1959).
634. R. L. Belford, A. E. Martell, and M. Calvin, *J. Inorg. Nucl. Chem.*, **2**, 11 (1956).
635. R. P. Dryden and A. Winston, *J. Phys. Chem.*, **62**, 635 (1958).
636. K. Nakamoto, Y. Morimoto, and A. E. Martell, *J. Phys. Chem.*, **66**, 346 (1962).
637. H. A. Kuska and M. T. Rogers, *J. Chem. Phys.*, **43**, 1744 (1965).
638. M. L. Morris, R. W. Moshier, and R. E. Sievers, *Inorg. Chem.*, **2**, 411 (1963).
639. J. J. Charette and P. Teyssié, *Spectrochim. Acta*, **16**, 689 (1960).
640. J. R. Allkins and P. J. Hendra, *J. Chem. Soc.*, **A,1967**, 1325; *Spectrochim Acta*, **22**, 2075 (1966); **23A**, 1671 (1967).
641. J. Chatt, L. A. Duncanson, and L. M. Venanzi, *Suomen Kemistilehti*, **B29**, 75 (1956); *Nature*, **177**, 1042 (1956).
642. K. Nakamoto, J. Fujita, R. A. Condrate, and Y. Morimoto, *J. Chem. Phys.*, **39**, 423 (1963).
643. F. A. Cotton and J. A. McCleverty, *Inorg. Chem.*, **3**, 1398 (1964).
644. F. W. Moore and M. L. Larson, *Inorg. Chem.*, **6**, 998 (1967).
645. M. L. Shankaranarayana and C. C. Patel, *Can. J. Chem.*, **39**, 1633 (1961).
646. L. H. Little, G. W. Poling, and J. Leja *Can. J. Chem.*, **39**, 745 (1961).
647. G. W. Watt and B. J. McCormick, *Spectrochim. Acta*, **21**, 753 (1965).
648. J. Fujita and K. Nakamoto, *Bull. Chem. Soc. Jap.*, **37**, 528 (1964).
649. S. H. H. Chaston and S. E. Livingston, *Aust. J. Chem.*, **20**, 1065 (1967); S. H. H. Chaston, S. E. Livingston, T. N. Lockyer, V. A. Pickles, and J. S. Shannon, *ibid.*, **18**, 673 (1965).
650. H. Shindo and T. L. Brown, *J. Amer. Chem. Soc.*, **87**, 1904 (1965).
651. C. A. McAuliffe, J. V. Quagliano, and L. M. Vallarino, *Inorg. Chem.*, **5**, 1996 (1966).
652. C. A. McAuliffe, *J. Chem. Soc.*, **A,1967**, 641 (1967).
653. A. Dyfverman, *Acta Chem. Scand.*, **17**, 1609 (1963).
654. H. Roszinski, R. Dautel, and W. Zeil ,*Z. Phys. Chem.*, **36**, 26 (1963).
655. I. R. Beattie and T. Gilson, *J. Chem. Soc.*, **1964**, 3528.
656. C. W. Heitsch and R. N. Kniseley, *Spectrochim. Acta*, **19**, 1385 (1963).
657. H. Gerding and H. Houtgraaf, *Rec. Trav. Chim.*, *Pays-Bas*, **74**, 15 (1955).
658. R. C. Taylor and C. L. Cluff, *Nature*, **182**, 390 (1958).
659. A. Derek, H. Clague, and A. Danti, *Spectrochim. Acta*, **23A**, 2359 (1967).
660. K. F. Purcell and R. S. Drago, *J. Amer. Chem. Soc.*, **88**, 919 (1966).
661. I. R. Beattie and T. Gilson, *J. Chem. Soc.*, **1964**, 2292.
662. R. C. Taylor and T. C. Bissot, *J. Chem. Phys.*, **25**, 780 (1956).
663. B. P. Susz and P. Chalandon, *Helv. Chim. Acta*, **41**, 1332, 697 (1958); B. P. Susz and A. Lachavanne, *ibid.*, **41**, 634 (1958); D. Cassimatis, P. Gagnaux, and B. P. Susz, *ibid.*, **43**, 424 (1960).
664. M. F. Lappert, *J. Chem. Soc.*, **1961**, 817.
665. A. D. E. Pullin and J. M. Pollock, *Trans. Faraday Soc.*, **54**, 11 (1959).
666. G. L. Vidale and R. C. Taylor, *J. Amer. Chem. Soc.*, **78**, 294 (1956).
667. G. S. Rao, *Z. Anorg. Allg. Chem.*, **304**, 176 (1960).
668. E. K. Plyler and R. S. Mulliken, *J. Amer. Chem. Soc.*, **81**, 823 (1959).
669. R. A. Zingaro and W. B. Witmer, *J. Phys. Chem.*, **64**, 1705 (1960).
670. F. Watari, *Spectrochim. Acta*, **23A**, 1917 (1967).
671. H. Luther, D. Mootz, and F. Radwitz, *J. Prakt. Chem.*, **5**, 242 (1958).

672. N. N. Greenwood and K. Wade, *J. Chem. Soc.*, **1966**, 1130.
673. K. Dehnicke, *Z. Anorg. Allg. Chem.*, **309**, 266 (1961).
674. G. S. Rao, *Z. Anorg. Allg. Chem.*, **304**, 351 (1960).
675. H. J. Coerver and C. Curran, *J. Amer. Chem. Soc.*, **80**, 3522 (1958).
676. W. Gerrard, M. F. Lappert, H. Pyszora, and J. W. Wallis, *J. Chem. Soc.*, **1960**, 2182.
677. T. L. Brown and M. Kubota, *J. Amer. Chem. Soc.*, **83**, 4175 (1961).
678. E. Allenstein and A. Schmidt, *Spectrochim. Acta*, **20**, 1451 (1964).
679. J. C. Sheldon and S. Y. Tyree, *J. Amer. Chem. Soc.*, **80**, 4775 (1958); **81**, 2290 (1959).
680. H. Gerding, J. A. Koningstein, and E. R. van der Worm, *Spectrochim. Acta*, **16**, 881 (1960).
681. I. R. Beattie and M. Webster, *J. Chem. Soc.*, **1964**, 3507.
682. M. F. Farona and J. G. Grasselli, *Inorg. Chem.*, **6**, 1675 (1967).
683. I. R. Beattie, G. P. McQuillan, L. Rule, and M. Webster, *J. Chem. Soc.*, **1963**, 1514.
684. S. S. Raskin, *Opt. Spektrosk.*, **1**, 516 (1956).
685. E. F. Gross and I. M. Ginzburg, *Opt. Spektrosk.*, **1**, 710 (1956).
686. L. W. Daasch, *Spectrochim. Acta*, **15**, 726 (1959); *J. Chem. Phys.*, **28**, 1005 (1958).
687. J. A. A. Ketelaar, C. Haas, and J. van der Elsken, *J. Chem. Phys.*, **24**, 624 (1956); J. A. A. Ketelaar and J. van der Elsken, *ibid.*, **30**, 336 (1959).
688. L. H. Jones and M. M. Chamberlain, *J. Chem. Phys.*, **25**, 365 (1956).
689. A. Strasheim and K. Buijs, *J. Chem. Phys.*, **34**, 691 (1961).
690. V. W. Meloche and G. E. Kalbus, *J. Inorg. Nucl. Chem.*, **6**, 104 (1958).
691. R. A. Buchanan and W. A. Bowen, *J. Chem. Phys.*, **34**, 348 (1961).
692. F. Vrátný, *J. Inorg. Nucl. Chem.*, **10**, 328 (1959).
693. V. C. Farmer, *Spectrochim. Acta*, **8**, 374 (1957).
694. R. A. Durie and J. Szewczyk, *Spectrochim. Acta*, **15**, 593 (1959).
695. R. P. Eischens, W. A. Pliskin, and S. A. Francis, *J. Chem. Phys.*, **22**, 1786 (1954); *J. Phys. Chem.*, **60**, 194 (1956).
695a. L. H. Little, "Infrared Spectra of Adsorbed Species," Academic Press, 1966.
696. C. W. Garland, *J. Phys. Chem.*, **63**, 1422 (1959); **65**, 617 (1961).
697. C. E. O'Neill and D. J. C. Yates, *Spectrochim. Acta*, **17**, 953 (1961); *J. Phys. Chem.*, **65**, 901 (1961); D. J. C. Yates, *ibid.*, **65**, 746 (1961).
698. H. L. Pickering and H. C. Eckstrom, *J. Phys. Chem.*, **63**, 512 (1959).
699. J. H. Taylor and C. H. Amberg, *Can. J. Chem.*, **39**, 535 (1961).
700. R. M. Hammaker, S. A. Francis, and R. P. Eischens, *Spectrochim. Acta*, **21**, 1295 (1965).
701. H. C. Eckstrom, *J. Phys. Chem.*, **70**, 594 (1966).
702. J. F. Harrod, R. W. Roberts, and E. F. Rissmann, *J. Phys. Chem.*, **71**, 343 (1967).
703. G. Blyholder, *J. Chem. Phys.*, **44**, 3134 (1966).
704. G. Blyholder and M. C. Allen, *J. Phys. Chem.*, **69**, 3998 (1965).
705. J. W. Ward and H. W. Habgood, *J. Phys. Chem.*, **70**, 1178 (1966).
706. J. E. Mapes and R. P. Eischens, *J. Phys. Chem.*, **58**, 1059 (1954); W. A. Pliskin and R. P. Eischens, *ibid.*, **59**, 1156 (1955).
707. M. Folman and D. J. C. Yates, *Proc. Roy. Soc.*, **A246**, 32 (1958); *J. Phys. Chem.*, **63**, 183 (1959).
708. N. W. Cant and L. H. Little, *Can. J. Chem.*, **42**, 802 (1964); **43**, 1252 (1965).
709. D. J. C. Yates, N. Sheppard, and C. L. Angell, *J. Chem. Phys.*, **23**, 1980 (1955).
710. M. Folman, *Trans. Faraday Soc.*, **57**, 2000 (1961).
711. N. Sheppard and D. J. C. Yates, *Proc. Roy. Soc.*, **A238**, 69 (1957).
712. W. A. Pliskin and R. P. Eischens, *Spectrochim. Acta*, **8**, 302 (1956), *J. Chem. Phys.*, **24**, 482 (1956).

713. L. H. Little, *J. Phys. Chem.*, **63**, 1616 (1959); *J. Chem. Phys.*, **35**, 342 (1961).
714. G. Kragounis and O. Peter, *Z. Elektrochim.*, **63**, 1120 (1959).
715. M. Folman and D. J. C. Yates, *J. Phys. Chem.*, **63**, 183 (1959).
716. J. K. A. Clarke and A. D. E. Pullin, *Trans. Faraday Soc.*, **56**, 534 (1960).
717. L. H. Little, H. E. Klauser, and C. H. Amberg, *Can. J. Chem.*, **39**, 421 (1961).
718. R. W. Hoffmann and G. W. Brindley, *J. Phys. Chem.*, **65**, 443 (1961).
719. W. A. Pliskin and R. P. Eischens, *Z. Phys. Chem.*, **24**, 11 (1960).
720. S. Matsushita and T. Nakata, *J. Chem. Phys.*, **36**, 665 (1962).
721. T. Ogawa, K. Kishi, and K. Hirota, *Bull. Chem. Soc. Jap.*, **37**, 1306 (1964).
722. D. K. Huggins and H. D. Kaesz, "Use of Infrared and Raman Spectroscopy in the Study of Organometallic Compounds," in "*Progress in Solid State Chemistry*, Vol. 1, H. Reiss., ed., Macmillan, 1964.
723. K. Nakamoto, "Characterization of Organometallic Compounds by Infrared Spectroscopy," in *Characterization of Organometallic Compounds*, Part I, M. Tsutsui, ed., Interscience, 1969.
724. J. Chatt and L. A. Duncanson, *J. Chem. Soc.*, **1953**, 2939.
725. J. A. Wunderlich and D. P. Mellor, *Acta Cryst.*, **7**, 130 (1954); **8**, 57 (1955); J. N. Dempsey and N. C. Baenziger, *J. Amer. Chem. Soc.*, **77**, 4984 (1955); G. B. Bokii and G. A. Kukina, *Kristallografiya*, **2**, 395 (1957).
726. D. B. Powell and N. Sheppard, *Spectrochim. Acta*, **13**, 69 (1958).
727. A. A. Babushkin, L. A. Gribov, and A. D. Gelman, *Dokl. Akad. Nauk. SSSR*, **123**, 461 (1958).
728. D. B. Powell and N. Sheppard, *J. Chem. Soc.*, **1960**, 2519.
729. J. Chatt, L. A. Duncanson, and R. G. Guy, *Nature*, **184**, 526 (1959).
730. M. J. Grogan and K. Nakamoto, *J. Amer. Chem. Soc.*, **88**, 5454 (1966).
731. J. Paradilla-Sorzano and J. P. Fackler, *J. Mol. Spectrosc.*, **22**, 80 (1967).
732. M. J. Grogan and K. Nakamoto, *J. Amer. Chem. Soc.*, **90**, 918 (1968).
733. H. B. Jonassen and J. E. Field, *J. Amer. Chem. Soc.*, **79**, 1275 (1957).
734. S. I. Shupack and M. Orchin, *Inorg. Chem.*, **3**, 374 (1964).
735. H. P. Fritz, *Chem. Ber.*, **94**, 1217 (1961).
736. R. A. Alexander, N. C. Baenziger, C. Carpenter, and J. R. Doyle, *J. Amer. Chem. Soc.*, **82**, 535 (1960).
737. J. R. Doyle and H. B. Jonassen, *J. Amer. Chem. Soc.*, **78**, 3965 (1956).
738. J. Chatt, L. M. Vallarino, and L. M. Venanzi, *J. Chem. Soc.*, **1957** 2496.
739. P. J. Hendra and D. B. Powell, *Spectrochim. Acta*, **17**, 909 (1961).
740. P. J. Hendra and D. B. Powell, *Spectrochim. Acta*, **17**, 913 (1961).
741. P. E. Slade and H. B. Jonassen. *J. Amer. Chem. Soc.*, **79**, 1277 (1957).
742. P. J. Hendra and D. B. Powell, *Spectrochim. Acta*, **18**, 1195 (1962).
743. M. J. Grogan and K. Nakamoto, *Inorg. Chim. Acta*, **1**, 228 (1967).
744. O. S. Mills and G. Robinson, *Proc. Chem. Soc.*, 421 (1960).
745. J. R. Doyle, J. H. Hutchinson, N. C. Baenziger, and L. W. Tresselt, *J. Amer. Chem. Soc.*, **83**, 2768 (1961).
746. B. Dickens and W. N. Lipscomb, *J. Amer. Chem. Soc.*, **83**, 4062 (1961); *J. Chem Phys.*, **37**, 2084 (1962).
747. R. T. Bailey, E. R. Lippincott, and D. Steele, *J. Amer. Chem. Soc.*, **87**, 5346 (1965).
748. J. Muller and E. O. Fischer, *J. Organometal. Chem.*, **5**, 275 (1966).
749. J. Chatt, G. A. Rowe, A. A. Williams, *Proc. Chem. Soc.*, 208 (1957); J. Chatt, R. G. Guy, and L. A. Duncanson, *J. Chem. Soc.*, **1961**, 827.
750. R. Nast and H. Schindel, *Z. Anorg. Allg. Chem.*, **326**, 201 (1963).
751. J. Chatt, R. G. Guy, L. A. Duncanson, and D. T. Thompson, *J. Chem. Soc.*, **1963**, 5170.

752. A. D. Allen and T. Theophanides, *Can. J. Chem.*, **44**, 2703 (1966).
753. T. J. Kealy and P. L. Pauson, *Nature*, **168**, 1039 (1951).
754. H. P. Fritz, "Infrared and Raman Spectral Studies of π-Complexes Formed between Metals and C_nH_n Rings," in *Advances in Organometallic Chemistry*, Vol. 1, F. G. A. Stone and R. West, eds., Academic Press, 1964.
755. G. Wilkinson, P. L. Pauson, and F. A. Cotton, *J. Am. Chem. Soc.*, **76**, 1970 (1954).
756. E. R. Lippincott and R. D. Nelson, *Spectrochim. Acta*, **10**, 307 (1958); *J. Chem. Phys.*, **21**, 1307 (1953); *J. Amer. Chem. Soc.*, **77**, 4990 (1955).
757. H. P. Fritz and R. Schneider, *Chem. Ber.*, **93**, 1171 (1960).
758. G. Wilkinson and T. S. Piper, *J. Inorg. Nucl. Chem.*, **2**, 32 (1956); **3**, 104 (1956).
759. E. Maslowsky and K. Nakamoto, *Inorg. Chem.*, **8**, 1108 (1969).
760. W. K. Winter, B. Curnutte, and S. E. Whitcomb, *Spectrochim. Acta*, **15**, 1085 (1959).
761. H. P. Fritz, *Chem. Ber.*, **92**, 780 (1959).
762. E. R. Lippincott, J. Xavier, and D. Steele, *J. Amer. Chem. Soc.*, **83**, 2262 (1961).
763. M. L. H. Green, L. Pratt, and G. Wilkinson, *J. Chem. Soc.*, **1959**, 989, 3753.
764. F. A. Cotton and L. T. Reynolds, *J. Amer. Chem. Soc.*, **80**, 269 (1958).
765. L. T. Reynolds and G. Wilkinson, *J. Inorg. Nucl. Chem.*, **2**, 246 (1956).
766. T. S. Piper, F. A. Cotton, and G. Wilkinson, *J. Inorg. Nucl. Chem.*, **1**, 165 (1955); F. A. Cotton, A. O. Liehr, and G. Wilkinson, *ibid.*, **1**, 175 (1955).
767. B. F. Hallam and P. L. Pauson, *Chem. Ind.* (*London*), 653 (1955).
768. T. S. Piper and G. Wilkinson, *Chem. Ind.* (*London*), 1296 (1955).
769. T. S. Piper and G. Wilkinson, *J. Inorg. Nucl. Chem.*, **3**, 104 (1956).
770. M. J. Bennett, F. A. Cotton, A. Davison, J. W. Faller, S. J. Lippard, and S. M. Morehouse, *J. Amer. Chem. Soc.*, **88**, 4371 (1966).
771. A. Davison, M. L. H. Green, and G. Wilkinson, *J. Chem. Soc.*, **1961**, 3172.
772. O. S. Mills, *Acta Cryst.*, **11**, 620 (1958).
773. F. A. Cotton, H. Stammreich, and G. Wilkinson, *J. Inorg. Nucl. Chem.*, **9**, 3 (1959); F. A. Cotton and G. Yagupsky, *Inorg. Chem.*, **6**, 15 (1967).
774. R. D. Fischer, A. Vogler, and K. Noack, *J. Organometal. Chem.*, **7**, 135 (1967); K. Noack, *J. Inorg. Nucl. Chem.*, **25**, 1383 (1963).
775. F. C. Wilson and D. P. Shoemaker, *J. Chem. Phys.*, **27**, 809 (1957).
776. G. Wilkinson, *J. Amer. Chem. Soc.*, **76**, 209 (1954).
777. H. P. Fritz and J. Manchot, *J. Organometal. Chem.*, **2**, 8 (1964).
778. R. B. King, *Inorg. Chem.*, **5**, 2227 (1966).
779. T. S. Piper and G. Wilkinson, *J. Inorg. Nucl. Chem.*, **2**, 38 (1956).
780. R. L. Cooper, M. L. H. Green, and J. T. Moelwyn-Hughes, *J. Organometal. Chem.*, **3**, 261 (1965).
781. M. P. Johnson and D. F. Shriver, *J. Amer. Chem. Soc.*, **88**, 301 (1966).
782. H. P. Fritz and E. O. Fischer, *J. Chem. Soc.*, **1961**, 547.
783. R. J. Haines, R. S. Nyholm, and M. H. B. Stiddard, *J. Chem. Soc.*, **A,1966**, 1606.
784. S. D. Robinson and B. L. Shaw, *J. Chem. Soc.*, **1965**, 1529.
785. D. W. McBride, E. Dudek, and F. G. A. Stone, *J. Chem. Soc.*, **1964**, 1752.
786. E. W. Abel, A. Singh, and G. Wilkinson, *J. Chem. Soc.*, **1960**, 1321.
787. R. T. Bailey and E. R. Lippincott, *Spectrochim. Acta*, **21**, 389 (1965).
788. H. E. Rubalcava and J. B. Thomson, *Spectrochim. Acta*, **18**, 449 (1962).
789. E. O. Fischer and U. Zahn, *Chem. Ber.*, **92**, 1624 (1959).
790. H. P. Fritz, J. F. W. McOmie, and N. Sheppard, *Tetrahedron Lett.*, **26**, 35 (1960).
791. R. P. Dodge and V. Schomaker, *Nature*, **186**, 798 (1960).
792. H. P. Fritz, *Z. Naturforsch.*, **B16**, 415 (1961).
793. E. O. Fischer and W. Hafner, *Z. Naturforsch.*, **B10**, 665 (1955).
794. E. Weiss and E. O. Fischer, *Z. Anorg. Allg. Chem.*, **286**, 142 (1956).

795. H. P. Fritz, W. Lüttke, H. Stammreich, and R. Forneris, *Chem. Ber.*, **92**, 3246 (1959); *Spectrochim. Acta*, **17**, 1068 (1961); H. P. Fritz and E. O. Fischer, *J. Organometal. Chem.*, **7**, 121 (1967).
796. R. G. Snyder, *Spectrochim. Acta*, **15**, 807 (1959).
797. E. O. Fischer and D. Seus, *Chem. Ber.*, **89**, 1809 (1956).
798. H. P. Fritz and J. Manchot, *Spectrochim. Acta*, **18**, 171 (1961).
799. R. E. Humphrey, *Spectrochim. Acta*, **17**, 93 (1961).
800. G. Winkhaus, L. Pratt, and G. Wilkinson, *J. Chem. Soc.*, **1961**, 3807.
801. R. D. Fischer, *Chem. Ber.*, **93** 165 (1960).

Appendices

APPENDIX I

POINT GROUPS AND THEIR CHARACTER TABLES

The following are the character tables of the point groups that appear frequently in this book. The species (or the irreducible representations) of the point group are labeled according to the following rules: A and B denote nondegenerate species (one-dimensional representation). A represents the symmetric species (character $= +1$) with respect to rotation about the principal axis (chosen as z axis), whereas B represents the antisymmetric species (character $= -1$) with respect to rotation about the principal axis; E and F denote doubly degenerate (two-dimensional representation) and triply degenerate species (three-dimensional representation), respectively. If two species in the same point group differ in the character of C (other than the principal axis), they are distinguished by subscripts 1, 2, 3, $\cdots$. If two species differ in the character of σ (other than σ_v), they are distinguished by $'$ and $''$. If two species differ in the character of i, they are distinguished by subscripts g and u. If these rules allow several different labels, g and u take precedence over 1, 2, 3, $\cdots$, which in turn take precedence over $'$ and $''$. The labels of species of point groups $\mathbf{C}_{\infty v}$ and $\mathbf{D}_{\infty h}$ (linear molecules) are exceptional and are taken from the notation for the component of the electronic orbital angular momentum along the molecular axis.

$\mathbf{C}_s$	I	$\sigma(xy)$		
A'	$+1$	$+1$	T_x, T_y, R_z	$\alpha_{xx}, \alpha_{yy}, \alpha_{zz}, \alpha_{xy}$
A''	$+1$	-1	T_z, R_x, R_y	α_{yz}, α_{xz}

$\mathbf{C}_2$	I	$C_2(z)$		
A	$+1$	$+1$	T_z, R_z	$\alpha_{xx}, \alpha_{yy}, \alpha_{zz}, \alpha_{xy}$
B	$+1$	-1	T_x, T_y, R_x, R_y	α_{yz}, α_{xz}

$\mathbf{C}_i$	i	I		
A_g	$+1$	$+1$	R_x, R_y, R_z	all components of α
A_u	$+1$	-1	T_x, T_y, T_z	

$\mathbf{C}_{2v}$	I	$C_2(z)$	$\sigma_v(xz)$	$\sigma_v(yz)$		
A_1	$+1$	$+1$	$+1$	$+1$	T_z	$\alpha_{xx}, \alpha_{yy}, \alpha_{zz}$
A_2	$+1$	$+1$	-1	-1	R_z	α_{xy}
B_1	$+1$	-1	$+1$	-1	T_x, R_y	α_{xz}
B_2	$+1$	-1	-1	$+1$	T_y, R_x	α_{yz}

$\mathbf{C}_{3v}$	I	$2C_3(z)$	$3\sigma_v$		
A_1	+1	+1	+1	T_z	$\alpha_{xx} + \alpha_{yy}, \alpha_{zz}$
A_2	+1	+1	−1	R_z	
E	+2	−1	0	$(T_x, T_y), (R_x, R_y)$	$(\alpha_{xx} - \alpha_{yy}, \alpha_{xy}), (\alpha_{yz}, \alpha_{xz})$

$\mathbf{C}_{4v}$	I	$2C_4(z)$	$C_4^2 \equiv C_2''$	$2\sigma_v$	$2\sigma_d$		
A_1	+1	+1	+1	+1	+1	T_z	$\alpha_{xx} + \alpha_{yy}, \alpha_{zz}$
A_2	+1	+1	+1	−1	−1	R_z	
B_1	+1	−1	+1	+1	−1		$\alpha_{xx} - \alpha_{yy}$
B_2	+1	−1	+1	−1	+1		α_{xy}
E	+2	0	−2	0	0	$(T_x, T_y), (R_x, R_y)$	$(\alpha_{yz}, \alpha_{xz})$

C_p^n (or S_p^n) denotes that C_p (or S_p) operation is carried out successively n times.

$\mathbf{C}_{\infty v}$	I	$2C_\infty^{\phi}$	$2C_\infty^{2\phi}$	$2C_\infty^{3\phi}$	...	$\infty\sigma_v$		
Σ^+	+1	+1	+1	+1	...	+1	T_z	$\alpha_{xx} + \alpha_{yy}, \alpha_{zz}$
Σ^-	+1	+1	+1	+1	...	−1	R_z	
Π	+2	$2\cos\phi$	$2\cos 2\phi$	$2\cos 3\phi$	...	0	$(T_x, T_y), (R_x, R_y)$	$(\alpha_{yz}, \alpha_{xz})$
Δ	+2	$2\cos 2\phi$	$2\cos 2\cdot 2\phi$	$2\cos 3\cdot 2\phi$	...	0		$(\alpha_{xx} - \alpha_{yy}, \alpha_{xy})$
Φ	+2	$2\cos 3\phi$	$2\cos 2\cdot 3\phi$	$2\cos 3\cdot 3\phi$	...	0		
...	...	...	...	...	...	...		

$\mathbf{C}_{2h}$	I	$C_2(z)$	$\sigma_h(xy)$	i		
A_g	+1	+1	+1	+1	R_z	$\alpha_{xx}, \alpha_{yy}, \alpha_{zz}, \alpha_{xy}$
A_u	+1	+1	−1	−1	T_z	
B_g	+1	−1	−1	+1	R_x, R_y	α_{yz}, α_{xz}
B_u	+1	−1	+1	−1	T_x, T_y	

$\mathbf{D}_3$	I	$2C_3(z)$	$3C_2$		
A_1	+1	+1	+1		$\alpha_{xx} + \alpha_{yy}, \alpha_{zz}$
A_2	+1	+1	−1	T_z, R_z	
E	+2	−1	0	$(T_x, T_y), (R_x, R_y)$	$(\alpha_{xx} - \alpha_{yy}, \alpha_{xy}), (\alpha_{yz}, \alpha_{xz})$

$\mathbf{D}_{2d} \equiv \mathbf{V}_d$	I	$2S_4(z)$	$S_4{}^2 \equiv C_2''$	$2C_2$	$2\sigma_d$		
A_1	+1	+1	+1	+1	+1		$\alpha_{xx} + \alpha_{yy}, \alpha_{zz}$
A_2	+1	+1	+1	−1	−1	R_z	
B_1	+1	−1	+1	+1	−1		$\alpha_{xx} - \alpha_{yy}$
B_2	+1	−1	+1	−1	+1	T_z	α_{xy}
E	+2	0	−2	0	0	$(T_x, T_y), (R_x, R_y)$	$(\alpha_{yz}, \alpha_{xz})$

$\mathbf{D}_{3d}$	I	$2S_6(z)$	$2S_6^2 \equiv 2C_3$	$S_6^3 \equiv S_2 \equiv i$	$3C_2$	$3\sigma_d$		
A_{1g}	+1	+1	+1	+1	+1	+1		$\alpha_{xx} + \alpha_{yy}, \alpha_{zz}$
A_{1u}	+1	−1	+1	−1	+1	−1		
A_{2g}	+1	+1	+1	+1	−1	−1	R_z	
A_{2u}	+1	−1	+1	−1	−1	+1	T_z	
E_g	+2	−1	−1	+2	0	0	(R_x, R_y)	$(\alpha_{xx} - \alpha_{yy}, \alpha_{xy}), (\alpha_{yz}, \alpha_{xz})$
E_u	+2	+1	−1	−2	0	0	(T_x, T_y)	

$\mathbf{D}_{4d}$	I	$2S_8(z)$	$2S_8^2 \equiv 2C_4$	$2S_8^3$	$S_8^4 \equiv C_2''$	$4C_2$	$4\sigma_d$		
A_1	+1	+1	+1	+1	+1	+1	+1		$\alpha_{xx} + \alpha_{yy}, \alpha_{zz}$
A_2	+1	+1	+1	+1	+1	−1	−1	R_z	
B_1	+1	−1	+1	−1	+1	+1	−1		
B_2	+1	−1	+1	−1	+1	−1	−1	T_z	
E_1	+2	$+\sqrt{2}$	0	$-\sqrt{2}$	−2	0	0	(T_x, T_y)	
E_2	+2	0	−2	0	+2	0	0		$(\alpha_{xx} - \alpha_{yy}, \alpha_{xy})$
E_3	+2	$-\sqrt{2}$	0	$+\sqrt{2}$	−2	0	0	(R_x, R_y)	$(\alpha_{yz}, \alpha_{xz})$

$\mathbf{D}_{2h} \equiv \mathbf{V}_h$	I	$\sigma(xy)$	$\sigma(xz)$	$\sigma(yz)$	i	$C_2(z)$	$C_2(y)$	$C_2(x)$		
A_g	+1	+1	+1	+1	+1	+1	+1	+1		$\alpha_{xx}, \alpha_{yy}, \alpha_{zz}$
A_u	+1	−1	−1	−1	−1	+1	+1	+1		
B_{1g}	+1	+1	−1	−1	+1	+1	−1	−1	R_z	α_{xy}
B_{1u}	+1	−1	+1	+1	−1	+1	−1	−1	T_z	
B_{2g}	+1	−1	+1	−1	+1	−1	+1	−1	R_y	α_{xz}
B_{2u}	+1	+1	−1	+1	−1	−1	+1	−1	T_y	
B_{3g}	+1	−1	−1	+1	+1	−1	−1	+1	R_x	α_{yz}
B_{3u}	+1	+1	+1	−1	−1	−1	−1	+1	T_x	

$\mathbf{D}_{3h}$	I	$2C_3(z)$	$3C_2$	σ_h	$2S_3$	$3\sigma_v$		
A_1'	+1	+1	+1	+1	+1	+1		$\alpha_{xx} + \alpha_{yy}, \alpha_{zz}$
A_1''	+1	+1	+1	−1	−1	−1		
A_2'	+1	+1	−1	+1	+1	−1	R_z	
A_2''	+1	+1	−1	−1	−1	+1	T_z	
E'	+2	−1	0	+2	−1	0	(T_x, T_y)	$(\alpha_{xx} - \alpha_{yy}, \alpha_{xy})$
E''	+2	−1	0	−2	+1	0	(R_x, R_y)	$(\alpha_{yz}, \alpha_{xz})$

$\mathbf{D}_{4h}$	I	$2C_4(z)$	$C_4^2 \equiv C_2''$	$2C_2$	$2C_2'$	σ_h	$2\sigma_v$	$2\sigma_d$	$2S_4$	$S_2 \equiv i$		
A_{1g}	+1	+1	+1	+1	+1	+1	+1	+1	+1	+1		$\alpha_{xx} + \alpha_{yy}, \alpha_{zz}$
A_{1u}	+1	+1	+1	+1	+1	−1	−1	−1	−1	−1		
A_{2g}	+1	+1	+1	−1	−1	+1	−1	−1	+1	+1	R_z	
A_{2u}	+1	+1	+1	−1	−1	−1	+1	+1	−1	−1	T_z	
B_{1g}	+1	−1	+1	+1	−1	+1	+1	−1	−1	+1		$\alpha_{xx} - \alpha_{yy}$
B_{1u}	+1	−1	+1	+1	−1	−1	−1	+1	+1	−1		
B_{2g}	+1	−1	+1	−1	+1	+1	−1	+1	−1	+1		α_{xy}
B_{2u}	+1	−1	+1	−1	+1	−1	+1	−1	+1	−1		
E_g	+2	0	−2	0	0	−2	0	0	0	+2	(R_x, R_y)	$(\alpha_{yz}, \alpha_{xz})$
E_u	+2	0	−2	0	0	+2	0	0	0	−2	(T_x, T_y)	

$\mathbf{D}_{5h}$	I	$2C_5(z)$	$2C_5^2$	σ_h	$5C_2$	$5\sigma_v$	$2S_5$	$2S_5^3$		
A_1'	+1	+1	+1	+1	+1	+1	+1	+1		$\alpha_{xx} + \alpha_{yy}, \alpha_{zz}$
A_1''	+1	+1	+1	−1	+1	−1	−1	−1		
A_2'	+1	+1	+1	+1	−1	−1	+1	+1	R_z	
A_2''	+1	+1	+1	−1	−1	+1	−1	−1	T_z	
E_1'	+2	2 cos 72°	2 cos 144°	+2	0	0	+2 cos 72°	+2 cos 144°	(T_x, T_y)	
E_1''	+2	2 cos 72°	2 cos 144°	−2	0	0	−2 cos 72°	−2 cos 144°	(R_x, R_y)	$(\alpha_{yz}, \alpha_{xz})$
E_2'	+2	2 cos 144°	2 cos 72°	+2	0	0	+2 cos 144°	+2 cos 72°		$(\alpha_{xx} - \alpha_{yy}, \alpha_{xy})$
E_2''	+2	2 cos 144°	2 cos 72°	−2	0	0	−2 cos 144°	−2 cos 72°		

$\mathbf{D}_{6h}$	I	$2C_6(z)$	$2C_6^2 \equiv 2C_3$	$C_6^3 \equiv C_2''$	$3C_2$	$3C_2'$	σ_h	$3\sigma_v$	$3\sigma_d$	$2S_6$	$2S_3$	$S_6^3 \equiv S_2 \equiv i$		
A_{1g}	+1	+1	+1	+1	+1	+1	+1	+1	+1	+1	+1	+1		$\alpha_{xx} + \alpha_{yy}, \alpha_{zz}$
A_{1u}	+1	+1	+1	+1	+1	+1	−1	−1	−1	−1	−1	−1		
A_{2g}	+1	+1	+1	+1	−1	−1	+1	−1	−1	+1	+1	+1	R_z	
A_{2u}	+1	+1	+1	+1	−1	−1	−1	+1	+1	−1	−1	−1	T_z	
B_{1g}	+1	−1	+1	−1	+1	−1	−1	−1	+1	+1	−1	+1		
B_{1u}	+1	−1	+1	−1	+1	−1	+1	+1	−1	−1	+1	−1		
B_{2g}	+1	−1	+1	−1	−1	+1	−1	+1	−1	+1	−1	+1		
B_{2u}	+1	−1	+1	−1	−1	+1	+1	−1	+1	−1	+1	−1		
E_{1g}	+2	+1	−1	−2	0	0	−2	0	0	−1	+1	+2	(R_x, R_y)	$(\alpha_{yz}, \alpha_{xz})$
E_{1u}	+2	+1	−1	−2	0	0	+2	0	0	+1	−1	−2	(T_x, T_y)	
E_{2g}	+2	−1	−1	+2	0	0	+2	0	0	−1	−1	+2		$(\alpha_{xx} - \alpha_{yy}, \alpha_{xy})$
E_{2u}	+2	−1	−1	+2	0	0	−2	0	0	+1	+1	−2		

$\mathbf{D}_{\infty h}$	I	$2C_\infty^{\phi}$	$2C_\infty^{2\phi}$	$2C_\infty^{3\phi}$	…	σ_h	∞C_2	$\infty\sigma_v$	$2S_\infty^{\phi}$	$2S_\infty^{2\phi}$	…	$S_2 \equiv i$		
Σ_g^+	+1	+1	+1	+1	…	+1	+1	+1	+1	+1	…	+1		$\alpha_{xx} + \alpha_{yy}, \alpha_{zz}$
Σ_u^+	+1	+1	+1	+1	…	−1	−1	+1	−1	−1	…	−1	T_z	
Σ_g^-	+1	+1	+1	+1	…	+1	−1	−1	+1	+1	…	+1	R_z	
Σ_u^-	+1	+1	+1	+1	…	−1	+1	−1	−1	−1	…	−1		
Π_g	+2	$2\cos\phi$	$2\cos 2\phi$	$2\cos 3\phi$	…	−2	0	0	$-2\cos\phi$	$-2\cos 2\phi$	…	+2	(R_x, R_y)	$(\alpha_{yz}, \alpha_{xz})$
Π_u	+2	$2\cos\phi$	$2\cos 2\phi$	$2\cos 3\phi$	…	+2	0	0	$+2\cos\phi$	$+2\cos 2\phi$	…	−2	(T_x, T_y)	
Δ_g	+2	$2\cos 2\phi$	$2\cos 4\phi$	$2\cos 6\phi$	…	+2	0	0	$+2\cos 2\phi$	$+2\cos 4\phi$	…	+2		$(\alpha_{xx} - \alpha_{yy}, \alpha_{xy})$
Δ_u	+2	$2\cos 2\phi$	$2\cos 4\phi$	$2\cos 6\phi$	…	−2	0	0	$-2\cos 2\phi$	$-2\cos 4\phi$	…	−2		
Φ_g	+2	$2\cos 3\phi$	$2\cos 6\phi$	$2\cos 9\phi$	…	−2	0	0	$-2\cos 3\phi$	$-2\cos 4\phi$	…	+2		
Φ_u	+2	$2\cos 3\phi$	$2\cos 6\phi$	$2\cos 9\phi$	…	+2	0	0	$+2\cos 3\phi$	$+2\cos 4\phi$	…	−2		
…	…	…	…	…	…	…	…	…	…	…	…	…		

$\mathbf{T}_d$	I	$8C_3$	$6\sigma_d$	$6S_4$	$3S_4^2 \equiv 3C_2$		
A_1	+1	+1	+1	+1	+1		$\alpha_{xx} + \alpha_{yy} + \alpha_{zz}$
A_2	+1	+1	−1	−1	+1		
E	+2	−1	0	0	+2		$(\alpha_{xx} + \alpha_{yy} - 2\alpha_{zz}, \alpha_{xx} - \alpha_{yy})$
F_1	+3	0	−1	+1	−1	(R_x, R_y, R_z)	
F_2	+3	0	+1	−1	−1	(T_x, T_y, T_z)	$(\alpha_{xy}, \alpha_{yz}, \alpha_{xz})$

$\mathbf{O}_h$	I	$8C_3$	$6C_2$	$6C_4$	$3C_4^2 \equiv 3C_2''$	$S_2 \equiv i$	$6S_4$	$8S_6$	$3\sigma_h$	$6\sigma_d$		
A_{1g}	+1	+1	+1	+1	+1	+1	+1	+1	+1	+1		$\alpha_{xx} + \alpha_{yy} + \alpha_{zz}$
A_{1u}	+1	+1	+1	+1	+1	−1	−1	−1	−1	−1		
A_{2g}	+1	+1	−1	−1	+1	+1	−1	+1	+1	−1		
A_{2u}	+1	+1	−1	−1	+1	−1	+1	−1	−1	+1		
E_g	+2	−1	0	0	+2	+2	0	−1	+2	0		$(\alpha_{xx} + \alpha_{yy} - 2\alpha_{zz}, \alpha_{xx} - \alpha_{yy})$
E_u	+2	−1	0	0	+2	−2	0	+1	−2	0		
F_{1g}	+3	0	−1	+1	−1	+3	+1	0	−1	−1	(R_x, R_y, R_z)	
F_{1u}	+3	0	−1	+1	−1	−3	−1	0	+1	+1	(T_x, T_y, T_z)	
F_{2g}	+3	0	+1	−1	−1	+3	−1	0	−1	+1		$(\alpha_{xy}, \alpha_{yz}, \alpha_{xz})$
F_{2u}	+3	0	+1	−1	−1	−3	+1	0	+1	−1		

APPENDIX II

General Formulas for Calculating the Number of Normal Vibrations in Each Species

These tables were quoted from G. Herzberg, Molecular spectra and Molecular Structure II (Ref. 8 of Part I).

A. Point groups including only nondegenerate vibrations

Point group	Total number of atoms	Species	Number of vibrations*
C_2	$2m + m_0$	A B	$3m + m_0 - 2$ $3m + 2m_0 - 4$
C_s	$2m + m_0$	A' A''	$3m + 2m_0 - 3$ $3m + m_0 - 3$
$C_i \equiv S_2$	$2m + m_0$	A_g A_u	$3m - 3$ $3m + 3m_0 - 3$
C_{2v}	$4m + 2m_{xz} + 2m_{yz} + m_0$	A_1 A_2 B_1 B_2	$3m + 2m_{xz} + 2m_{yz} + m_0 - 1$ $3m + m_{xz} + m_{yz} - 1$ $3m + 2m_{xz} + m_{yz} + m_0 - 2$ $3m + m_{xz} + 2m_{yz} + m_0 - 2$

$\mathbf{C}_{2h}$	$4m + 2m_h + 2m_2 + m_0$	A_g	$3m + 2m_h + m_2 - 1$
		A_u	$3m + m_h + m_2 + m_0 - 1$
		B_g	$3m + m_h + 2m_2 - 2$
		B_u	$3m + 2m_h + 2m_2 + 2m_0 - 2$
$D_{2h} \equiv \mathbf{V}_h$	$8m + 4m_{xy} + 4m_{xz} + 4m_{yz} + 2m_{2x} + 2m_{2y} + 2m_{2z} + m_0$	A_g	$3m + 2m_{xy} + 2m_{xz} + 2m_{yz} + m_{2x} + m_{2y} + m_{2z}$
		A_u	$3m + m_{xy} + m_{xz} + m_{yz}$
		B_{1g}	$3m + 2m_{xy} + m_{xz} + m_{yz} + m_{2x} + m_{2y} - 1$
		B_{1u}	$3m + m_{xy} + 2m_{xz} + 2m_{yz} + m_{2x} + m_{2y} + m_{2z} + m_0 - 1$
		B_{2g}	$3m + m_{xy} + 2m_{xz} + m_{yz} + m_{2x} + m_{2z} - 1$
		B_{2u}	$3m + 2m_{xy} + m_{xz} + 2m_{yz} + m_{2x} + m_{2y} + m_{2z} + m_0 - 1$
		B_{3g}	$3m + m_{xy} + m_{xz} + 2m_{yz} + m_{2y} + m_{2z} - 1$
		B_{3u}	$3m + 2m_{xy} + 2m_{xz} + m_{yz} + m_{2x} + m_{2y} + m_{2z} + m_0 - 1$

* Note that m is always the number of sets of equivalent nuclei not on any element of symmetry; m_0 is the number of nuclei lying on all symmetry elements present; m_{xy}, m_{xz}, m_{yz} are the numbers of sets of nuclei lying on the xy, xz, yz plane, respectively, but not on any axes going through these planes; m_2 is the number of sets of nuclei on a two-fold axis but not at the point of intersection with another element of symmetry; m_{2x}, m_{2y}, m_{2z} are the numbers of sets of nuclei lying on the x, y, or z axis if they are two-fold axes, but not on all of them; m_h is the number of sets of nuclei on a plane σ_h but not on the axis perpendicular to this plane.

B. Point groups including degenerate vibrations

Point group	Total number of atoms	Species	Number of vibrations**
$\mathbf{D_3}$	$6m + 3m_2 + 2m_3 + m_0$	A_1	$3m + m_2 + m_3$
		A_2	$3m + 2m_2 + m_3 + m_0 - 2$
		E	$6m + 3m_2 + 2m_3 + m_0 - 2$
$\mathbf{C_{3v}}$	$6m + 3m_v + m_0$	A_1	$3m + 2m_v + m_0 - 1$
		A_2	$3m + m_v - 1$
		E	$6m + 3m_v + m_0 - 2$
$\mathbf{C_{4v}}$	$8m + 4m_v + 4m_d + m_0$	A_1	$3m + 2m_v + 2m_d + m_0 - 1$
		A_2	$3m + m_v + m_d - 1$
		B_1	$3m + 2m_v + m_d$
		B_2	$3m + m_v + 2m_d$
		E	$6m + 3m_v + 3m_d + m_0 - 2$
$\mathbf{C_{\infty v}}$	m_0	Σ^+	$m_0 - 1$
		Σ^-	0
		Π	$m_0 - 2$
		$\Delta, \Phi \cdots$	0
$\mathbf{D_{2d} \equiv V_d}$	$8m + 4m_d + 4m_2 + 2m_4 + m_0$	A_1	$3m + 2m_d + m_2 + m_4$
		A_2	$3m + m_d + 2m_2 - 1$
		B_1	$3m + m_d + m_2$
		B_2	$3m + 2m_d + 2m_2 + m_4 + m_0 - 1$
		E	$6m + 3m_d + 3m_2 + 2m_4 + m_0 - 2$

$\mathbf{D}_{3d}$	$12m + 6m_d + 6m_2 + 2m_6 + m_0$	A_{1g}	$3m + 2m_d + m_2 + m_6$
		A_{1u}	$3m + m_d + m_2$
		A_{2g}	$3m + m_d + 2m_2 - 1$
		A_{2u}	$3m + 2m_d + 2m_2 + m_6 + m_0 - 1$
		E_g	$6m + 3m_d + 3m_2 + m_6 - 1$
		E_u	$6m + 3m_d + 3m_2 + m_6 + m_0 - 1$
$\mathbf{D}_{4d}$	$16m + 8m_d + 8m_2 + 2m_8 + m_0$	A_1	$3m + 2m_d + m_2 + m_8$
		A_2	$3m + m_d + 2m_2 - 1$
		B_1	$3m + m_d + m_2$
		B_2	$3m + 2m_d + 2m_2 + m_8 + m_0 - 1$
		E_1	$6m + 3m_d + 3m_2 + m_8 + m_0 - 1$
		E_2	$6m + 3m_d + 3m_2$
		E_3	$6m + 3m_d + 3m_2 + m_8 - 1$
$\mathbf{D}_{3h}$	$12m + 6m_v + 6m_h + 3m_2 + 2m_3 + m_0$	A_1'	$3m + 2m_v + 2m_h + m_2 + m_3$
		A_1''	$3m + m_v + m_h$
		A_2'	$3m + m_v + 2m_h + m_2 - 1$
		A_2''	$3m + 2m_v + m_h + m_2 + m_3 + m_0 - 1$
		E'	$6m + 3m_v + 4m_h + 2m_2 + m_3 + m_0 - 1$
		E''	$6m + 3m_v + 2m_h + m_2 + m_3 - 1$
$\mathbf{D}_{4h}$	$16m + 8m_v + 8m_d + 8m_h + 4m_2 + 4m_2' + 2m_4 + m_0$	A_{1g}	$3m + 2m_v + 2m_d + 2m_h + m_2 + m_2' + m_4$
		A_{1u}	$3m + m_v + m_d + m_h$
		A_{2g}	$3m + m_v + m_d + 2m_h + m_2 + m_2' - 1$
		A_{2u}	$3m + 2m_v + 2m_d + m_h + m_2 + m_2' + m_4 + m_0 - 1$
		B_{1g}	$3m + 2m_v + m_d + 2m_h + m_2 + m_2'$
		B_{1u}	$3m + m_v + 2m_d + m_h + m_2'$
		B_{2g}	$3m + m_v + 2m_d + 2m_h + m_2 + m_2'$
		B_{2u}	$3m + 2m_v + m_d + m_h + m_2$
		E_g	$6m + 3m_v + 3m_d + 2m_h + m_2 + m_2' + m_4 - 1$
		E_u	$6m + 3m_v + 3m_d + 4m_h + 2m_2 + 2m_2' + m_4 + m_0 - 1$

Point group	Total number of atoms	Species	Number of vibrations**
$\mathbf{D}_{5h}$	$20m + 10m_v + 10m_h + 5m_2 + 2m_5 + m_0$	A_1'	$3m + 2m_v + 2m_h + m_2 + m_5$
		A_1''	$3m + m_v + m_h$
		A_2'	$3m + m_v + 2m_h + m_2 - 1$
		A_2''	$3m + 2m_v + m_h + m_2 + m_5 + m_0 - 1$
		E_1'	$6m + 3m_v + 4m_h + 2m_2 + m_5 + m_0 - 1$
		E_1''	$6m + 3m_v + 2m_h + m_2 + m_5 - 1$
		E_2'	$6m + 3m_v + 4m_h + 2m_2$
		E_2''	$6m + 3m_v + 2m_h + m_2$
$\mathbf{D}_{6h}$	$24m + 12m_v + 12m_d + 12m_h + 6m_2 + 6m'_2 + 2m_6 + m_0$	A_{1g}	$3m + 2m_v + 2m_d + 2m_h + m_2 + m_2' + m_6$
		A_{1u}	$3m + m_v + m_d + m_h$
		A_{2g}	$3m + m_v + m_d + 2m_h + m_2 + m_2' - 1$
		A_{2u}	$3m + 2m_v + 2m_d + m_h + m_2 + m_2' + m_6 + m_0 - 1$
		B_{1g}	$3m + m_v + 2m_d + m_h + m_2'$
		B_{1u}	$3m + 2m_v + m_d + 2m_h + m_2 + m_2'$
		B_{2g}	$3m + 2m_v + m_d + m_h + m_2$
		B_{2u}	$3m + m_v + 2m_d + 2m_h + m_2 + m_2'$
		E_{1g}	$6m + 3m_v + 3m_d + 2m_h + m_2 + m_2' + m_6 - 1$
		E_{1u}	$6m + 3m_v + 3m_d + 4m_h + 2m_2 + 2m_2' + m_6 + m_0 - 1$
		E_{2g}	$6m + 3m_v + 3m_d + 4m_h + 2m_2 + 2m_2'$
		E_{2u}	$6m + 3m_v + 3m_v + 2m_h + m_2 + m_2'$

$\mathbf{D}_{\infty h}$	$2m_\infty + m_0$	Σ_g^+	m_∞
		Σ_u^+	$m_\infty + m_0 - 1$
		Σ_g^-, Σ_u^-	0
		Π_g	$m_\infty - 1$
		Π_u	$m_\infty + m_0 - 1$
		$\Delta_g, \Delta_u,$	0
		$\Phi_g, \Phi_u, \cdots$	0
$\mathbf{T}_d$	$24m + 12m_d$ $+6m_2 + 4m_3 + m_0$	A_1	$3m + 2m_d + m_2 + m_3$
		A_2	$3m + m_d$
		E	$6m + 3m_d + m_2 + m_3$
		F_1	$9m + 4m_d + 2m_2 + m_3 - 1$
		F_2	$9m + 5m_d + 3m_2 + 2m_3 + m_0 - 1$
$\mathbf{O}_h$	$48m + 24m_h + 24m_d$ $+12m_2 + 8m_3$ $+6m_4 + m_0$	A_{1g}	$3m + 2m_h + 2m_d + m_2 + m_3 + m_4$
		A_{1u}	$3m + m_h + m_d$
		A_{2g}	$3m + 2m_h + m_d + m_2$
		A_{2u}	$3m + m_h + 2m_d + m_2 + m_3$
		E_g	$6m + 4m_h + 3m_d + 2m_2 + m_3 + m_4$
		E_u	$6m + 2m_h + 3m_d + m_2 + m_3$
		F_{1g}	$9m + 4m_h + 4m_d + 2m_2 + m_3 + m_4 - 1$
		F_{1u}	$9m + 5m_h + 5m_d + 3m_2 + 2m_3 + 2m_4 + m_0 - 1$
		F_{2g}	$9m + 4m_h + 5m_d + 2m_2 + 2m_3 + m_4$
		F_{2u}	$9m + 5m_h + 4m_d + 2m_2 + m_3 + m_4$

** Note that m is the number of sets of nuclei not on any element of symmetry; m_0 is the number of nuclei on all elements of symmetry; m_2, m_3, m_4, $\cdots$ are the numbers of sets of nuclei on a two-fold, three-fold, four-fold. $\cdots$ axis but not on any other element of symmetry that does not wholly coincide with that axis; m_2' is the number of sets of nuclei on a two-fold axis called C_2' in the previous character tables; m_v, m_d, m_h are the numbers of sets of nuclei on planes σ_v, σ_d, σ_h, respectively, but not on any other element of symmetry.

APPENDIX III

DERIVATION OF EQUATION 11.3 (PART I)

Using the rectangular coordinates, we write the kinetic energy as

$$2T = \tilde{\dot{\mathbf{X}}}\mathbf{M}\dot{\mathbf{X}} \tag{1}$$

where

$$\mathbf{X} = \begin{bmatrix} x_1 \\ y_1 \\ z_1 \\ x_2 \\ \vdots \\ z_N \end{bmatrix} \quad \text{and} \quad \mathbf{M} = \begin{bmatrix} m_1 & & & & & \\ & m_1 & & & & \\ & & m_1 & & & \\ & & & m_2 & & \\ & & & & \ddots & \\ & & & & & m_N \end{bmatrix}$$

By definition, the momentum p_{x_1} conjugated with x_1 is given by

$$p_{x_1} = \frac{\partial T}{\partial \dot{x}_1} = m_1 \dot{x}_1$$

$p_{y_1} \cdots p_{z_N}$ take similar forms. Using the conjugate momenta, we write T as

$$\begin{aligned} 2T &= \frac{1}{m_1} p_{x_1}{}^2 + \frac{1}{m_1} p_{y_1}{}^2 + \cdots + \frac{1}{m_N} p_{z_N}{}^2 \\ &= \tilde{\mathbf{P}}_x \mathbf{M}^{-1} \mathbf{P}_x \end{aligned} \tag{2}$$

where

$$\mathbf{P}_x = \begin{bmatrix} p_{x_1} \\ p_{y_1} \\ \vdots \\ p_{z_N} \end{bmatrix} \quad \text{and} \quad \mathbf{M}^{-1} = \begin{bmatrix} \mu_1 & & & \\ & \mu_1 & & \\ & & \ddots & \\ & & & \mu_N \end{bmatrix}$$

The column matrix $\mathbf{P}_x$ can be expressed as

$$\mathbf{P}_x = \mathbf{M}\dot{\mathbf{X}} \tag{3}$$

Define a set of conjugate momenta **P** associated with internal coordinates, **R**. As is shown at the end of this Appendix, we have

$$\mathbf{P}_x = \tilde{\mathbf{B}}\mathbf{P} \tag{4}$$

Equations 3 and 4 give

$$\mathbf{M}\dot{\mathbf{X}} = \tilde{\mathbf{B}}\mathbf{P} \tag{5}$$

Equation 11.8 in the text gives

$$\mathbf{R} = \mathbf{B}\mathbf{X} \quad \text{and} \quad \dot{\mathbf{R}} = \mathbf{B}\dot{\mathbf{X}} \tag{6}$$

By inserting Eq. 5 into Eq. 6, we obtain

$$\dot{\mathbf{R}} = \mathbf{B}\mathbf{M}^{-1}\tilde{\mathbf{B}}\mathbf{P} \tag{7}$$

Using Eq. 4, we write Eq. 2 as

$$2T = \tilde{\mathbf{P}}\mathbf{B}\mathbf{M}^{-1}\tilde{\mathbf{B}}\mathbf{P} \tag{8}$$

If we define

$$\mathbf{G} = \mathbf{B}\mathbf{M}^{-1}\tilde{\mathbf{B}} \qquad \text{(11.7), text}$$

Eq. 8 is written as

$$2T = \tilde{\mathbf{P}}\mathbf{G}\mathbf{P} \tag{9}$$

If Eq. 11.7 is combined with Eq. 7, we obtain

$$\dot{\mathbf{R}} = \mathbf{G}\mathbf{P}$$

or

$$\mathbf{G}^{-1}\dot{\mathbf{R}} = \mathbf{G}^{-1}\mathbf{G}\mathbf{P} = \mathbf{P} \tag{10}$$

Using Eq. 10, Eq. 9 can be written

$$\begin{aligned} 2T &= \tilde{\dot{\mathbf{R}}}\tilde{\mathbf{G}}^{-1}\mathbf{G}\mathbf{G}^{-1}\dot{\mathbf{R}} \\ &= \tilde{\dot{\mathbf{R}}}\mathbf{G}^{-1}\dot{\mathbf{R}} \qquad \text{(11.3), text} \end{aligned}$$

Derivation of Eq. 4

The momentum p_{R_k} conjugated with the internal coordinate, R_k, is given by

$$p_{R_k} = \frac{\partial T}{\partial \dot{R}_k} \qquad k = 1, 2, \cdots, s$$

If we denote the coordinates corresponding to the translational and rotational motions of the molecule by R_j^0 and its conjugate momentum by $p_{R_j^0}$

$$p_{R_j^0} = \frac{\partial T}{\partial \dot{R}_j^0} \qquad j = 1, 2, \cdots, 6$$

Then the momentum, p_{x_1}, in terms of rectangular coordinates is written as

$$\begin{aligned} p_{x_1} &= \frac{\partial T}{\partial \dot{x}_1} = \sum_k^s \frac{\partial T}{\partial \dot{R}_k}\frac{\partial R_k}{\partial x_1} + \sum_j^6 \frac{\partial T}{\partial \dot{R}_j^0}\frac{\partial R_j^0}{\partial x_1} \\ &= \sum_k^s p_{R_k} B_{k, x_1} + \sum_j^6 p_{R_j^0}\frac{\partial R_j^0}{\partial x_1} \end{aligned}$$

The second term becomes zero since the momenta corresponding to the translational and rotational motions are zero. Thus, we have

$$p_{x_1} = \sum_k^s p_{R_k} B_{k,x_1}$$

$$p_{y_1} = \sum p_{R_k} B_{k,y_1}$$

$$\vdots \qquad \vdots$$

$$p_{z_N} = \sum p_{R_k} B_{k,z_N}$$

In a matrix form, this is written as

$$\mathbf{P}_x = \tilde{\mathbf{B}}\mathbf{P} \tag{4}$$

APPENDIX IV

THE G AND F MATRIX ELEMENTS OF TYPICAL MOLECULES

In the following tables, F represents F matrix elements in the GVF field, whereas F^* denotes those in the UBF field. In the latter, $F' = -\frac{1}{10}F$ was assumed for all cases, and the *molecular tension* (see Ref. I-37) was ignored.

(1) Bent XY_2 Molecules (C_{2v})

A_1 species—infrared and Raman active

$$G_{11} = \mu_y + \mu_x(1 + \cos\alpha)$$

$$G_{12} = -\frac{\sqrt{2}}{r}\mu_x \sin\alpha$$

$$G_{22} = \frac{2}{r^2}[\mu_y + \mu_x(1 - \cos\alpha)]$$

$$F_{11} = f_r + f_{rr}$$

$$F_{12} = (\sqrt{2})\, r f_{r\alpha}$$

$$F_{22} = r^2 f_\alpha$$

$$F^*_{11} = K + 2F\sin^2\frac{\alpha}{2}$$

$$F^*_{12} = (0.9)(\sqrt{2})\, rF \sin\frac{\alpha}{2}\cos\frac{\alpha}{2}$$

$$F^*_{22} = r^2\left[H + F\left\{\cos^2\frac{\alpha}{2} + (0.1)\sin^2\frac{\alpha}{2}\right\}\right]$$

B_2 species—infrared and Raman active

$$G = \mu_y + \mu_x(1 - \cos\alpha)$$

$$F = f_r - f_{rr}$$

$$F^* = K - (0.2)F\cos^2\frac{\alpha}{2}$$

(2) Pyramidal XY_3 Molecules (C_{3v})

A_1 species—infrared and Raman active

$$G_{11} = \mu_y + \mu_x(1 + 2\cos\alpha)$$

$$G_{12} = -\frac{2}{r}\frac{(1 + 2\cos\alpha)(1 - \cos\alpha)}{\sin\alpha}\mu_x$$

$$G_{22} = \frac{2}{r^2}\left(\frac{1 + 2\cos\alpha}{1 + \cos\alpha}\right)[\mu_y + 2\mu_x(1 - \cos\alpha)]$$

$$F_{11} = f_r + 2f_{rr}$$

$$F_{12} = r(2f_{r\alpha} + f'_{r\alpha})$$

$$F_{22} = r^2(f_\alpha + 2f_{\alpha\alpha})$$

$$F^*_{11} = K + 4F\sin^2\frac{\alpha}{2}$$

$$F^*_{12} = (1.8)rF\sin\frac{\alpha}{2}\cos\frac{\alpha}{2}$$

$$F^*_{22} = r^2\left[H + F\left\{\cos^2\frac{\alpha}{2} + (0.1)\sin^2\frac{\alpha}{2}\right\}\right]$$

E species—infrared and Raman active

$$G_{11} = \mu_y + \mu_x(1 - \cos\alpha)$$

$$G_{12} = \frac{1}{r}\frac{(1 - \cos\alpha)^2}{\sin\alpha}\mu_x$$

$$G_{22} = \frac{1}{r^2(1 + \cos\alpha)}[(2 + \cos\alpha)\mu_y + (1 - \cos\alpha)^2\mu_x]$$

$$F_{11} = f_r - f_{rr}$$

$$F_{12} = r(-f_{r\alpha} + f'_{r\alpha})$$

$$F_{22} = r^2(f_\alpha - f_{\alpha\alpha})$$

$$F^*_{11} = K + \left[\sin^2\frac{\alpha}{2} - (0.3)\cos^2\frac{\alpha}{2}\right]F$$

$$F^*_{12} = -(0.9)rF\sin\frac{\alpha}{2}\cos\frac{\alpha}{2}$$

$$F^*_{22} = r^2\left[H + F\left\{\cos^2\frac{\alpha}{2} + (0.1)\sin^2\frac{\alpha}{2}\right\}\right]$$

Here $f_{r\alpha}$ denotes interaction between Δr and $\Delta\alpha$ having a common bond (e.g., Δr_1 and $\Delta\alpha_{12}$ or $\Delta\alpha_{31}$); $f'_{r\alpha}$ denotes interaction between Δr and $\Delta\alpha$ having no common bonds (e.g., Δr_1 and $\Delta\alpha_{23}$); see Fig. I-9*c*.

(3) Planar XY_3 Molecules (D_{3h})

A'_1 species—Raman active

$$G = \mu_y$$

$$F = f_r + 2f_{rr}$$

$$F^* = K + 3F$$

A''_2 species—infrared active

$$G = \frac{9}{4r^2}(\mu_y + 3\mu_x)$$

$$F = r^2 f_\theta$$

E' species—infrared and Raman active

$$G_{11} = \mu_y + \frac{3}{2}\mu_x$$

$$G_{12} = \frac{3\sqrt{3}}{2}\mu_x$$

$$G_{22} = \frac{3}{2r^2}(2\mu_y + 3\mu_x)$$

$$F_{11} = f_r - f_{rr}$$

$$F_{12} = r(f'_{r\alpha} - f_{r\alpha})$$

$$F_{22} = r^2(f_\alpha - f_{\alpha\alpha})$$

$$F^*_{11} = K + 0.675F$$

$$F^*_{12} = -(0.9)\frac{\sqrt{3}}{4}rF$$

$$F^*_{22} = r^2(H + 0.325F)$$

The symbols $f_{r\alpha}$ and $f'_{r\alpha}$ are defined in (2), f_θ denotes the force constant for the out-of-plane mode (see Fig. I-9*f*).

(4) Tetrahedral XY_4 Molecules (T_d)

A_1 species—Raman active

$$G = \mu_y$$

$$F = f_r + 3f_{rr}$$

$$F^* = K + 4F$$

E species—Raman active

$$G = \frac{3\mu_y}{r^2}$$

$$F = r^2(f_\alpha - 2f_{\alpha\alpha} + f'_{\alpha\alpha})$$

$$F^* = r^2(H + 0.37F)$$

F_2 species—infrared and Raman active

$$G_{11} = \mu_y + \tfrac{4}{3}\mu_x$$

$$G_{12} = -\frac{8}{3r}\mu_x$$

$$G_{22} = \frac{1}{r^2}(\tfrac{16}{3}\mu_x + 2\mu_y)$$

$$F_{11} = f_r - f_{rr}$$

$$F_{12} = (\sqrt{2})r(f_{r\alpha} - f'_{r\alpha})$$

$$F_{22} = r^2(f_\alpha - f'_{\alpha\alpha})$$

$$F^*_{11} = K + \tfrac{6}{5}F$$

$$F^*_{12} = \tfrac{3}{5}rF$$

$$F^*_{22} = r^2(H + \tfrac{1}{2}F)$$

where $f_{\alpha\alpha}$ denotes interaction between two $\Delta\alpha$ having a common bond; $f'_{\alpha\alpha}$ denotes interaction between two $\Delta\alpha$ having no common bond.

(5) Square-Planar XY_4 Molecules (D_{4h})

A_{1g} species—Raman active

$$G = \mu_y$$

$$F = f_r + 2f_{rr} + f'_{rr}$$

$$F^* = K + 2F$$

B_{1g} species—Raman active

$$G = \mu_y$$

$$F = f_r - 2f_{rr} + f'_{rr}$$

$$F^* = K - 0.2F$$

B_{2g} species—Raman active

$$G = \frac{4\mu_y}{r^2}$$

$$F = r^2(f_\alpha - 2f_{\alpha\alpha} + f'_{\alpha\alpha})$$

$$F^* = r^2(H + 0.55F)$$

E_u species—infrared active

$$G_{11} = 2\mu_x + \mu_y$$

$$G_{12} = -\frac{2\sqrt{2}}{r}\mu_x$$

$$G_{22} = \frac{2}{r^2}(\mu_y + 2\mu_x)$$

$$F_{11} = f_r - f'_{rr}$$

$$F_{12} = (\sqrt{2})r(f_{r\alpha} - f'_{r\alpha})$$

$$F_{22} = r^2(f_\alpha - f'_{\alpha\alpha})$$

$$F^*_{11} = K + 0.9F$$

$$F^*_{12} = -(\sqrt{2})r(0.45)F$$

$$F^*_{22} = r^2(H + 0.55F)$$

The symbol f_{rr} denotes interaction between two Δr perpendicular to each other; f'_{rr} denotes interaction between two Δr on the same straight line. In addition, a square-planar XY_4 molecule has two out-of-plane vibrations in the A_{2u} and B_{2u} species.

(6) Octahedral XY_6 Molecules (O_h)

A_{1g} species—Raman active

$$G = \mu_y$$

$$F = f_r + 4f_{rr} + f'_{rr}$$

$$F^* = K + 4F$$

E_g species—Raman active

$$G = \mu_y$$
$$F = f_r - 2f_{rr} + f'_{rr}$$
$$F^* = K + 0.7F$$

F_{1u} species—infrared active

$$G_{11} = \mu_y + 2\mu_x$$
$$G_{12} = -\frac{4}{r}\mu_x$$
$$G_{22} = \frac{2}{r^2}(\mu_y + 4\mu_x)$$
$$F_{11} = f_r - f'_{rr}$$
$$F_{12} = 2rf_{r\alpha}$$
$$F_{22} = r^2(f_\alpha + 2f_{\alpha\alpha})$$
$$F^*_{11} = K + 1.8F$$
$$F^*_{12} = 0.9rF$$
$$F^*_{22} = r^2(H + 0.55F)$$

F_{2g} species—Raman active

$$G = \frac{4\mu_y}{r^2}$$
$$F = r^2(f_\alpha - 2f'_{\alpha\alpha})$$
$$F^* = r^2(H + 0.55F)$$

F_{2u} species—inactive

$$G = \frac{2\mu_y}{r^2}$$
$$F = r^2(f_\alpha - 2f_{\alpha\alpha})$$
$$F^* = r^2(H + 0.55F)$$

The symbol f_{rr} denotes interaction between two Δr perpendicular to each other, whereas f'_{rr} denotes those between two Δr on the same straight line; $f_{\alpha\alpha}$ denotes interaction between two $\Delta\alpha$ perpendicular to each other, whereas $f'_{\alpha\alpha}$ denotes those between two $\Delta\alpha$ on the same plane. Only the interaction between two $\Delta\alpha$ having a common bond was considered.

APPENDIX V

GROUP FREQUENCY CHARTS

The data cited in this book were used in the preparation of the following group frequency charts. Each section of Part III gives a number of group frequencies that are not included here. For the physical meaning of group frequency, see Sec. I-18.

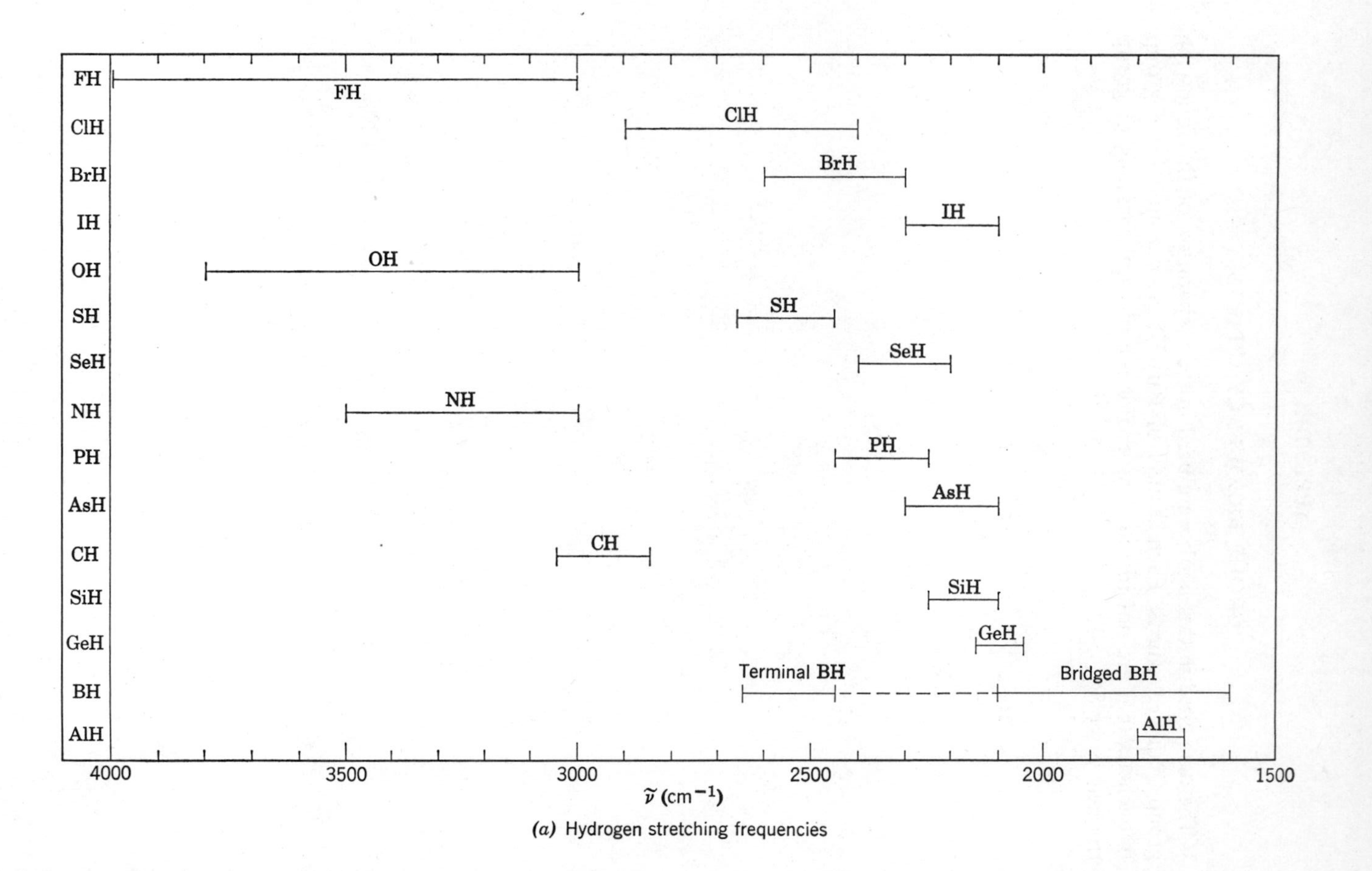

(*a*) Hydrogen stretching frequencies

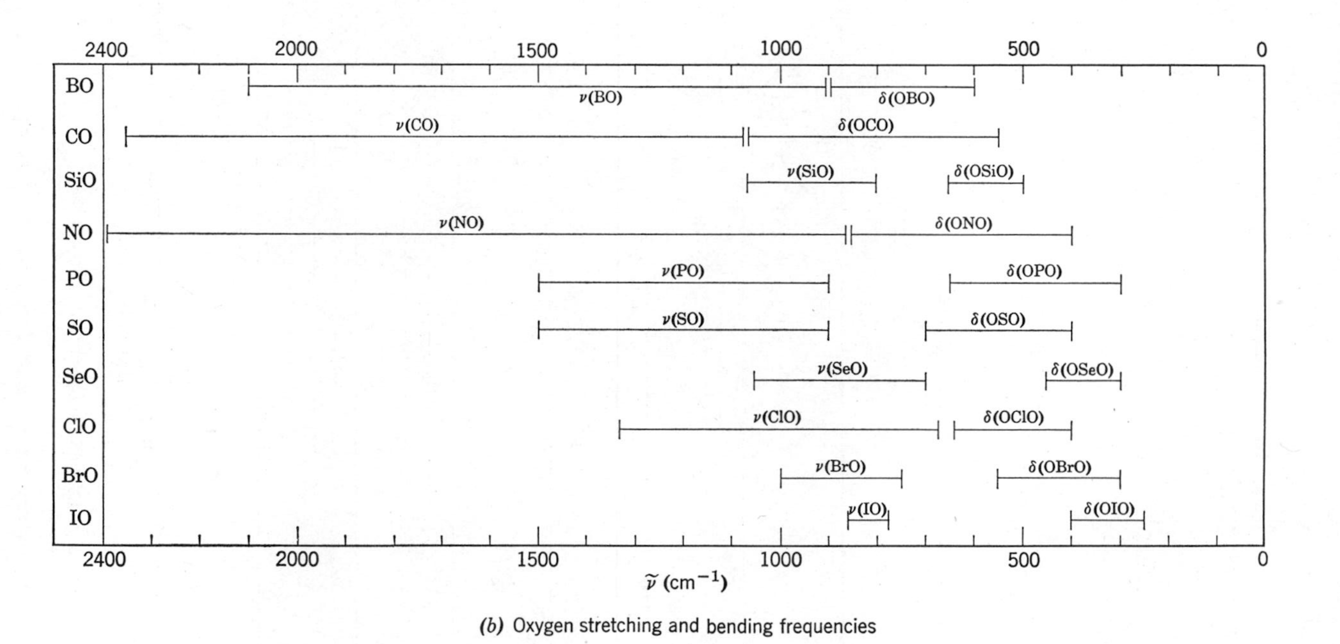

(*b*) Oxygen stretching and bending frequencies

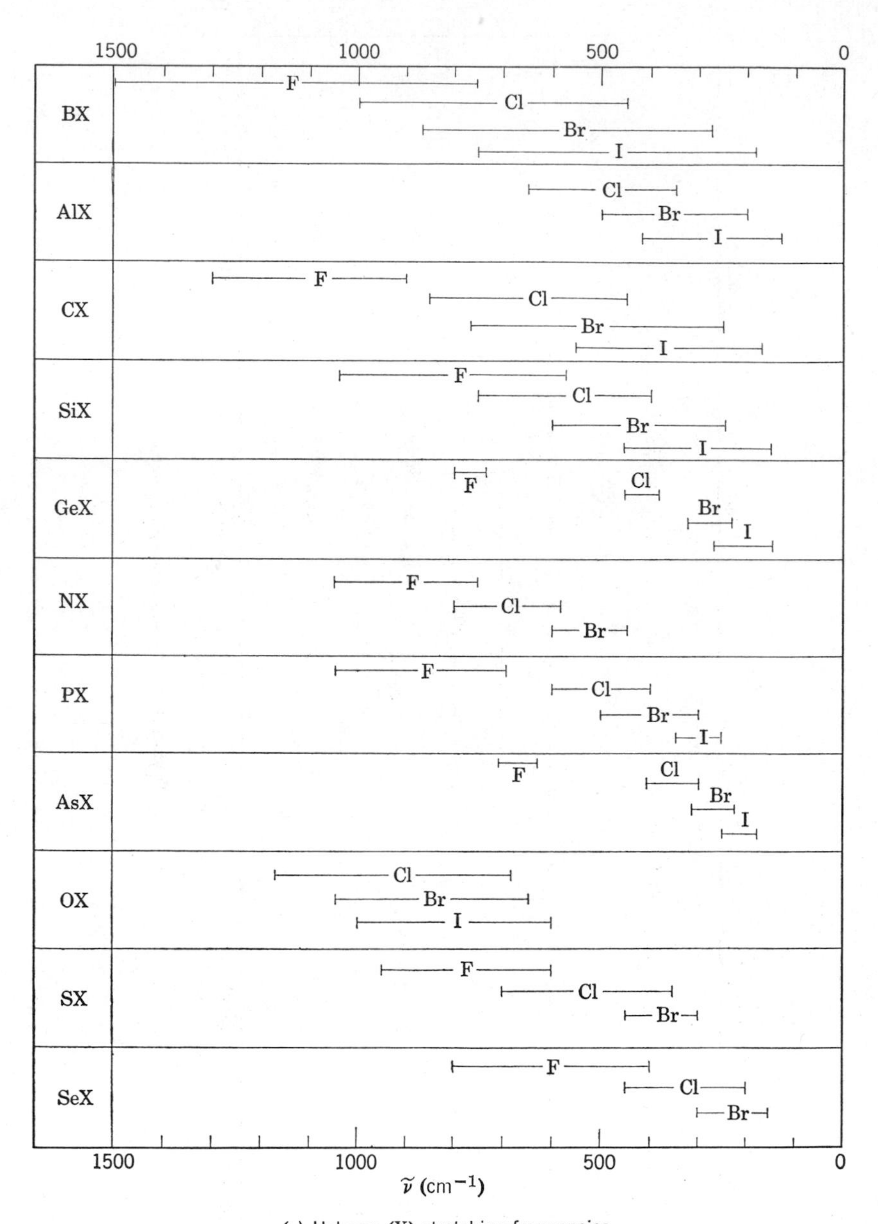

(c) Halogen (X) stretching frequencies

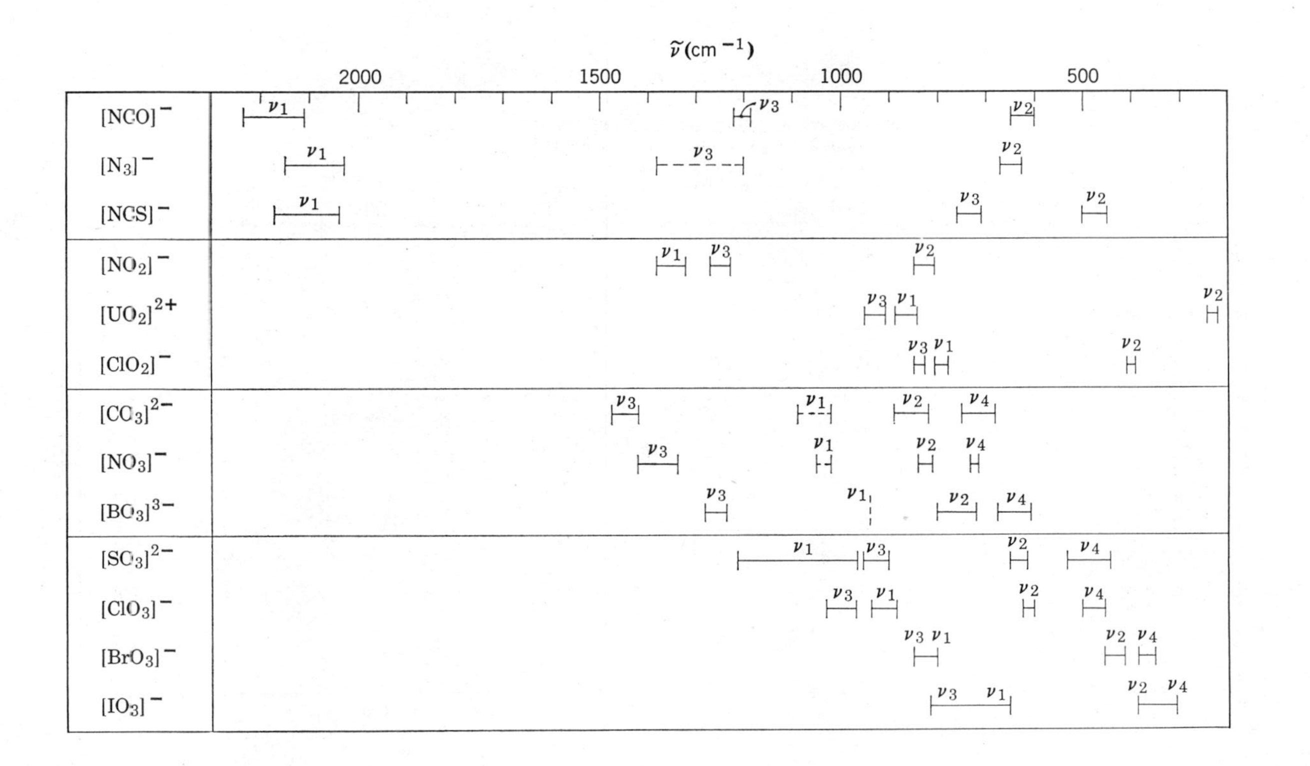

$\tilde{\nu}$ (cm^{-1})
2000
1500
1000
500
$[NCO]^-$
$[N_3]^-$
$[NCS]^-$
$[NO_2]^-$
$[UO_2]^{2+}$
$[ClO_2]^-$
$[CO_3]^{2-}$
$[NO_3]^-$
$[BO_3]^{3-}$
$[SO_3]^{2-}$
$[ClO_3]^-$
$[BrO_3]^-$
$[IO_3]^-$
ν_1
ν_2
ν_3
ν_4

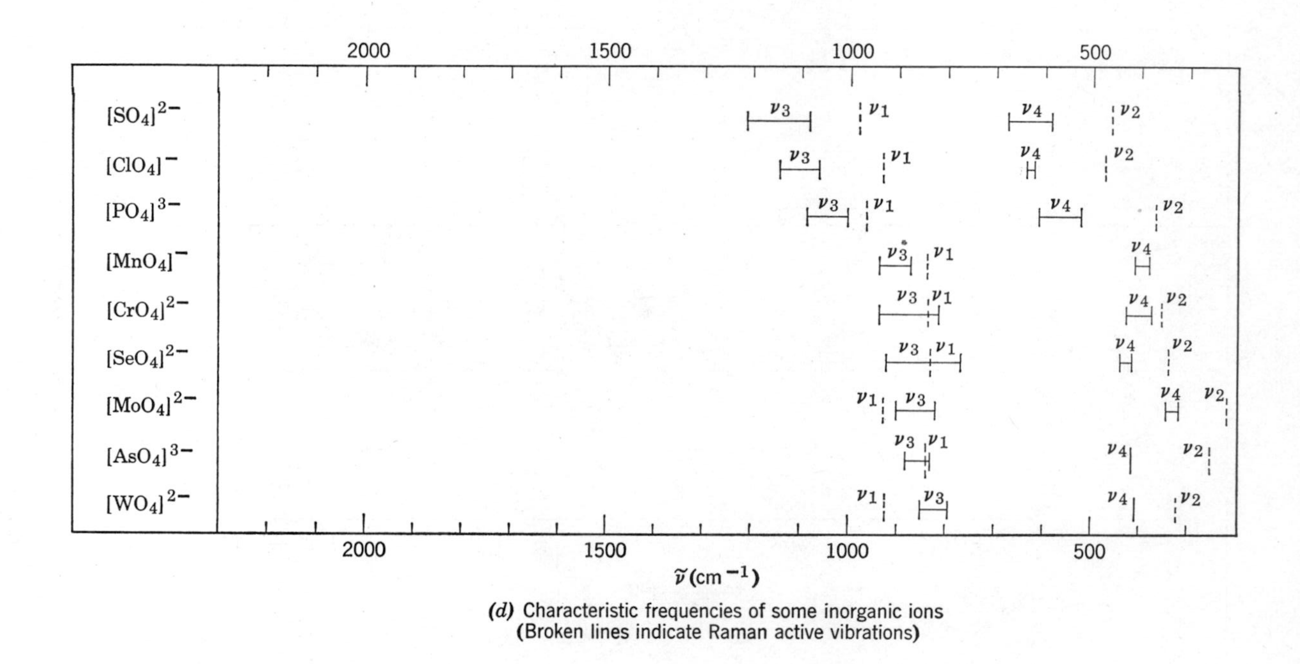

(*d*) Characteristic frequencies of some inorganic ions
(Broken lines indicate Raman active vibrations)

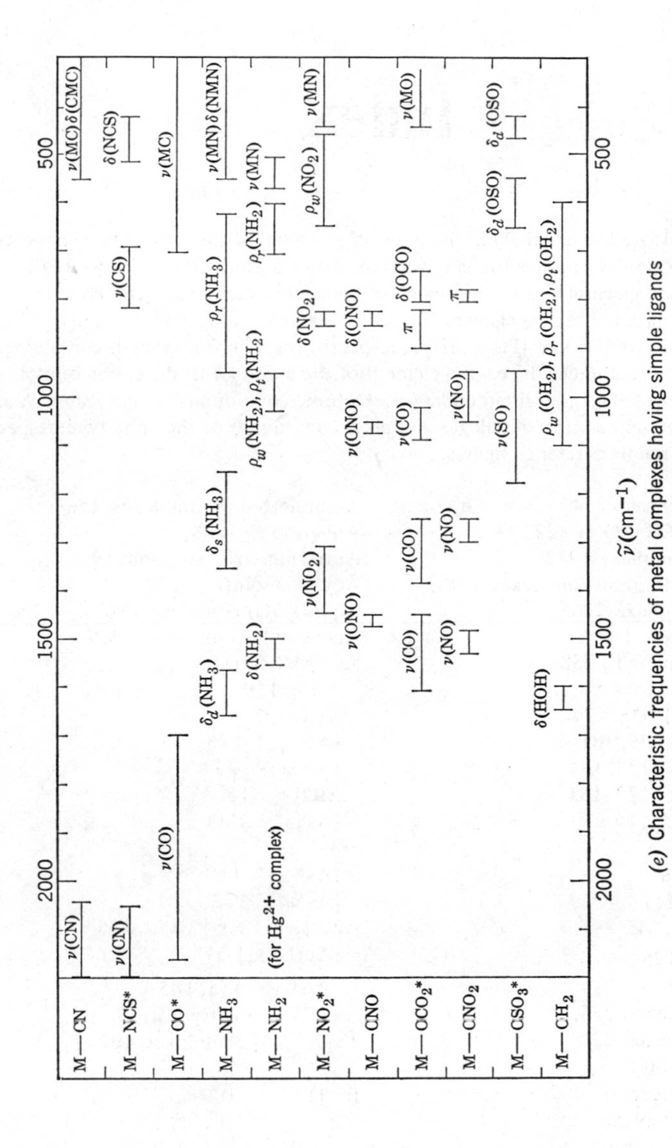

(e) Characteristic frequencies of metal complexes having simple ligands

(Frequency ranges include bidentate and bridged complexes for the ligands marked by an asterisk)

Index

The compounds are listed alphabetically under the symbol of the first element appearing in the chemical formulas as conventionally written, with the following exceptions: (1) For certain inorganic species (mainly oxy-acids and their ions), chemical formulas have been rearranged to place the central element first. For example, H_2O, H_2SO_4 and $[H_2PO_4]^-$ are written OH_2, $SO_2(OH)_2$ and $[PO_2(OH)_2]^-$, respectively. (2) Compounds containing complex anions are listed under the central element of the anion (with the cation omitted). For example, $Na_3[Co(NO_2)_6]$ is listed under Co. Isotopic compounds are not listed. X and R are used to represent a series of halogen and alkyl compounds of the same type, respectively. Boldface page numbers refer to figures.